Isaac Newton

ISAAC
NEWTON
ADVENTURER
IN THOUGHT

A. RUPERT HALL

CAMBRIDGE
UNIVERSITY PRESS

Published by the Press Syndicate of the University of Cambridge
The Pitt Building, Trumpington Street, Cambridge CB2 1RP
40 West 20th Street, New York, NY 10011-4211, USA
10 Stamford Road, Oakleigh, Melbourne 3166, Australia

First published 1992 by Blackwell Publishers Oxford
Reissued by Cambridge University Press 1996

Printed in Great Britain by Biddles Ltd, Guildford & King's Lynn

British Library Cataloguing in Publication Data available

Library of Congress Cataloging-in-Publication Data available

ISBN 0 521 56221 X hardback

ISBN 0 521 56669 X paperback

Contents

Figures

General Editor's Preface

Our society depends upon science, and yet to many of us what scientists do is a mystery. The sciences are not just collections of facts, but are ordered by theory, and this is where Einstein's famous phrase about science being a free creation of the human mind comes in. Science is a fully human activity; the personalities of those who practise it are important in its progress and often interesting to us. Looking at the lives of scientists is a way of bringing science to life.

By the time of his death in 1727 Newton stood as the representative figure of modern science. His name was something to conjure with, evoking ideas of the absent-minded professor, the solitary genius, and the power of mathematical and experimental science to answer questions about the world. He became a key figure in the Enlightenment of the eighteenth century, and his work correspondingly aroused unease among Romantics who saw his science as inhuman and reductive.

A mythical Newton, a new Adam born on Christmas Day and nourished by an apple from the tree of knowledge, came to obscure the real man who had worked in dynamics, astronomy and optics, and less successfully in chemistry, to synthesize the work of great predecessors such as Kepler, Galileo, Descartes and Boyle. Rupert Hall comes to this biography after editing the correspondence of both Newton himself and also Henry Oldenburg, the first secretary of the

Royal Society and the editor of its journal. He has also edited some of Newton's unpublished papers. His interpretation of Newton thus has the authority brought by the easy familiarity with the documents which a lifetime's research can yield. Since many of Newton's papers were not accessible to scholars until the twentieth century, we can now perhaps meet Newton the man more easily than at any time since his death.

Newton was a secretive man, as Hall shows, easier to admire than to like, and lonely in his eminence at the end of his long life. But Hall is sceptical about seeing him as a psychological case, or as a magus whose real interests were in alchemy and in the interpretation of the more obscure biblical prophecies. The great virtue of this biography is that we feel that Newton is understandable. We see extraordinary ability and application in mathematics and experiment; we see Newton's mind gradually gaining upon the dark, but also making mistakes and sometimes from our perspective retreating from a better to a worse understanding. And yet Hall does not forget that Newton was a man of his time, involved in the political and religious issues of his day, making friends and allies and falling out with them.

As an introduction to an extraordinary man, and through him to that combination of experimental and mathematical reasoning which is modern science, this biography would be hard to beat. Readers will remember the formidable and reserved man who, for all sorts of reasons discussed by Hall, became that representative figure for his successors.

David Knight
University of Durham

Foreword

I have endeavoured here to write an account of the greatest mind in British history. If any names of Englishmen survive into the remote future, those of Shakespeare and Newton will surely be among them. The latter was above all things a mathematician and a natural philosopher, but he also gave deep scholarship and profound thought to ancient history, especially to the early history of Christianity, the unravelling of sacred prophecy and even to monetary theory and practice. On all these topics he left vast accumulations of manuscript material.

In this book attention is chiefly directed to Newton the mathematician and philosopher. As such he worked his great transformation in human thought. Even so, a volume of modest size permits no very technical treatment of his researches in mathematics and mechanics, and in experimental optics. For such treatment the reader may turn to D. T. Whiteside's epoch-making edition of Newton's *Mathematical Papers*, the parallel volumes of *Optical Papers* (in progress) edited by Alan E. Shapiro, and very many specialist studies by these and other scholars. As for Newton's daily life and personal pursuits, for all the huge amount of material by and concerning Newton now accessible to us, these are shadowy at all periods of his life. I have not tried to emulate here Frank E. Manuel's psychological analysis of Newton in terms of theories whose validity seems to be doubtful. In general, I believe it imprudent to try to interpret Newton's life and writings in terms of single factors, whether these be his infantile experiences, his reading of the strange books of the alchemists, his faith in God or even his confidence in number and measure. There is no single key to understanding Newton, no single source or stream of knowledge to which he applied his unique mental powers.

In less specialized formulae than those of Manuel, the student of Newton soon discovers that his temperament was volatile, sensitive and egocentric. He could also be steady, resolute and generous. His life shows little affection for people or places and no pretence of it. Outside his immediate family and possibly (for a year or two) his friend Nicolas Fatio de Duillier, Newton demonstrated attachment to no one; in later life he abandoned Lincolnshire and Cambridge without apparent regret. He was free from all sentimentality, some might say from warmth. He had little mercy for the human foibles of others, and held up no mirror to discern his own. Yet he gave away large sums and pushed the careers of young men of whose ability he was convinced. Geniality was not his obvious characteristic, but the few admitted to friendship found him a lively and hospitable companion. His ideas of principle and duty were deep-seated and inflexible; when these were at stake he was bold, intransigent, cunning. His threshold of anger and resentment was low. But to paint Newton as a psychopathic genius is to take as romantic a view of his life as painting him as a saintly genius. Such romance is the biographer's creation. Newton was a man; Robert Hooke, Gottfried Wilhelm Leibniz and John Flamsteed were not angels either.

To observe Newton's transcendence of the intellectual bounds of his age, while at the same time never questioning some of its entrenched certainties, is a fascinating exercise. Yet his thinking was consistent and all of a piece, just as he believed the universe of matter and spirit to be. To Newton an investigation of God's workmanship in the solar system by no means seemed as distant from an analysis of the language of biblical prophecy as it does to us. (The former we can understand, at the latter we only marvel.) Modern scholars like Manuel have rediscovered what the close successors of Newton knew: to Newton's mind his study of human experience was all one, whether it was experience of Nature, of human history or of the actions of divine mercy. In a sense, thinking man is as much the hero of the *Principia* as sinful man is of *Paradise Lost*. Newton was the last polymath, the last mind to believe that all knowledge was within its grasp. Only sensual and aesthetic experiences were denied to

Newton: food and drink meant nothing to him, verse was an antic game with words, music a tedious jumble of sounds. I think he never spoke of any picture.

The documented, analytical biography of Newton was begun by Sir David Brewster more than 150 years ago; his book, however dated his attitudes, is useful still. Since 1980 all who are fascinated by Newton have been indebted to its successor, the large, scholarly and immensely detailed account of his life and work by Richard S. Westfall. Not the least of the merits of *Never at Rest* is a bibliographical essay of almost eight pages, with which it would be pointless to compete; for reference, Peter and Ruth Wallis's *Newton and Newtoniana, 1672–1975* is an invaluable resource. Much useful information is compiled accessibly in Derek Gjertsen's *Newton Handbook*.

I should apologize to readers for the length of my notes, arising from a wish to simplify the text. They may also provide some guidance to the extensive corpus of Newtonian studies.

I beg to thank my wife for encouragement, criticism and forbearance. To Tom Whiteside goes my gratitude for his devoting his immense knowledge to the emendation of some chapters, and to the Librarians of the Royal Society and the Bodleian Library for facilitating my use of their collections. Mrs Sheila Edwards and her predecessors have cheerfully met my requests during more than forty-five years. Members of the Wellcome Institute for the History of Medicine have continued to be my friends, and to them also go my thanks.

Throughout this book I have restricted double quotation marks to Newton's words. As usual, square brackets denote editorial insertions. I have copied original spelling and punctuation, but make no apology for not perpetuating obsolete writing contractions such as y^e and y^n. In an Appendix I have outlined very briefly the history of Newton's manuscript legacy, and of his library.

A. Rupert Hall
Tackley

Acknowledgement

This new edition is issued without any change, except that I have corrected a mistake on p. 74. I owe this correction to Dr Alan E. Shapiro.

A. Rupert Hall

1

The Hopeful Youth, 1642–1664

Early childhood

The Newton family belonged to the yeoman class descended from the most modest of the free landholders of manorial England. Socially beneath the esquires and knights, some members of this class had prospered greatly after the fourteenth-century decline of feudalism. Later the dissolution of the monasteries, sheep-farming and inflation had given some yeoman families, like the Newtons, means to enter the gentry class. As small landowners they lived in solid houses of brick or stone, sometimes adjacent to the barns and farmyard, like Newton's birthplace at Woolsthorpe in Lincolnshire. Yeomen, who could not prosper by idleness, constituted a great reservoir of ambition and talent, supplying Church, universities and the law, as well as commerce and industry. They wrote plays for the stage and music for the nobility; they staffed the empire. The father whom Isaac Newton never knew, also Isaac, could not sign his name but his mother and half-sisters were literate after the phonetic style also practised by the greatest ladies. When the posthumous child was three years old his mother, Hannah, remarried Barnabas Smith, rector of nearby North Witham. Smith owned a small library of theological books, works of the Fathers and so on, which passed to his stepson. Thus Isaac was born into the lower limit of landed property and learning alike. When he was old and famous he took pains to satisfy the College of Heralds of his common descent with an established armigerous gentleman, Sir John Newton, who was glad to bring so famous a man among his own kin. Sir Isaac Newton for his part responded to many begging letters from *his* poor relations.

The connected families – Newtons, Babingtons, Blyths, Chapmans, Smiths and Ayscoughs (Hannah's family) – were spread all over the East Midlands.[1] Apart from any undocumented direction, spiritual or scholastic, given to young Isaac by his stepfather, we know of two close connections, his uncle William Ayscough and Humphrey Babington, who were members of Trinity College, Cambridge, the latter being a Fellow of the College when Newton went into residence there. Each has been credited with the chief responsibility for sending Newton to the University, and it was certainly at Babington's rectory at Boothby Pagnell (another village close by) that Isaac Newton, in the autumn of 1665, evaluated a hyperbolic area to "two and fifty figures": the computation survives to this day.[2]

The impulse to thrive, strong in the yeoman class and evident among Newton's forebears, was certainly possessed by Isaac himself. After an unpromising start in life – for he was to go up to Cambridge as a sizar or student servitor who might act as his tutor's copyist – Isaac far surpassed socially all his family except Sir John. Early life revealed in him practical talents in mechanics, skill with his hands that matured into experimentation; later life brought out his capacity for administration and public business. If Newton had not achieved great intellectual distinction, he would have made his mark in some other way. And he had it in him to excel as scholar or theologian if he had not first taken to mathematics and natural philosophy.

Isaac's family had been settled in Woolsthorpe-by-Colsterworth in Lincolnshire, hardly more than a hamlet, since at least the mid-sixteenth century. Whether, as Newton claimed, they were entitled to the status of lord of the manor of Woolsthorpe seems rather doubtful, since it is not clear that a legal manor distinct from Colsterworth existed. One of Isaac's great-uncles did very well; that line produced his contemporary, the Sir John Newton already mentioned, and married into the great family of Coke of Norfolk, later Earls of Leicester. On a different level, when all the lands at Woolsthorpe came into his possession they seem to have been worth in rents and dues at least £200 per annum to Isaac Newton: no despicable sum in those days, approximating to the salary of Newton's professorship at Cambridge. Newton was therefore never poor and, after his mother's death in 1679, could have lived in Cambridge, if he had wished, in a state of some affluence. He was well able to keep up the dignity of his fellowship and University offices. By the time he was writing the *Principia* ('The Mathematical Principles of Natural Philosophy') in the 1680s Newton was a wealthy if eccentric senior professor, a man of some authority. In the last phase of his life he was to become a very rich man indeed.

Isaac Newton's father died in October 1642, six months after wedding Hannah Ayscough and almost three months before his son's birth on Christmas Day. His will defined his flourishing estate: besides the 'manor' and its houses he owned property worth £460, including a large flock of sheep, cattle, and ample stores of hay and grain. With his father dead before his birth and his mother remarried, the boy was brought up at Woolsthorpe by his grandmother, Margery Ayscough. His grandfather, James, left nothing to young Isaac Newton in his will, made when Isaac was ten years old, nor did Isaac ever refer to him (though he did recall his grandmother as his foster-mother and otherwise). Hannah's inheritance under James Ayscough's will came to Newton after his mother's death. He also profited (after coming of age on 25 December 1663) from the settlements made at the time of his mother's marriage to the deceased Barnabas Smith.

The mature and opulent Newton took pride in his landed status and defended his (dubious) rights as lord of the manor to control the pasturage of animals on the common lands of Woolsthorpe and the felling of timber. Until after his mother's death he returned there at regular intervals and it was in Lincolnshire that Newton spent that marvellous eighteen months when he was "in the prime of" his "age for invention in mathematics and natural philosophy". His county retained its hold on his affections to the end of his days and he relished the opportunity to chat with younger men from Grantham and thereabouts, like the future antiquary William Stukeley. Though never setting foot in his village during the latter half of his life, Newton kept in touch with its affairs through his reeve or agents, insisted upon his rents and performed his charitable duties in repair of the church and so forth. Late in life Newton was generous to his relatives. One wonders whether Humphrey Newton, who came up to Trinity College from Grantham in 1685 and served for five years as Isaac's copyist – the printer's manuscript of the *Principia* is in his hand – was not one of these, though all writers except Stukeley deny the connection; the rather pompous name of Humphrey was also borne by Dr Babington. After Newton's death Humphrey, then a physician in the town of his birth, was solicited for his recollections of the great man with whom he had lived intimately. Humphrey replied with invaluable accounts of Newton's habits and behaviour without eluci- dating the nature of his genius, except to say that he was absent- minded and eccentric.

Isaac Newton was particularly fond of his half-sister Hannah, named Barton by her marriage. At the end of his life he provided for the children of her son Robert who (as an army colonel) had been drowned in the course of a disastrous expedition to French Canada in 1711, buying them a country estate. Hannah's daughter Catherine was the dearest of Newton's young relatives. Since she lived with him in London from 1696 until her late marriage in 1717 she will reappear in the biography; it was she with her husband, John Conduitt, and their descendants who transmitted Newton's unpub- lished intellectual legacy to posterity.

The elder Hannah's first baby was a tiny infant, presumably premature, which no one expected to live. As Newton himself quaintly put it, he could be fitted into a quart pot – a strange experiment! Whether he was put out to nurse, what was his life in petticoats, we do not know. As might be expected, his mother's departure to North Witham when Isaac was three afflicted the small child grievously. We do not know that James Ayscough disliked children, or that he gave young Isaac his first lessons, or that he whipped him for his mistakes; but the treatment of children in the seventeenth century is not generally regarded as having been soft and kindly. 'Spare the rod and spoil the child' was an injunction ubiquitously accepted as true, and no doubt young Isaac received his share of this blessing. A fictional picture of Newton's childhood may perhaps have been written by Samuel Butler in imagining an upbring- ing early in the last century:

> Before Ernest could well crawl he was taught to kneel; before he could well speak he was taught to lisp the Lord's prayer, and the general confession. How was it possible that these things could be taught too early? If his attention flagged or his memory failed him, here was an ill weed which would grow apace, unless it were plucked out immedi- ately, and the only way to pluck it out was to whip him, or shut him up in a cupboard, or dock him of some of the small pleasures of childhood. Before he was three years old he could read and, after a fashion, write. Before he was four he was learning Latin, and could do rule of three sums.[3]

It seems strange that (unlike Butler, and Somerset Maugham in his finest novel) Newton's biographers have generally spent little pity on his (probably) loveless and lonely childhood, while expatiating on the twists that this may have introduced into his personality. Virtually orphaned for eight years, he lacked siblings to provide comfort or divert wrath. It is unsurprising that Frank E. Manuel has taken the Freudian theory of the Oedipus complex to explain the whole future development of Newton's character, distorted by the loss of his father and the absence of his mother during infancy.

In the summer of 1662, perhaps when home for the long vacation from Cambridge, Newton experienced a fit of religious fervour (of which Ernest's father in Butler's novel would surely have approved) inducing him to make a record of his private sins in Shelton's shorthand, which he had just mastered. Presumably the twenty-year-old used this script, as Samual Pepys did, as much for secrecy as for speed. In all his mass of papers, it seems, Newton only once employed shorthand again, about a couple of years later for an entirely unprivate record. Since this bizarre document is unique, we cannot tell whether the undergraduate Newton was usually, or on any other occasion whatever, so punctilious in accounting to God for his peccadilloes, but probably not. Several of the sins indicate normal observance of formal religious obligations: they include breaches of the sabbath, such as "Making a mousetrap on Thy day" or "Idle discourse on Thy day and at other times" and "Twisting a cord on Sunday morning". Others record a more general impiety: "Not loving Thee for Thyself" or "Not desiring Thy ordinances" and "A breaking again of my covenant renewed in the Lord's Supper" and "Missing Chapel [at College]". A few hint at the normal sexual pressures upon adolescent boys and young bachelors: "Having unclean thoughts, words and actions and dreams", "A relapse", again "A relapse" and possibly even more directly, "Using unlawful means to bring us out of distresses". Some may take this to refer to magical practices, but this seems really unlikely. Most interesting in the list of forty-nine sins committed before Whitsuntide 1662 are those that give a glimpse of Newton's family life, indicative of that sort of bad temper that mars us all from time to time: "Falling out with the servants" or "Peevishness with my mother" or "Refusing to go to the close [field] at my mother's command" and "Punching my sister" – younger than he! Other casual acts of violence recorded are ordinary enough with schoolboys, and naughty playfulness appears again in "Squirting water on Thy day" or "Robbing my mother's box of plums and sugar" and "Calling Dorothy Rose a jade". The sporting tastes of older boys seem to be implied in the following: "Denying a crossbow to my mother and grandmother though I knew of it" – surely forbidden indulgence in fowling with another boy's sporting weapon? (Would Newton have been reassured by knowing that that great and pious man of science, Robert Boyle, also went fowling with a crossbow?) Many of the sins already quoted plainly belong to a childhood long before the ripe age of twenty, to which may be added "Peevishness at Mr Clarks for a piece of bread and butter" (he had lodged with the Clarks when at school in Grantham) or eating an apple in church, stealing cherry cobs from Edward Storer, swimming in a kimnel [tub] on the Lord's day, so that we may imagine these trivial offences being recollected as the major crimes of past years. At Cambridge, besides missing chapel (where daily attendance was required of under-graduates) and also sermons at Great St. Mary's church, Newton had done such terrible things as using a fellow-student's towel "to spare my own" and "Deceiving my chamber-fellow of the knowledge of him that took him for a sot". Gluttony is also recorded more than once, a sin conspicuously absent from Newton's mature life.[4]

As the notes above indicate, Newton's stepfather had died in 1653 and from that date until his departure for Cambridge in 1661 the boy lived in a family consisting of his grandmother, mother, half-brother and two half-sisters, as well as (one may suppose) three or four servants. The eight years of isolation were over. To those years we may confidently assign the most extraordinary of the sins Newton listed: "Threatening my father and mother Smith to burn them and the house over them" to which may be related "Wishing death and hoping it to some". Newton was not the first child nor the last to nourish furious anger against parents or those *in loco parentis* and the setting of houses afire by quite small children is not unheard of. That Newton remembered for ten years such impotent fury and then recorded it solemnly is a measure of the damage that he had undergone. The other sin may be a personal death-wish, as well as a wish for the deaths of others. Did Isaac hope that Barnabas Smith and James Ayscough might soon die?

Whatever the black misery in his heart, we have nevertheless a little evidence that Isaac Newton lived as little boys do live in the country, playing, fighting, robbing orchards, quarrelling with his sisters, persecuting small birds, eating sweets when he could get them. We need not imagine that he never laughed – though Newton was not given to mirth in later life – nor that he was always peevish. At a time when even quite small boys were expected to take some share in the farm-work, and the Newtons were not so grand that the heir to the estate might count himself exempt from this necessity, Isaac clearly disliked such tasks. He would not take a message to the men in the field, and one anecdote tells of his tendency to read a book in the hedge-bottom rather than watch for the straying of the sheep. Another states that he once came back from Grantham with only a halter in his hand, having become quite oblivious of the horse that had slipped it. Yet another sin recorded in 1662 was "Setting my heart on money, learning [and] pleasure more than Thee". Few boys would couple learning with money and pleasure as one of the great besetting temptations of life. Do the compass-scratchings and the sketch of Grantham church adorning the Woolsthorpe walls – if Newton really made them – indicate boredom?[5]

Grantham School

Even in remote Lincolnshire – not of course the fens, but what Newton's young friend Stukeley thought a country 'exceeded by none [in all England], for a fine air and for pleasantness; being most agreeably diversify'd with open heaths, rich meadow and inclosure, woods and parks; the most beautiful cornfields, springs, brooks and rivers'[6] – the eldest son of a family such as the Newtons now were was expected to have more learning than could be acquired in a village dame-school, such as those in which Isaac learned his letters. Five miles away in Grantham town was a Free Grammar School, ancient and esteemed. A pupil earlier in the century had been Henry More (1614–87), now a senior Fellow of Christ's College, Cambridge, and a distinguished academic author. Isaac probably encountered More at Grantham, since upon occasional visits to his home town he stayed with the Clarks, in whose house Isaac lodged; they were certainly acquainted at Cambridge later. Isaac attended the Grantham Free School from the age of eleven or twelve until he was eighteen or nineteen, for he entered Cambridge at a greater age than was then usual, when upper-class boys still arrived with their private tutors at the tender ages of fourteen or fifteen. Contemporary anecdote accounts for this by the story that Newton's mother took him from the school to manage the estate, then was persuaded to allow him to return and prepare for Cambridge. It is sad that only generalities can be offered about his instruction there; Latin certainly (in this language Newton, like all educated men and women, was always fluent on paper, and he may well have conversed in Latin with some visitors from the Continent); the elements of Greek, again used by Newton in his scholarly activities but no one would call him a Greek scholar;

arithmetic but perhaps no geometry; no modern languages.[7] D. T. Whiteside emphasizes that Isaac was no calculating boy-prodigy;[8] throughout his life he was liable to commit unnoticed arithmetical errors. The biographer, however, cannot exclude the possibility that either Stokes, the 'Master' of the Free School, or Mr Clark his 'usher' (brother of the apothecary with whom Newton lodged and a former pupil of Henry More at Cambridge) privately took the able boy well beyond the usual limits of grammar-school mathematics. Both were able men. Again, anecdote has it that it was Stokes, together with her brother William Ayscough, who persuaded Mrs Hannah Smith 'what a loss it was to the world, as well as a vain attempt, to bury so extraordinary a talent in rustic business' and that he should go to the University.[9] Moreover, Newton, during his brief return to school in 1660, lodged with Stokes, as though for private tuition, and certainly if the anecdote is true *some* work must have been done at the school that convinced his teachers of Newton's great mental powers. Another anecdote tells of his dismissing Euclid's *Elements* with contempt during his early Cambridge days; if he had already mastered some of its propositions while at school this might account for his putting on airs about the book which was new to most of his fellow-undergraduates.

Stukeley's *Memoirs* are also the only source for there having been a touch of romance in Newton's early life. Apothecary Clark had married a widow, Mrs Storer, with two sons and a daughter; Arthur and Edward are positively recorded in Newton's own manuscripts. As for the nameless Miss Storer, who was also a niece by marriage of Dr Humphrey Babington of Trinity College, Cambridge, Newton 'was said' to have 'entertaind a passion for her when they grew up', a story that Mrs Vincent (as she became) in extreme old age was vain enough to confirm to Stukeley, who continues:

> 'Tis certain he always had a great kindness for her. He visited her whenever in the country, in both her husbands days, and gave her, at a time when it was useful to her, a sum of money. She is a woman but of a middle stature, of a brisk eye, and without difficulty we may discern she has been very handsom.[10]

No other woman outside Newton's own family is known to have aroused any sentiment in him, and he is confidently supposed to have remained a chaste bachelor till the end. I agree with Louis Trenchard More that the letter printed by Brewster as a proposal of marriage from Newton to Lady Norris – a widow with whom he was indeed acquainted – is very unlikely to be genuine.[11] It is not in his hand or his style and it is hardly likely that he would figure as a bridegroom for the first time in his sixties.

Mrs Vincent describes Isaac as a 'sober, silent, thinking lad' who wasted little time on play with other boys. There may be cause to doubt that this common sombre recollection of a great man is exactly true, though Newton may have grown more studious and reserved as he passed into adolescence. She also recalled Isaac's skill in copying engravings with the pen (I suppose this is what was meant); a sketch of Grantham church survives on the wall at Woolsthorpe and numerous little sketches in Newton's scientific manuscripts witness to some slight skill in pen and pencil drawing. Those who recalled Newton's childhood also dwelt on his mechanical bent; at some period of his life he must have spent many hours fabricating models of wood. Stukeley wrote of older people marvelling at a windmill constructed not far from Grantham, the first in a district hitherto content with its ample streams for the grinding of corn: 'Newton's innate fire was soon excited. He penetrated beyond the superficial view of the thing . . . He obtain'd so exact a notion of the mechanism of it, that he made a true and perfect model of it in wood; and it was said to be as clean a piece of workmanship as the original.'[12] He also made a wooden clock driven by weights in the usual way, and also two water-clocks working on different principles. Like other boys, he made kites and flew them at night to take lighted lanterns aloft, and contrived sundials 'So that Isaacs dyals, when the sun shined, were the common guide of the family and neighbourhood.'[13] Dialling involves some sense of geometry and even astronomy; a method is copied into a notebook dated 1659.[14] The boy exercised his mechanical skill not only at home but in his free hours at school in Grantham.

Indirect confirmation of this artisanal skill is given in Newton's later life by his handiwork in practical optics and chemistry. He knew how to turn wood on a lathe, to melt metals in a furnace, to undertake simple blacksmith's work, to cut screw-threads, to lay bricks, as well as engaging in carpentry. This was more than a century before Louis XVI amused himself with lathe-work and lock-making, and Tolstoy's eccentric Prince Bolkonsky released his psychological tensions by engine-turning. The gentlemen of the Royal Society were of course familiar with firearms and they carefully collected accounts of many trades from agriculture to textiles, but few of them had actually acquired and practised manual dexterity – Robert Hooke was one of the exceptions. Designs for new scientific instruments and apparatus were normally turned over to the professional instrument-maker, as James Gregory did with his design for a reflecting telescope, and both Hooke and Christiaan Huygens did with improved mechanisms for clocks and watches. Newton was highly exceptional in his mastery of craft skills for scientific ends.

Akin to young Newton's interest in mechanisms and perhaps not unconnected with his lodging above an apothecary's shop was his concern for pigments and receipts. In a notebook seemingly bought (for $2\frac{1}{2}$d) in 1659 Isaac entered notes about drawing, painting, pigments (including "Colours for naked pictures"!), gilding and all sorts of practical tips and wrinkles – "A bait to catch fish", "A Salve for all sores", "A Water to clear the sight", "To Cut a Glasse". These, like some of the ideas for mechanisms, were copied from one edition or another of John Bate's *The Mysteries of Nature and Art* – the book is not one which Newton is known to have owned – and he may also have drawn on John Wilkins's *Mathematical Magic* (1648).[15] Into the same notebook, probably a Cambridge undergraduate now, Newton entered astronomical tables and an ecclesiastical calendar (beginning in 1662), three pages on the Copernican system, and notes upon plane and spherical trigonometry. This notebook, in the Pierpont Morgan Library in New York, clearly extended from Isaac's late schooldays into his undergraduate period. Into it he also copied lists of words from Francis Gregory's *Nomenclatura brevis anglo-latino*, about 2400 in all, interpolating words of his own into the sequences. So he made *Offender* follow *Orphan*, making his own 'self-accusatory association', as Frank Manuel has it. Manuel also finds testimony to Newton's puritanical, grim, guilty frame of mind in the English sentences he made up for rendering into Latin: for example, "What else is to dance but to play the foole?"[16]

From this slight evidence, whether or not it casts light upon Newton's psychological constitution, it is hard now to determine if he was well prepared for the University in comparison with other boys, or where the tinge of genius may have been disclosed. He cannot have had that fierce drilling in the classical tongues and their literature that Dr Busby's Westminster gave, nor had his speech, tastes and graces been refined on the Grand Tour by which the aristocracy already polished its scions. What he had clearly acquired, and would exploit in his mature life, was a tremendous capacity for teaching himself; all the anecdotes insist upon his autodidact knowledge and skills. He had also acquired the habit of steady toil. No man has ever possessed a greater capacity for intense, concentrated, protracted intellectual effort than Isaac Newton. Such a habit of regular daily work is not acquired in adult life.

Cambridge, 1661

With his mother's permission won and his final coaching at Grantham completed, Newton was formally admitted at Trinity College, Cambridge, on 5 June 1661, having satisfied examiners of his fitness for admission. The journey down the Great North Road had taken three days – though it could be ridden in two – one night being spent at his mother's property at Sewstern, the other at Stilton in Leicestershire. The college assigned him a tutor – Benjamin Pulleyn – and put him in a room, or set of rooms, which he shared with a chamberfellow. Newton's accounts show that he at once bought a lock for his desk, ink, a notebook, candles and a chamber-pot. Richer undergraduates decorated and furnished their rooms in style, but Newton was presumably content with the college provision of furniture and fittings, together with whatever he may have brought with him. Anecdote relates that he was unhappy with his first room-mate, until he accidentally met John Wickins, '& thereupon [they] agreed to shake off their present disorderly Companions & Chum together, which they did as soon as conveniently they could, & so continued as long as my Father staid at College.'[17] Newton made as close a friend of Wickins as of anyone; the latter sometimes acted as Newton's copyist and assisted in some of his optical experiments.

Frank E. Manuel has built his portrait of Newton upon the hypothesis that he suffered from a frustrated fixation upon his mother, the chief direct evidence for this being his ungrudging attendance upon her deathbed, hardly a peculiar act of filial devotion. Newton seems to have been eager to get to Cambridge, and though homesick once he got there (as many undergraduates are who have not endured a boarding-school) he was within a year or two content to foresee a career in the University for himself. He never considered taking a living in Lincolnshire. His mother, from a unique letter to Newton surviving (because of his optical notes on the back), was undemonstrative towards him and certainly kept him on a short leash. Westfall's computation is that Newton was allowed £10 a year for personal expenses; his mother, out of her income of some £700 per annum, very comfortable indeed for those days, had also to find his university and college fees of about £10 or £15 a year. Perhaps £20 or £25 annually was thought quite adequate for the education of a sizar, but it is far from clear why so wealthy a widow entered her eldest son in so lowly a class of students. Westfall makes the plausible suggestion that he was Babington's sizar and that it was this little act of patronage that opened Cambridge to him. As Babington resided in college only occasionally, any work that Newton may have done for him cannot have been onerous. Sizars, according to a literal reading of the statues which is not necessarily safe, were set lower than other groups of undergraduates in a variety of ways, and had the duty of supplying the scholars and Fellows with their allowance of beer and commons from the buttery. In practice, it seems that this was often done by the women servants, 'bedmakers', in whose recruitment to be ill-favoured was an important advantage. If Newton ran a few errands for Babington and copied his sermons, this is a very different thing for 'fagging' for *all* students and emptying their chamber-pots. Recent historians of Balliol College, Oxford, have suggested that this kind of student was so disgusting by reason of his poverty and grovelling eagerness to make his way in the world (most became parish priests) that he became 'a social pariah with whom men of ordinary good sense and good feeling hardly cared to be seen walking and conversing in public.'[18] Indeed, it is a stereotype that the 'swell' always looks down on the 'swot'. Whether Isaac Newton suffered any such scorn in his day we cannot know now, but it seems that in any case he associated with pensioners like Wickins rather than with fellow-sizars; to some of these he regularly lent small sums of money.[19] Whether that made him seem a 'good fellow' or a usurer cannot be determined either, but it seems that sympathy for Newton's ostracism on social grounds may be wasted. And if we censure Newton's mother for depressing him to the sizar class, we can hardly blame him for a snobbish desire to rise from it.

An undergraduate who sticks to his books, who may be intent upon impressing his teachers, who is devout and tender of conscience, who is poor and lives quietly, is not likely to be either celebrated or popular. We may take it without argument that young Isaac would in our own time have been as unlikely to be President of the Union as a Blue, and very likely would not have been chosen as an Apostle. We need not necessarily assume, however, that money-lending and solitary walks furnished Newton's only relaxations from his desk. A few years later his accounts show that Newton enjoyed the civilized pastime of bowls, that he played cards and drank at a tavern; but we have no evidence for such diversions in the years of his sizarship, when (as we have already given reason to believe) Newton was preoccupied with religious observance and may (for all we know of his books at this time) have thought of making a mark in religious scholarship and theology – as indeed, in a sense, he was to do. Perhaps some further light may be thrown on Newton's early studies and intellectual tastes by considering what little we know of his early book purchases. But first we must recall that Newton's mother had presumably brought to Woolsthorpe his stepfather Smith's collection, that Newton presumably found books at school and at the Clarks', and that he may have read in the chained library of St Wulfram's church in Grantham, which he would have passed every day on his way to and from school.[20] Further, we can only know when Newton bought a book if he made a note of it in his accounts or recorded the date (and often, price) of purchase in the book itself. Both these records are very rare. Moreover, Newton's name and date in a book are not proof that he himself bought it; some books so marked might have been taken to Cambridge by Newton from Barnabas Smith's library, and so inscribed to protect his ownership there. (Presumably this might be taken as a sign of some interest in the book.) Above all, it can be confidently stated of few books whose ownership has been ascribed to Newton that he actually read them, even casually; many books in his library bear no trace of having been read. Of course he must in addition have read many borrowed books. Conventional learning dominates those few titles we can assign to Newton's early years of book acquisition. As a sixteen-year-old he bought the *Metamorphoses* of Ovid and the *Pindar* already mentioned. Newton's next known purchases, early in 1661, were in aid of his study of Greek and especially of the Greek New Testament, of which he bought a recent edition. They were Pasor's Greek–Latin lexicon for New Testament students and Beza's *Annotationes maiores* upon it.[21] Newton's first purchases at Cambridge included another aid to biblical study, Trelcatius's book of commonplaces in the Holy Scriptures (1608) and, less conventionally perhaps, Calvin's *Institutes* (1561).[22] Concern for classical literature is manifest in a small bilingual edition of the *Iliad* and the *Odyssey* – a volume given to Humphrey Newton in 1688 – Stephanus's *Dictionarium nominum propriorum virorum, mulierum, populorum . . .* (1576) and a book on the philosophy of nature reconciling Plato and Aristotle.[23] To these he added another lexicon (Greek–Latin, Latin–Greek) and Robert Sanderson's *Logicae artis compendium* in an edition of 1631.[24] Newton had been advised by Babington that this would be his first set book at the university, and so, it seems, it was. A widening of interests in classical scholarship is indicated by his buying Johann Sleiden, *De quatuor monarchiis*, though the copy now among Newton's books is not the one he bought in 1661.[25]

We cannot tell what other books Newton may have acquired by (say) 1662 without entering a date in them. Nor can we tell whether Newton bought books to read immediately, or for some possible future use. With limited funds, could he have afforded any unnecessary books? But for what they are worth the few titles indicate great seriousness of purpose, and a readiness to go far beyond what could be required of any undergraduate as part of his normal course of study. As we shall see in a moment, this is precisely true of his mathematical reading too; but for the present any hint of interest in mathematics or science is conspicuously absent – and it would have been very unusual had the case been otherwise.

Newton's Academic Progress

To return to the plain facts of Newton's life: he was elected scholar on 28 April 1664 after eight terms of residence.[26] Presumably sizars and pensioners were promoted to exhibitions and scholarships then, as now, as a reward for studious excellence (subject to the curious caprice of the geographical and family restrictions then prevailing). This promotion, together with the fact that Newton proceeded BA in the normal way in January 1665 (in a batch of twenty-six Trinity men), seems to belie the anecdotes of his incompetence or eccentricity as an undergraduate. The most circumstantial is told by John Conduitt, husband of Newton's favourite niece Catherine, as a story from Newton's own mouth:

> When he stood to be scholar of the house his tutour sent him to Dr Barrow then mathematical professor to be examined, the Dr examined him in Euclid which Sir I. had neglected & knew little or nothing of, & never asked him about Descartes's Geometry which he was master of [.] Sir I. was too modest to mention it himself & Dr Barrow could not imagine that any one could have read that book without being first master of Euclid, so that Dr Barrow conceived then but an indifferent opinion of him but however he was made scholar of the house.[27]

Similarly Stukeley says in his recollections: 'I have heard it as a tradition, whilst I was a student at Cambridge, that when Sir Isaac stood for Bachelor of Arts degree, he was put to second posing, or lost his groats, as they term it: which is look'd upon as disgraceful.'[28] However, the same John Conduitt also assured Fontenelle (the Secretary of the Académie Royale des Sciences in Paris, to whom it fell to write an obituary of Newton) that he had 'always informed himself before hand of the books his tutor intended to read [i.e. lecture upon], and when he came to the lectures, he found he knew more of them than his tutor: the first books he read for that purpose were Saunderson's Logic and Kepler's Optics.'[29] Presumably this too came from Newton, and it is quite reasonable that the great man should have told two kinds of stories about his early years. Scorn has been poured upon the idea that Benjamin Pulleyn could have read Kepler's *Optics* to anyone; but (as we shall see) something is to be derived from that part of the story when we realize that Newton – by inadvertence or lapse of memory – left such hearers as Conduitt and Stukeley with the mistaken impression that Isaac Barrow had been his tutor.

Specifically, the Elizabethan (1570) statutes of the university still operative in Newton's day required no undergraduate to study or have studied Euclid. They prescribed rhetoric for the first year, logic for the second and third, philosophy for the fourth. However, they also prescribed mathematics (along with logic and moral or natural philosophy) as providing the matter of the third-year disputations.[30] It seems to have been the theory, at least, that the colleges should provide teaching in mathematics as well as other subjects; elementary mathematics was meant, but this included astronomy. In the early seventeenth century some tutors, like Joseph Mede at Christ's College, expected their pupils to buy mathematical books and to work at them. It may well be that there had been some decline towards the middle of the century, though (again at Christ's) Henry More took pupils through Descartes's *Dioptrique*, a book hardly accessible to the innumerate. No doubt there was much variation from tutor to tutor, and the reality of instruction certainly departed from the model. Some undergraduates seem to have gained their degrees after paying virtually no heed to the statutory reading and exercises.

In theory, an undergraduate was supposed to acquire a thorough knowledge of the writings of Aristotle on logic, ethics, physics and cosmology, and to dispute systematically problems concerning these subjects in the mould of the Aristotelian syllogism. If the student notebook (Cambridge University Library MS Add. 3996) in Newton's collection is really in his own hand – and if not, it was studied by him – he preferred textbooks to original authorities, like most under-graduates before and since. He copied sentences in Greek from the *Nicomachaean Ethics* of Aristotle, and notes on his *Organon* (logic) followed by other Greek passages from Porphyry. Another book annotated was Eustachius on St Paul, *Ethica sive summae moralis disciplinae*, of which an edition had been published at Cambridge in 1654. There are also metaphysical notes upon being and non-being, action and potential, and so forth. With Gerard Vossius, *Rhetorices contractae*, he plunged into rhetoric and oratory. None of these books is now in the reconstructed 'Newton's library', though he owned and read several other books by the celebrated Vossius, possibly even in his undergraduate days. Another book he did not retain, if he ever owned it, though it may have been of some significance in his intellectual development, was the 'Peripatetic Physiology' (that is, physics) of Johannes Magirus; Newton carefully summarized, in Latin of course, the chapters of this lengthy exposition of Aristotle's *Physics* and *De Caelo* without taking notice of Magirus's own lengthier commentaries. All these books and some others were read only in part, according to this record – perhaps he was so instructed.[31]

If the load seems light and the achievement small one must recollect that the university was entering upon its slow decline – indirectly to be a little alleviated by Newton himself through the rising importance of the Tripos and of his own works within it – that was to last until the nineteenth century. The restoration of the Stuart monarchy during Newton's last year at school had, if anything, deferred rather than advanced hopes of its renewal. In official encomia the high intellectual life of the university might be praised, but one wonders how genuine was its distinction in oriental and modern languages, mathematics and natural philosophy, botany and chemistry, as recited by Isaac Barrow in 1654.[32] The studies and examinations of the university, like those of the schools preparing boys for it, were out of step not only with the outside world but with contemporary literature, scholarship and science. The naturalist John Ray, in a college lecture of about 1660, lamented that 'so little account [is] made of real Experimental Philosophy in this University and that those ingenious sciences of the Mathematics are so much neglected by us; and therefore [I] do earnestly exhort those that are young, especially Gentlemen, to set upon these studies, and take some pains in them.'[33]

Other well-known evidence from the generation before Newton's at Cambridge points to the difficulty of finding scholars able to give keen young men an entrance to mathematics and natural philosophy, not to say such specialized sciences as botany and chemistry.[34] (And certainly Newton himself did not change things: he himself had few auditors and fewer students and was for a time a non-resident professor employing a deputy.) The mathematician John Wallis said in his autobiography:

> I did thenceforth prosecute [mathematics] . . . not as a formal Study, but as a pleasing Diversion, at spare hours; as books of *Arithmetick*, or others *Mathematical* fel occasionally in my way. For I had none to direct me, what books to read, or what to seek, or in what method to proceed. For Mathematicks, (at that time, with us) were scarce looked upon as *Accademical* studies, but rather *Mechanical*; as the business of *Traders, Merchants, Seamen* . . . And amongst more than Two hundred Students (at that time) in our College, I do not know of any Two (perhaps not any) who had more of *Mathematicks* than I, (if so much) which was then but little.[35]

Wallis, recalling this state of affairs at the age of eighty in 1697, went on to a Savilian professorship at Oxford (1649), where he was joined by Seth Ward, another budding mathematician who had sought instruction outside Cambridge. There is, however, almost as much testimony on the other side. Wallis credited Cambridge with his first introduction to 'the Principles of what they now call the *New Philosophy*' and claimed that he had been the first undergraduate to defend the idea of circulation of the blood in a public disputation. Ward had been made a college lecturer in mathematics at Sidney Sussex in 1642. As Feingold has demonstrated, there was never a time when no college taught the subject, nor did the university ever completely lack competent mathematicians. It supplied five effective professors of astronomy and geometry to Gresham College in London (most of the rest were Oxford men).[36] Despite the lively intellectual activity associated at Oxford with the names of John Wilkins, Thomas Willis and John Wallis, as well as many younger men destined to become celebrated, Joseph Glanvill when a student there believed 'the new philosophy and art of philosophizing' to be better developed at Cambridge.[37] This university is generally held to have been the first locus of Cartesianism in England, and Henry More is considered to have been a leader in teaching it.[38]

However, the rather meagre evidence in favour of a revival of mathematics and science at Cambridge during the Interregnum has only an indirect bearing on the question of the character of the intellectual environment in which Newton found himself there, in the year following the Restoration. That the curriculum was outmoded and the instructional programme chaotic no one doubts. Boys were at their tutor's mercy. If Newton had fallen into the hands of Joseph Mede at Christ's College a generation earlier, we might have found him buying Keckermann's *Systema compendiosa* of mathematics, Thomas Digges's *Pantometria* and Sebastian Munster's *Cosmographia* upon his tutor's advice.[39] As it is, we know nothing at all of how Benjamin Pulleyn handled his pupils, Newton among them.

The Church and medicine were the only professions for which the University of Cambridge might be said to prepare its undergraduates, though it taught no medicine and little theology. (The Regius Professor of Medicine, Francis Glisson, did desert his lucrative London practice for a few days in most years in order to examine candidates for medical degrees.) Nor did Cambridge train or encourage undergraduates to become men of intellect and scholarship. The university course was merely a ritual of a peculiarly boring kind, a fact few tried to disguise. As rituals should not be allowed to interfere with important realities, university statutes were regularly defeated by neglect or nullified by exemption. A budding physician, by a short excursion to Angers or Leiden, could acquire an MD that Cambridge would recognize; royal letters patent gave degrees, fellowships, even headships of colleges to those unqualified for these positions, the Crown exercising a prerogative power which, in time, was to confirm Newton's curious privilege as, at the same time, a layman, a Fellow of Trinity and a professor. The incompetence of the university and the indifference of the colleges had, however, the negative merit that an undergraduate eager to write poetry, to hunt for plants in the countryside around Cambridge, to dissect small animals or to read mathematics was free to do so. The fact that the Copernican system was officially anathema – as was brought out in a specimen academic disputation arranged for the visit of the arch-conservative Cosimo de'Medici in 1669 – was totally irrelevant to what was read and believed. In the same university Henry More's pupils at Christ's College were reading Descartes's *Principia philosophiae* and *Dioptrique*. Cambridge in Newton's day was, in brief, a place of idleness or of intellectual liberty: the young man in search of anything might find it, provided he was resolute enough to overcome the obstacles that the university put in his way. The merit of this freedom was that young men did not have to devote all their efforts to becoming merchants, lawyers or civil servants, or much into becoming priests.

Newton takes to Mathematics

Against this background the foundation of a professorship of mathematics by Henry Lucas, and the translation to it in 1663 (as its first occupant) from the chair of Greek of Isaac Barrow, a Fellow of Newton's own college, seems almost providential as far as he was concerned. It is conceivable that otherwise he might have remained as relatively obscure in rural isolation as Jeremiah Horrocks or Richard Towneley.

We may be certain that the chaotic undergraduate system of teaching and examining gave Newton time for reading of which we are ignorant. We have reason to infer that John Ray as an under-graduate was already intent upon plants and animals, but nothing during Newton's first years in Cambridge connects him with mathem-atics and physics. One may guess that in them he plunged into theology and the history of Christianity (on the slight evidence of his known purchases of books), interests that would emerge into the light only much later. There is no hint of his life's work, unless one is found in Stokes's confidence in him, until near the close of his undergraduate days, when we find him taking full advantage of the laxity of the system and his own immense autodidact capacity. According to his own record:

July 4th 1699. By consulting an accompt of my expenses at Cambridge in the years 1663 & 1664 I find that in the year 1664 a little before Christmas I being then senior Sophister, I bought Schooten's Miscel-lanies & Cartes's Geometry (having read this Geometry & Oughtred's Clavis above half a year before) & borrowed Wallis's works and by consequence made these Annotations out of Schooten & Wallis in winter between the years 1664 & 1665. At which time I found the method of Infinite series. And in summer 1665 being forced from Cambridge by the Plague I computed the area of the Hyperbola at Boothby in Lincolnshire to two & fifty figures by the same method.
Is. Newton[40]

This oft-quoted record was written in a 'mathematical' notebook now in the University Library at Cambridge (MS Add. 4000) begun at the turn of the year 1663–4; the account-book mentioned in it (coming in order between the Trinity College notebook and the Fitzwilliam Museum notebook) has vanished. Another memorandum of a much later date takes the story further than we need go at present:

In the beginning of the year 1665 I found the Method of approximating series and the rule for reducing any dignity of any Binomial into such a series. The same year in May I found the method of Tangents of Gregory and Slusius, & in November had the direct method of fluxions, & the next year in January had the Theory of Colours & in May following I had entrance into the inverse method of fluxions . . . All this was in the two plague years 1665 & 1666 for in those days I was in the prime of my age for invention and minded Mathematics and Philosophy more than at any time since.[41]

The omitted sentences relate the origin of Newton's physical discoveries, to which we shall return later.

The notebook entry, wholly borne out by the extant "Annotations" which D. T. Whiteside has printed,[42] raises extraordinary problems. William Oughtred's *Clavis mathematicae* of 1631 (available later in English) possessed a 'clean workmanlike notation allied to concise exposition of basic methods', qualities giving it a long life as a textbook of arithmetic and algebra.[43] But what is one to make of Newton's unaided attack upon Descartes's *La Géométrie* (1637) – he used a Latin version, of course – and "Schooten's Miscellanies"? This was to leap at once virtually to the frontiers of mathematical research. Reading this literally, and bearing in mind that we have no certain knowledge of Newton's receiving any previous training in algebra and geometry, one can but remark that only the highest genius could have done it. Anecdote confuses as much as it helps to explain. After Newton's death the mathematician Abraham de Moivre, when asked by John Conduitt for an account of Newton's early mathematical development, told the following story (which is here paraphrased):

In [autumn] 1663 Newton bought a book of astrology at Stourbridge Fair (held at Cambridge) where he came across mathematics that defeated him. A book on trigonometry was equally obscure to him because of its geometry. When he looked into the *Elements*, Euclid's theorems seemed trivially obvious at first. 'Changed his mind when he read that Parallelograms upon the same base & between the same Parallels are equal' and the 'theorem of Pythagoras'. Mastering these forced him to work carefully through the book again from the beginning. Then he began Oughtred's *Clavis*, again encountering matters that baffled him. He also took up Descartes's *Géométrie* and by repeatedly working through from the beginning whenever he was baulked by reasoning that he could not follow, he mastered the whole of that book. He then went through Euclid once more, and after that Descartes again. Then he started on Wallis's *Algebra*.[44]

This indirect reminiscence, supplemented but not contradicted by any extant original document, again emphasizes the autodidactic character of Newton's mathematical study.

Other mathematical books from which Newton took notes were omitted from these accounts: Descartes's *La Dioptrique* (again read in

Latin), and François Viète's collected works edited by Frans van Schooten (1646). But it does seem that his most important sources (before Wallis) were Oughtred's *Clavis mathematicae*, read in the third Latin edition of 1652, and that great two-volume compendium of modern mathematics, the second edition of van Schooten's *Geometria, à Renato des Cartes Anno 1637 Gallicè edita* (1659–61) together with book V, a miscellany of thirty problems, of his earlier *Exercitationum Mathematicarum Libri Quinque* (Leyden, 1657). Each of van Schooten's books contained important works by other mathematicians, among them (in the *Exercitationes*) a tract by Christiaan Huygens, whose scientific life was to be interwoven with Newton's.[45]

Though the notes are quite extensive (sixty-three printed pages in Whiteside's edition) they are far from covering systematically every detail of the books Newton read, and they are not wholly summaries, for what he read led him into explorations of his own (sometimes mistaken). This smooth passage from copying to original work is typical of Newton's early notebooks. While giving him a firm grounding in many areas of algebra and geometry – though many topics of seventeenth-century mathematics seem to be missing from these early materials – these annotations do not lead anywhere immediately. They simply prove the acquisition of a necessary competence. With the notes on John Wallis's *Operum Mathematicorum Pars Altera* (Oxford, 1656) the case is different. Newton was brought to the inception of his great mathematical discoveries, to be treated in the next chapter.

Obviously as an undergraduate Newton had contact with at least one individual interested in mathematics, since he borrowed Schooten's *Geometria* (in its second edition of 1659–61) and also "Wallis's Works". These were far from being ordinary or cheap books. The individual might have been someone unknown to history, like William Sherlock of Peterhouse, from whom (or some owner after him) Newton acquired his own copy of the 1649 edition of the same book. And it is very likely that other books read by Newton were borrowed, for example Boyle's *History of Colours* (1664) of which he never owned a copy.[46]

Newton, Barrow and Descartes

We have no evidence from these early Cambridge years of any acquaintance between Isaac Newton and Isaac Barrow, appointed to the Lucasian chair in 1663 and a senior Fellow of Newton's own college, and it is the merest presumption that he *might* have lent Newton mathematical and scientific books (he owned a large library) and *might* have encouraged Newton's enthusiasm for reading them. There is slightly dubious evidence that, a little before Newton's election as a scholar, he began to attend Barrow's mathematical lectures, which were of a general, introductory character.[47] Conduitt's account of Barrow's interviewing the sizar before the scholarship election cannot be right, but if Barrow had happened to notice an intent young man from his own college (identifiable by his gown) among his undoubtedly small audience, might he not have asked Pulleyn to send Newton to see him? And if, as the anecdote has it, Barrow then found Newton only slightly acquainted with Euclid, perhaps to the extent of being unaware of Barrow's own 1655 edition of that text, might not Barrow have expatiated to him on the importance of the logical and historical foundations of mathematics, so inspiring Newton to the careful study of Barrow's Euclid that we know he made?[48] Barrow was no fool, nor was he idle; he had been reckoned in his day (with John Ray) one of the most brilliant undergraduates to come to the university.[49] He was also a major figure in the modernization of Cambridge philosophy, though, like Henry More, his attitude to Descartes was far from being one of complete approbation.[50] Since Barrow was unquestionably Newton's patron in the late 1660s, might he not have assumed this role as early as 1664? It was not utterly unusual for young men to be thus taken up by seniors in their colleges.[51] Westfall plausibly suggests that Newton's college career went so smoothly because he had a patron among the Fellows; it is as likely to have been Barrow as Humphrey Babington, who rarely resided there.[52] I am far from reviving the notion that Barrow taught Newton mathematics, but I do suggest that he might have helped Newton to learn them for himself, and that he had a permanent influence upon Newton's ideas. Pursuing this hypothesis, one may wonder whether it was not Barrow, rather than Pulleyn, his tutor, who advised Newton to study "Kepler's Optics". Barrow must have had the course of lectures on geometrical optics that he was to deliver in 1668 in his mind some time before. He was to bring Newton into this work as his assistant in publication. What more natural than that he should have suggested Newton's looking into this branch of applied mathematics? For Barrow, Kepler's more classical treatment of the subject may have had more appeal than that of Descartes in *La Dioptrique*, whether or not he was aware that Newton had already gone through that book.

Modern scholars have discovered from Newton's notebooks and his published writings the potent influence upon him of Descartes, whose initial impact was made in the first half of 1664.[53] This influence was by no means restricted to mathematics, where it was great; powerful Cartesian ideas about Nature were to retain their hold on Newton almost to the time of his writing the *Principia*. Such an interest in Descartes was not unusual in Cambridge then.[54] Newton acquired, and carefully studied, with many markings of his attention, a copy of the Amsterdam, 1656, Latin edition of Descartes's *Opera philosophica* containing the *Meditations, Discourse on Method, Meteors* and *Dioptrics, Principia philosophiae* and *Passions of the Soul* (but not the *Geometry*).[55] Some of Descartes's influence upon Newton was, it must be said, negative, for Newton from the first reacted strongly against many of Descartes's views, especially in epistemology, metaphysics and the theory of matter.

A Philosophical Notebook

This becomes evident from a group of notes, sketches and essays put together under the heading *Quaestiones quaedam philosophiae* ('Some Problems in Philosophy') in yet another notebook.[56] The material is arranged under forty-five headings or catchwords; in some cases nothing was ever written under the heading (as with "Of Odours & Vapours"), under others Newton wrote a small essay. The headings are general ideas like matter, place, time, motion, or they deal with the cosmos, as "Of the Sunn Starrs & Planetts & Comets". Newton observed comets in December 1664 and in January 1665 (without realizing that these were the same object) and also in early April 1665. We may guess that his use of the notebook extended over the greater part of these two years. Under "Of Water & Salt" Newton wrote about the moon's influence upon the tides; it is curious that this author of the first quantified theory of the tides probably never saw the sea! Under "Of minerals" he noted the composition of a complex arsenical bronze suitable for casting mirrors. Under "Attraction Magneticall" he sketched ideas for perpetual motion, and "Of Gravity & Levity" suggested others:[57]

Try whither the weight of a body may be altered by heate or cold, by dilatation or condensation, beating, powdering, transferring to severall places or several heights, or placing a hot or heavy body over it or under it or by magnetisme, whither lead or its dust spread abroade, whither a plate flat ways or edg ways is heaviest. Whither the rays of gravity may be stopped by reflecting them if so a perpetuall motion may be made one of these two ways.

Two sketches indicate how a 'current' of gravity might be harnessed like a flow of water; evidently at this point Newton could contemplate a purely mechanical gravitation without embarrassment. Over the whole compilation he inscribed the Latin tag *Amicus Plato amicus Aristotelis sed magis amica veritas*[58] (Plato is a friend, Aristotle is a friend, but truth is a greater friend).

Some of the material sprang from Newton's own mind, notably his various remarks of criticism or approbation of the authors he read, or his suggestions for exploring their treatment further. Not infrequently he suggests an experimental trial:[59]

> To try whither the Moon pressing the Atmosphere causes the flux and reflux of the sea [by means of a barometer]
> Whither magneticall rays will blow [out] a candle . . .
> To try whither the weight of a body may be altered by heate or cold . . .
> To try if two prisms the one casting blew upon the other's red doe not produce a white.

In the case of optics Newton actually went on to try the experiments. For the most part, however, the *Quaestiones* are heavily dependent upon a wide range of books read: those by Robert Boyle, Joseph Glanvill, Kenelm Digby, Henry More, Galileo and (surprisingly) Thomas Hobbes, as well as Walter Charleton and Descartes.[60] From these books he must have become acquainted with two major aspects of seventeenth-century natural philosophy. He would have encountered the experimentalism associated with the very new Royal Society of London (and here he would have learned more from Robert Hooke's *Micrographia* and the early numbers of the *Philosophical Transactions*, both also annotated by Newton). Further, he would have grasped the import of the 'mechanical philosophy of Nature' that was rising into ascendancy over the whole of Europe. Boyle, Charleton, Galileo, Glanvill, Hobbes, Hooke, even Henry More were all in their varying styles mechanical philosophers, all critical of the qualitative, irreducible philosophy of Aristotle.

The Mechanical Philosophy

The mechanical philosophy of the mid-seventeenth century had grown from the seeds of ancient atomism sown by Epicurus (*c*.300 BC). Aristotle had been a harsh, persistent and effective critic of the atomists, whom he had (long after his own death) succeeded in burying for a thousand years until Lucretius's poem *De rerum natura* ('On the nature of things') was rediscovered in 1417.[61] The antitheism and especially the anticreationism of the atomic doctrine (which were its original *raison d'être*) had of course to be suppressed in Christian Europe; there remained an explanation of Nature in material or structural terms rather than in teleological or organismic terms. Atomism made it possible to think of all changes in Nature (whether cyclical or not, even such qualitative changes as the insipid being made sweet or sour) as reducible to the movements of bodies or the invisible particles composing visible bodies. Just as alternations of day and night or the displacements of the seasons are attributed to the movements of the heavenly bodies, so the mechanical philosopher looked for possible ways in which the motions of invisible particles might create other effects. Sound was readily traced to a vibration of the particles of bodies; light might be the effect of particles radiating outwards from the flame; changes of a chemical type could be accounted for as alterations in the patterns of arrangement of particles of different kinds or sizes.

In Newton's undergraduate days two schools of mechanical philosophers, both of French origin, were in contention. A Christianized Epicurean atomism was powerfully advocated by Pierre Gassendi, whose vast formulations had recently been published in Lyons. He had his English champion in Walter Charleton. He taught that the ultimate constituents of matter were true unbreakable atoms and that between and besides these there existed nothing but absolute vacuum. Gassendi took his exemplification – detailed explanations of effects in Nature – straight from Lucretius. René Descartes, though he agreed with the atomists that all action and change must arise from the direct impact of matter upon matter, utterly rejected atoms and the void. He regarded the notion of empty space as a logical contradiction. If there is nothing between two bodies A and B, this can only mean that they are in contact. Hence for Descartes the universe is a plenum, it contains nowhere any gap or void unfilled by matter, and in consequence matter itself must be amorphous. There cannot be atoms endowed with constancy of shape and size. Though

Descartes for convenience divided his material particles into three types, each is in principle reducible by attrition to the next below.

Finding a plenum logically essential, Descartes realized it by supposing an amorphous aether to fill all the spaces between the particles of ordinary matter. Each star, of which the sun is one, was placed at the centre of a swirling whirlpool (or vortex) of this aether. Because it is everywhere compressed by similar vortices, Descartes held that the aether of each must press inwards and so cause gravity. It could also transmit a vibration, which is light, but he also wrote as though light were a transmission of aetherial particles. Proceeding in this way Descartes imagined a tightly interlocking series of mechanisms to account for the major part of the phenomena of Nature known to him, including the tides, boiling and freezing, the rainbow, the formation of metals and minerals and magnetism (caused by peculiar screw-like particles able to act upon the threaded pores of magnetic bodies). In his separate book, *On Man*, he explained the chief phenomena of physiology in an analogous manner: the pulsation of the heart, the digestion of food, sensation and the motor action of nerves. Though Descartes set out to explain in great detail how his mechanisms (or models of reality) were capable of reproducing all our real experience of things, they were all (like his theory of matter which was the root of the whole system) wholly speculative. Indeed, very little in the physical world reconstructed theoretically by Descartes could be put to any experimental test, since the active agents were too small to be experienced separately. Descartes recognized the importance of systematic enquiry into how things actually are constituted, of which experimentation is a major branch but (as a deductive philosopher) he preferred to follow reason, experimenting little, if at all, in the physical sciences. He did carry out some anatomical dissections, especially of the eye, and Newton followed his example.

Young Newton paid more attention to the general ideas upon which Descartes had founded his reconstruction of the universe than to the detailed mechanisms. (At this time, c.1665, it was still not generally realized even that Descartes's formal rules prescribing the partition of motion by impact between particles were incorrect. Newton later came to believe that Descartes's principle of the conservation of motion in the universe – all cause being a movement – was formally false.) When some forty years ago Newton's papers began to be examined, it came as a surprise to many to discover how much time Newton had devoted to Descartes, to his metaphysics and epistemology as well as to his science. The whole of Cartesian physics, and the essential choice between Descartes's and Gassendi's ideas, turned upon Descartes's rejection of atoms and the void, matters which Newton examined with great care in the *Quaestiones*. He was unequivocally on the side of the atomists, arguing strongly for indivisible least particles and the necessity for the void; it is impossible that motion should take place in the absence of empty spaces.[62] Newton also refused to identify space with material substance. In his atomism he was more definite than some of his authors, like Boyle and Hooke, who were more eclectic in their attitudes to the two rival schools. Newton preferred to follow Charleton and More, who were keen atomists:

Of Attomes

It remaines therefore that the first matter must be attoms And that Matter may be so small as to be indiscerpible the excellent Dr. Moore in his booke of the soules immortality hath proved beyond all controversie yet I shall use one argument to shew that it cannot be divisible in infinitum & that is this: Nothing can be divided into more parts than it can possibly be constituted of. But matter (i.e. finite) cannot be constituted of infinite parts.[63]

Whether Newton at this stage actually looked into Gassendi's great folios is uncertain; Westfall thinks that he did.[64] Later Newton, who was as much scholar as experimenter, carefully read Lucretius, Gassendi's authority.

Newton differed from Descartes on many more general issues of metaphysics and philosophy. Like Henry More – who may have been one of his mentors though there is no direct evidence of this – Newton believed that Descartes had gone too far in eliminating God from the creation and governance of the universe. Newton, like More but in opposition to Descartes and Hobbes, believed that empty space can not only be filled by material substance but also by spiritual substance; in fact, God is omnipresent "but He being a spirit and penetrating all matter, can be no obstacle to the motion of matter; no more than if nothing were in its way."[65] From this conception Newton never budged. He also rejected such Cartesian conceptions as the definitions of motion and rest: motion could not be properly understood as merely relative to other bodies, nor could it be rest (absence of motion) that causes matter to cohere into hard bodies.

Something of the set of Newton's mind emerges from the *Quaestiones*. We see him accepting the mechanical philosophy in preference to that of Aristotle yet being critical of its most effective and persuasive modern exponent, Descartes. He was conscious of the defects in the arguments and evidence offered by both of the rival schools. 'With ruthless disregard for mere verbal solutions he insisted upon probing theories to their final consequences, which he held up for examination.'[66] Newton was not prepared to follow any line of thought that seemed to banish the deity from the universe: complete mechanism could not provide the answer to all problems of Nature. With this proviso, his programme for natural science favoured the reduction of theories to factually testable propositions, though as yet he had attempted nothing of the sort himself. His own ideas were unformed and inconsistent; sometimes he was dissatisfied with what he wrote and struck it out. But, as Westfall has remarked, the *Quaestiones* reveal 'both unmistakable marks of his genius and suggestions of the path it would follow'.[67] Only one element – perhaps the most important of all – is evidently still missing: the concept that natural philosophy is *mathematical*. How this entered Newton's philosophy of nature we may see in the next chapter.

Trinity College before 1661

Before concluding this first chapter, it might be useful to add a little more about Isaac Barrow, because of the significant role assigned to him in Newton's development by most scholars (with the exception of D. T. Whiteside). Admittedly, some of this significance was based on a false view of the relationship between the professor and the undergraduate in Trinity College.

Barrow's writings contain important information about intellectual life in Cambridge (and Trinity in particular) ten years or less before Newton came into residence. He was a senior member of a group within the college which – in the same manner as better-known groups in Oxford and London – took an interest in medicine, natural history, mathematics and experimental philosophy rather than religion and affairs of state. Of this group, which was completely informal, the most scientifically distinguished members besides Barrow himself, all subsequently Fellows of the Royal Society, were the naturalists John Ray and Walter Needham, but a Fellow of Trinity, John Nidd, has often been taken as its central figure.[68] The experimenter and microscopist Henry Power, though of Christ's College, was linked to this group, writing to a Trinity friend in 1654: 'What

Hapynesse I enjoyed at Cambridge I could never properly tell till now that I am removed from it. I can now well understand the greatnesse of the losse I sustaine in the disenjoyment of that worthy Societie there, which used to entertaine mee with such excellent & noble discourses, Physicall, mathematicall & Anatomicall.' Needham was a close friend of John Ray, as was also the mathematician Francis Jessop, who like Needham came to Trinity in 1654 and was a pupil of the same tutor. Jessop spent all his later life in Sheffield and was never FRS, but he had much contact with the Royal Society. He possessed real talent, receiving the praise of Barrow in his *Geometrical Lectures*.[69]

Evidently there was a vigorous intellectual life in Trinity College during Barrow's first period there under the Commonwealth, and the college was by no means unique in Cambridge in this respect. But of this group none save Barrow himself and John Ray (who left Cambridge in 1662) survived into Newton's time from 1661 onwards. Nidd was dead, others like Power and Needham had long pursued professional careers elsewhere. The Restoration of the monarchy which brought Barrow back to Cambridge drove out Ray, who would not falsify his oath to the Presbyterian confession. Any connection between the maturing Newton of the post-plague years and this Cambridge activity of nearly fifteen years before must necessarily be slight or indeed non-existent.

2

"The prime of my age for invention", 1664–1667

"The same year I found the method of Tangents . . ."

The origins of Newton's principal innovations in mathematics and science are to be found in the records of his reading as an undergraduate; immediately after these studies came "the prime of my age for invention" which Newton placed in 1665 and 1666. In mathematics and optics the transition from reading to original investigation can be perceived (though not exactly dated). In mechanics, on the other hand, we find Newton's earliest discovery on the first page of his "Waste Book" (Cambridge University Library MS Add. 4004), where the *tenth* page is dated "Jan 20th 1664" (that is, 1665 in our reckoning). No known annotation from Galileo or any other likely source antedates the "Waste Book". From this and other dates in the notebooks it is evident that Newton's great epoch of creative scientific work began some months before the start of the year 1665, while he continued to mine the rich veins that he had opened as late as (probably) 1668. The two years of Newton's reminiscences stretch more accurately to three or more, of which 1665 and 1666 were central and critical.

That a large portion of this creative epoch was passed in Lincoln-shire – not by any means altogether at home in Woolsthorpe, but probably also at Grantham and certainly at Boothby Pagnell – was a matter of chance. Not for the first time, Cambridge University was closed by plague. Bubonic plague had been endemic in Britain since the great pandemic of the mid-fourteenth century; the last major outbreak had been in 1625. Since opinion viewed all gatherings of people and especially all urban crowds as swift means of diffusing the disease, those who could fled the cities where playhouses were closed, bull-baitings prohibited, preaching abandoned and schools and universities dismissed. Newton had graduated BA in January 1665 when he was already deep in mathematical discovery. On 6 May his unique surviving letter from his mother was sent to him in Cambridge,[1] but he had already left the college before it was officially closed in August, because of the plague. We may imagine that in June or July 1665 he had packed up his notebooks and his little library and ridden home. Samuel Pepys first came across the plague infection in London on 7 June and the number of deaths began to increase rapidly thereafter, but it might have been the excessively hot weather as much as the fear of infection that sent Newton away. He returned on 20 March 1666 (after the usual winter reduction of mortality) only to leave once more in June when a renewed outbreak occurred in Cambridge. This time he did not come back until April 1667.[2] Thus it seems likely that his total rustication to Lincolnshire amounted to some eighteen or nineteen months. Newton would not be so long absent from Cambridge again until he ceased to reside there thirty years later.

Newton claimed that this optical experimentation had been ob-structed by the Cambridge plague: "Amidst these thoughts [about light] I was forced from Cambridge by the Intervening Plague, and it was more than two years before I proceeded further."[3] His work in mathematics, also begun in Cambridge, was not similarly interrupted to any significant extent. The surviving annotations from Wallis's *Opera mathematica* of 1656 were made by Newton's own account in the winter of 1664–5 and from them he went straight to "the method of Infinite series".

In January 1663/4 Newton dated his extracts from Wallis's proce-dures for squaring the parabola and hyperbola by means of indivi-sibles; a little later he was annotating Wallis's treatment of series in

Arithmetica infinitorum and applying these procedures also to quadratures.[4] All the time he was thinking and inventing, for example: "Thus Wallis doth it, but it may bee done thus . . ."[5] The study of series brought him to the problem of how to interpolate terms. The point of interpolation – that is, finding the value of a new term, inserted between other terms in a series whose values were already established – was that by this method Newton knew he would be able to 'square' the general sector of the circle, for example, or the segment under an hyperbola, in terms of the two limiting ordinates. In algebra, Wallis had got so far as discovering the limits of π between two infinitely extendable series.[6] He had also established generally that the area under any curve $y = x^n$ was

$$\frac{x^{n+1}}{n+1}$$

Now the equation for the circle is $y = (1 - x^2)^{1/2}$ which can be put in Wallis's form as $y = z^{1/2}$. Wallis did not know how to deal with this fractional exponent, but Newton now supplied the answer by interpolation. Following Wallis, Newton wrote down a regular arithmetical progression thus:

$$1.\ \square\ .\ \frac{2}{3}\ .\ *\ .\ \frac{8}{15}\ .\ *\ .\ \frac{48}{105}\ .\ *\ .\ \frac{384}{945}$$

"Twixt which termes [he continued] if the intermediate termes $\square$ [and] * can bee found the 2d $\square$ will give the area of the line $y = \sqrt{aa - xx}$, the circle."[7]

In effect, working (as Wallis, Pascal and others had done before) with arrays of tabulated values of coefficients, Newton arrived at a version of the bounding series used by Wallis:

$$\frac{1}{2} \times \frac{1}{4} \times \frac{3}{6} \times \frac{5}{8} \times \frac{7}{10} \times \frac{9}{12} \times \frac{11}{14} \times \ \ldots$$

Multiplying each successive fraction into the preceding product yields another series of numbers:

$$1, \frac{1}{2}, \frac{1}{8}, \frac{1}{16}, \frac{5}{128}, \frac{7}{256}, \frac{21}{1024}, \frac{33}{2048}$$

Applying these fractions as coefficients to the series of even powers Newton produced the equivalence:

$$\frac{x}{2}\sqrt{1 - x^2} = \frac{x}{2} - \frac{x^3}{4} - \frac{x^5}{16} - \frac{x^7}{32} - \frac{5x^9}{256} - \cdots \qquad ^8$$

From this expression the quadrature of the circle can be derived, the number of terms being extendable to the limits of the calculator's patience. Newton worked out similar power series for $\sin^{-1}x$ and the area under the hyperbola, computing the latter at Boothby Pagnell to 51 decimal places.

Archimedes had long ago shown how to square the parabola, an achievement that remained unique till the seventeenth century. Much effort was wasted on futile attempts to square the circle by a geometrical construction, a problem we now know to be insoluble and so regarded in Newton's time by a growing number of mathematicians. Indeed, in 1667 James Gregory was to outline what he thought could be made into a rigorous proof that it was so. In place of geometrically exact quadratures, converging series (the term is Gregory's also) promised the hope of effecting any desired quadrature to an increasingly accurate approximation, according to the number of terms considered. (Thomas Harriot had, unknown to everyone, made great progress of the same kind at the opening of the century.) James Gregory, Newton's slightly older contemporary, was in particular to employ converging series to obtain quadratures; later still his younger contemporary, Gottfried Wilhelm Leibniz, would master this technique. But it was Newton who was first (in private) to greatly extend the usefulness of series by his methods of expansion, and especially by the binomial expansion which he discovered in Lincolnshire and first applied to the circle series given above.

About twelve years later, in June 1676, Newton stated the binomial expansion in his first letter to Leibniz, as an easy way of expanding roots;[9] in his second letter to Leibniz he relaxed enough to explain how he had come upon it:[10]

At the beginning of my mathematical studies, wherein I chanced upon the works of our [fellow-countryman] Mr Wallis, in considering the series by the interpolation of which he shows the area of the circle and the hyperbola, namely, that in a series of curves of which x is the common base or axis, and the [various] ordinates are

$$(1 - x^2)^{\frac{0}{2}}, (1 - x^2)^{\frac{1}{2}}, (1 - x^2)^{\frac{2}{2}}, (1 - x^2)^{\frac{3}{2}}, (1 - x^2)^{\frac{4}{2}}$$

[It occurred to me that] if [between] the alternate [corresponding] areas which are

$$x, \text{———}, x - \tfrac{1}{3}x^3, \text{———}, x - \tfrac{2}{3}x^3 + \tfrac{1}{5}x^5, \cdots$$

we could interpolate [the missing terms] we would have the intermediate areas of which the first, [corresponding to the term]$(1 - x^2)^{1/2}$, is the circle. In order to effect this interpolation I noticed that in all the first term was x and that the second terms $\frac{0}{3} \times^3, \frac{1}{3} \times^3, \frac{2}{3} \times^3, \frac{3}{5} \times^3$ etc. were in arithmetic progression and hence that the two first terms of the series [plural] to be intercalated ought to be

$$x - \frac{1}{2}\frac{x_3}{3}, \quad x - \frac{x_3}{3}, \quad x - \frac{5}{2}\frac{x_3}{3} \text{ etc}$$

In order to interpolate the rest I considered that the denominators 1, 3, 5, 7, were in arithmetic progression and so only the numerical coefficients of the numerators remained to be investigated. Now these in the given alternate areas were the figures of powers of the number 11, namely $(11)^0$, $(11)^1$, $(11)^2$, $(11)^3$, $(11)^4$, that is, first 1, then 1,1, thirdly 1,2,1, fourthly 1, 3, 3, 1, fifthly 1, 4, 6, 4, 1. Accordingly I inquired how in these series the remaining figures could be derived from the two first that are given, and I found that calling the second figure m, the remainder might be produced by the continued multiplication of the terms of this series

$$\frac{m - 0}{1} \times \frac{m - 1}{2} \times \frac{m - 2}{3} \times \frac{m - 3}{4} \times \ldots$$

This rule for the coefficients is of course identical with the modern rule expressing them as the series

$$m, \frac{m(m - 1)}{1.2}, \frac{m(m - 1)(m - 2)}{1.2.3} \text{ etc.}$$

Evidently, though in 1676 Newton correctly expressed to Leibniz the *content* of his process of arriving at the binomial expansion he did not, despite the reference to his notebooks which he told Leibniz he had made, represent the details of the process quite accurately; to Leibniz he made the process appear more deductive than it had been historically. He omitted, in Whiteside's words, to 'hint at the effort it [had] cost him to lay out the extended Wallisian interpolation schemes he had found necessary in the winter of 1664/5 to deduce the binomial expansion.'[11] Further, the proof of this expansion, also attributed to 1664/5 in the letter to Leibniz, is in fact first encountered in Newton's *De Analysi* ('On analysis', 1669). Like most of us, when looking back on his earlier self Newton wrote with the advantage of hindsight and set himself in the best possible light.

The important and certain truth is that at this early stage of his mathematical development Newton had forged the key to open one wide door to his inventions. This door would later admit him to the vast chambers of the integral calculus (in our language; to Newton it was the *method of flowing quantities, or fluents*). It would enable him to achieve at least the approximate quadrature of any mathematically definable curve and to integrate a very large number of algebraic expressions.

The key to the counterpart set of chambers, those of the differnetial calculus (or, in Newton's language, the *calculus of fluxions*) he found only a little later, and this method he developed more rapidly by concentrated application. "The same year [1665] I found the method of Tangents of Gregory and Slusius, & in November had the direct method of fluxions" (see p. 19 above). The problem of determining the position of the tangent to a curve at a specified point upon it was a very old one: for example, in geometrical optics it has to be solved in order to ascertain the place to which a ray of light will be reflected from the surface of a spherical mirror. Descartes's invention of analytical geometry made the problem more pressing; Newton came across it in the van Schooten edition of *La Géométrie*. In Cartesian terms, the curve being AB, D the point to which the tangent is to be drawn and CD its ordinate, the problem is to find the position of E, the meet of the tangent and the base-line. Not even the beginnings of a good method for drawing the tangent to an algebraic or simple transcendental curve were known till the late 1630s, when Descartes, Pierre de Fermat and Gilles Personne de Roberval each furnished his own solution. In its generality the problem is equivalent to finding at the point of tangency what Newton called the fluxions and Leibniz the differentials.

Passing over Newton's excursions in the wake of Descartes into analytical geometry (which were 'unfamiliar to Newton's contempo-

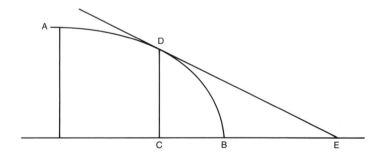

FIGURE 2.1   The angle of tangency and the subtangent CE

raries and unsuspected by later scholars[;] they have remained unknown till the present day'),[12] Newton spent months working his way to the concept and practice of the fluxion before, during and after the plague period. None of the mathematicians (Fermat, Johann Hudde, René François de Sluse) who preceded Newton in publishing improved tangent methods before Newton's was known could have influenced him, because when his method was devised in 1664–5 none of their methods was printed. Newton's starting-point was rather in a cumbersome method of Descartes's, improved by Hudde;[13] by the late autumn of 1664 he was gaining considerable mastery over the algebraic treatment of curves, recognizing (for example) that the problem of ascertaining the "crookednesse" (radius of curvature) of a line at any point involves finding the fluxion of the tangent at that point (late 1664 to May 1665).[14] On 20 May 1665 he headed a new page in the "Waste Book" that contains so much of this early mathematical work "A Method for finding theorems concerning Quaestiones de Maximis et Minimis", echoing a tract by Johann Hudde in the van Schooten edition of Descartes.[15] Returning to tangents after a few pages, Newton stated a modification of Hudde's multipliers which converted the terms of the given equation into those of an equation defining its sub-tangent (the interval CE in the last figure). The next day Newton went back to the "crookednesse of lines" with even greater success. There he introduced a cursive capital X to denote an arbitrary implicit function of x and y, and set points to either side of it to represent its first and second partial derivatives with respect to x and y.[16] By the summer of 1665 Newton was tabulating derivatives such as

$$\frac{x^6}{a + bx} = y$$

and their related "areas" (integrals). All this was done without introducing any new, formally defined mathematical concepts or anything much in the way of explanation of his processes, until at last in the late summer or early autumn of 1665 we encounter a statement in which Newton for the first time introduced the idea of a "motion" of a variable (with respect to some time uniformly elapsing, it is understood):

> If two bodys c, d describe the streight lines ac, bd, in the same time, (calling $ac = x$, $bd = y$, $p =$ motion of c, $q =$ motion of d) & if I have an equation expressing the relation of $ac = x$ & $bd = y$ whose termes are all

put equall to nothing. I multiply each terme of that equation by so many times py or $\frac{p}{x}$ as x hath dimensions in it. & also by soe many times qx or $\frac{q}{y}$ as y hath dimensions in it. The summe of these products is an equation expressing the relation of the motions of c & d. Example if $ax^3 + a^2yx - y^3x + y^4 = 0$ then $3apx^2 + a^2py - py^3 + a^2qx - 3qy^2x + 4qy^3 = 0$.[17]

By "motion" it will be evident that Newton here means speed of motion. He then writes of the converse procedure: "If an equation expressing the relation of their motions bee given, tis more difficult & sometimes Geometrically impossible, thereby to find the relation of the spaces described by these motions." At this stage Newton understood the variation of a quantity between two values as a motion or flow (hence *fluxion* and *fluent*) and perceived that the flowing of one quantity (x) could be compared with the flowing of a second quantity (y). The motions are in relation to a time which itself always flows uniformly and constantly. From this point he was able to handle not only 'geometrical' curves but such other relatively tractable curves as the spiral and the cycloid.

The underlying *raison d'être* of his method he set out in a fundamental paper (dated 13 November 1665) as follows:

If two bodys $\frac{A}{B}$ move uniformely the $\overset{\text{one}}{\text{other}}$ from $\frac{a}{b}$ to $\frac{c}{d}, \frac{e}{f}, \frac{g}{h}$, &c in the same time. Then are the lines $\frac{ac}{bd}$ & $\frac{ce}{df}$ & $\frac{eg}{fh}$ &c as their velocitys $\frac{p}{q}$. And though they move not uniformly yet are the infinitely little lines which each moment they describe as their velocitys are which they have while they describe them.[18]

Paraphrasing lightly, Newton continues: If the first body with velocity p describes the infinitely little line o ($= ce$) in one moment, in that moment the second body with the velocity q will describe the line $o.\frac{q}{p}$ ($= df$) . . . So that if the lines described be x ($= ac$) and y ($= bd$) in one moment, they will be $(x + o) = ae$ and $(y + oq/p) = bf$ in the next. Then, in an equation given relating x and y we may substitute $(x + o)$ for x

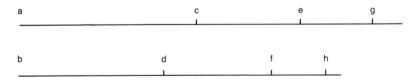

FIGURE 2.2   Increments of lines as velocities

and $(y + q/p.o)$ for y to obtain the incremented function. Putting all terms equal to zero, after appropriate manipulation in which the original equation is subtracted, "those terms in which o is [being] infinitely less than those in which o is not" they are blotted out to leave a new equation involving p and q. D. T. Whiteside points out that though Newton may *appear* to restrict himself to equations of the form $y = f(x)$ this is by no means the case. The character of the infinitesimal o is such that we have $f(x, y) = 0$, whence $f(x + o, y + q/p.o) = 0$ also. It therefore becomes evident that Newton's ratio q/p is exactly equivalent to the later Leibnizian ratio dy/dx, or to $-f_x/f_y$.

Apart from a short paper on finding tangents and curvature written in the following mid-May, Newton seems to have done little or nothing further to advance his study of calculus until almost a year later, when in October 1666 he compiled and refined the essence of all his discoveries to date in a thick tract – it makes forty-nine printed pages in Whiteside's edition – bearing the heading "To resolve Problems by Motion these following Propositions are sufficient."[19] It constitutes the first systematic treatise on calculus. Like the later *De Analysi* (1669) – the mathematical essay which Newton first allowed any other person to read (in July 1669 he put it into the hands of Isaac Barrow, who soon sent it off to John Collins in London) – the October 1666 tract was far ahead of its time. We do not know what else of Newton's Barrow may have seen at this time, if anything, but Collins at any rate before the end of 1669 had copied not only this tract but the notebook (CUL 4000) containing Newton's earliest mathematical notes and discoveries. If the importance of the steps taken by him in initiating a new branch of mathematics had been sharply perceived by these few early readers, and if (a great assumption!) Newton had acceded to their demand for a wider circulation of his work, perhaps both his own life and the development of mathematics would have taken a different course.

The essay (with numerous defects of spelling or writing) is composed in English, and is not perhaps an ideal exposition of such highly novel ideas as it contained, but it is fully intelligible (far more so than Leibniz's first published exposition of *his* calculus). Although it is more than a sketch, being elaborately worked out from its basic ideas with many examples and the necessary diagrams, we cannot presume that Newton actually prepared the tract for other eyes than his own. However that may be, the form of a communicable text is present, and there can be no question of the clarity of the expression of his basic ideas and the method of utilizing them. The obvious feature (to our eyes) missing from Newton's early calculus essays is a particular notation for the derivative. From the beginning Leibniz was to conceive of the symbolism 'dx' (originally '$d\overline{x}$'), indicating the infinitesimal difference between two adjacent values of x, to which he attached enormous importance, as being universally applicable: for, (as he was to tell Oldenburg in his 'calculus' letter of 11 June 1676) 'not only can [my new method] be applied where there are more indeterminate letters than y and x (which is often highly advantageous) but it is useful also when irrational quantities intervene, since [the method] is not at all obstructed thereby . . .' for they are always outside the irrational vinculum (*extra vinculum irrationale*).[20] Leibniz rightly judged this symbolic manipulation to facilitate the application of his method in what he now called 'differential equations'. Newton introduced into his expositions of calculus no regular notation for fluxions and fluents until the early 1690s. Though he had earlier used letters marked (in various ways) with dots, to denote homogenized derivatives or partial derivatives, this device did not pass into his first 'public' tract, *De Analysi*, 'perhaps the most celebrated of all Newton's mathematical writings', and its successors.[21] In defence of his own priority over Leibniz, Newton always denied that there was any special merit in the Leibnizian symbolism.

After explaining how to obtain the 'velocity ratio' p/q ($= dx/dy$) much as he had done already in the November 1665 paper, Newton plunged into the inverse problem of integration:

> If two Bodys A & B, by their velocitys p & q describe the lines x and y. & an Equation bee given expressing the relation twixt one of the lines x, & the ratio q/p of their motions q & p: To find the other line y.
> Could this ever bee done all problems whatever might bee resolved. But by the following rules it may very often bee done.[22]

The rules are elaborate, occupying thirteen pages: the treatment is far more detailed and specific than in anything Newton had composed before. (There are, by my account, fifty-two examples of how integrals may be obtained; in some Newton uses $\square$ as a symbol for the integral.) The simplest process is, obviously, an inversion of 'differentiation':

$$\text{As if } ax^{m/n} = q/p. \text{ Then is } \frac{na}{m+n}x^{\frac{m+n}{n}} = y.$$

Newton understood that where an easy integration like this could not be effected, even after algebraic manipulations, a result could be obtained "mechanically" by means of tables of logarithms, but he did not show how this might be done. Some integrations could proceed by finding an expression that could be interpreted as representing an area beneath a central conic, which could again be evaluated mechanically by means of tables of sines and tangents. Newton did not, however, fully execute his plan for the tract, leaving blanks for further discussion.[23] This would come later in *Methodus fluxionum et serierum* (1671).

After this detailed treatment of integration, Newton continued with applications of his new calculus to the determination of the tangents of curves, of their points of inflection, of their maximum and minimum curvatures, the quadrature of curves and their rectification. The text breaks off abruptly without providing an example of the final rectification process, though a blank space remains on the page.

The tract shows that by October 1666 it was no longer Newton's success in tackling particular problems in mathematics which most mattered, but the fact that he had discovered a well-spring of innovation, a new concept of the handling of changing quantities that could resolve so many of the problems puzzling the last generation – problems of tangents, quadratures, rectifications – treating them as particular cases to be dealt with straightforwardly by the application of the new concept. From the autumn of 1666 Newton had in his possession a document demonstrating that this new concept and method of his could be set out in such a way as to be understood and utilized by others. But it was a document that he would let few see, even in the last years of his life.[24] By then all that it contained had long been overtaken. It is true that a modern mathematical logician might find its arguments less than fully rigorous, but Newton made no assumptions not generally accepted in his time. We cannot tell whether it was desire for enhanced perfection, a natural secretiveness, a reluctance to enter into disputes or fears of attacks upon his own originality that caused Newton to admonish his friends: "Pray let none of my mathematical papers be printed without my special licence."[25] The injunction was to cause him more trouble and distress than he could ever have imagined, quite apart from the fact that it delayed the widespread introduction of calculus into mathematics for almost a generation.

"The next year in January [1666] had the Theory of Colours"

The origin of Newton's experimental study of light is well enough documented, but the course of events thereafter can hardly be established with certainty. Newton's own autobiographical remarks relating to optics omit points of great interest, and the anecdotes recorded by others are unreliable. John Conduitt wrote down what he gathered from Newton during the last months of his life (in August 1726):

> In August 1665 Sir Isaac who was then not 24 bought at Sturbridge Fair a prism to try some experiments upon Descartes book of colours & when he came home he made a hole in his shutter & darkened the room & put his prism between that & the wall found instead of a circle the light made [a spectrum] with strait sides & circular ends &c which convinced him immediately that Descartes was wrong & he then found out his own Hypothesis of colours tho he could not demonstrate it for want of another prism for which he staid till next Sturbridge fair & then proved what he had before found out.[26]

Newton may well have mistaken recollected dates so long in the past. There was no Stourbridge Fair outside Cambridge in either 1665 or 1666 because of the plague, or it may be that he acquired his first prism elsewhere. Descartes did not write a 'book of colours' and it is certain that Newton's experiments follow from Boyle's book of this title. Newton did not begin his experiments by boring a hole in his window-shutter, either at home or in Cambridge. Finally, there is a fundamental historical improbability in the idea of even Newton's developing a complete theory of light and colours, within a few weeks, on the basis of an experiment or two made with prisms.

It seems likely that the story of Newtonian optics began in Cambridge, when Newton was reading a great many books on mathematics, natural philosophy and experimental science. One of the sources first drawn upon by him, from which he took notes in his *Quaestiones quaedam philosophiae*, was Walter Charleton's *Physiologia Epicuro-Gassendo-Charltoniana* (1654). Though Charleton's Lucretian account of light and colour is of no great merit, Newton drew some material from it, for example: "Colours arise either from Shaddows intermixed with light or stronger & weaker reflections or parts of the body mixed with & carried away by light." Possibly from the same source Newton indicated further awareness of the possibility of treating light mechanically: "That darke colours seeme further of[f] than light ones may be from hence that the beames lose little of their force in reflecting from a white body because they are powerfully assisted thereby but a darke body be reason of the looseness of its parts give[s] some admission to the light & reflects it but weakly."[27] Far more significant was Newton's study of Robert Boyle's *Experiments & Considerations touching Colours*, first issued at the end of 1663 though dated 1664. It is quite likely that Newton read this book fairly soon after its publication; at any rate, material derived from it appeared under the heading "Of Colours" in the *Quaestiones quaedam philosophiae*. This is a major part of that compilation. Further, in another notebook now in the Cambridge University Library (MS Add. 3975) Newton made other notes on Boyle's book, being especially interested in the fact that a substance may transmit light of one colour but reflect light of a different colour (a fact presumably known to the Romans, since it is a striking property of the Lycurgus Cup, now in the British Museum). These notes are immediately followed by his own prismatic experiments.[28]

But to revert to the *Quaestiones*, which were probably written earlier. Besides having knowledge of Charleton and Boyle, it is evident that Newton had also read Descartes's ideas about light and colour, without forming any inclination to adopt them. Under "Of Light" he wrote (*contra* Descartes):

> Light cannot be by pressure &c for wee should see in the night as well or better than in the day[;] we should see a bright light above us because we are pressed downards [by the aether] . . . ther could be no refraction since the same matter cannot press two ways . . . A man goeing or running would see in the night . . . Also the Vortex is Ellipticall therefore light cannot come from the sunn directly.[29]

(Note the other implication that Newton had learned of the ellipticity of the planetary orbits, a point not taken into account by Descartes.) Experiment with the prism, so vital to Newton's later achievement, first appears under "Of Colours":

> No colours will arise out of the mixture of pure black and white . . .
> Try if two prisms the one casting blew upon the other's red doe not produce a white.

In curious conformity with Conduitt's anecdote, Newton, having imagined an experiment requiring *two* prisms, went on to perform (for we must believe that what follows next describes real experiments) a set of experiments requiring only *one*, thereby rendering this glass toy one of the most potent of all scientific instruments. Reversing the path of light in which coloured patches formed by a prism are seen by the eye, Newton painted coloured patches which he examined by placing a prism before his eye (see figure 2.3). Since he could readily vary the colours of the two patches juxtaposed, he could with only a single prism mimic the effect of mixing light from a pair of prisms. He carefully tabulated the enigmatic results, *eocd* and *cdpq* being areas of coloured fringe seen either side of the dividing line between the patches:

If *abdc* be white & *cdsr* black then *eodc* is read
If *abdc* be black & *cdsr* white then *eodc* is blew . . .

and so on. These two observations we, as Newton did later, can easily interpret; but we can only guess why he did not actually record a result for patches of blue and red, as originally conceived. Newton then recorded a distinctly Cartesian reflection: "The more uniformly the globuli move the optick nerves the more bodys seeme to be coloured red yellow blew greene &c but the more variously they move them the more bodys appear white black or Greys."[30] The globuli are Descartes's light particles; radiating from sun or flame they impinge on the nerves in the eyes. Colour, according to him, is our sensation arising from rotation (spin) in the individual globuli. Newton was saying that if all the particles striking the nerves possess

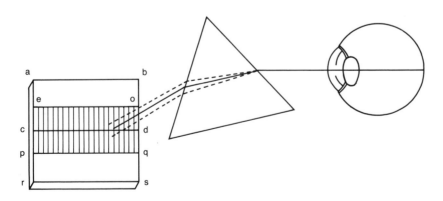

FIGURE 2.3   Colours examined through a prism; Newton's sketch in MS
3996 redrawn

a single strong spin we should perceive a single clear colour; if the globuli are endowed with different spins then the colours perceived will be weak, confused, or reduced to white (the allusion to *black* seems to be a conceptual aberration). Note that white is thought to result from a heterogeneity of physical effect, colour from a uniformity of the physical effect. Soon Newton was to make this idea clearer.

Meanwhile he went on with a series of numbered paragraphs, examining the physical properties to be associated with colour:

> 1. Note that slowly moved rays are refracted more then swift ones. 2ndly. If *abcd* be shaddow and *cdsr* white then the slowly moved rays coming from *cdqp* will be refracted as if they had come from *eodc* soe that the slowly moved rays being separated from the swift ones by refraction there arise 2 kinds of colours viz. from the slow ones blew, sky colours, & purple[;] from the swift ones red yellow[31] & from them which are neither moved very swift nor slow greene but from the slow & swiftly moved rays mingled ariseth white grey and black.[32]

This is a different theory from Descartes's. Now Newton is saying that it is speed of motion that distinguishes the reddish group of colours (swift) from the bluish group of colours (slow); evidently also the former resist refraction better than the latter do. White is again an aggregate of colours. The distinction as to speed of motion was a natural conjecture, but the distinction as to refraction was a matter of fact, as Newton goes on to record:

> 3rdly That the rays which make blew are refracted more than the rays which make red appears from this experiment. If one halfe of the thred abc be blew & the other red & a shade or black body be put behind it then looking on the thred through a prism one halfe of the thred shall appear higher than the other, not both in one direct line, by reason of unequall refractions in the differing colours.[33]

Exactly the same experiment is the first recorded by Newton in MS 3975, the so-called 'Chemical Notebook', written immediately after his annotations from Boyle's book on colours (p. 42 above) and finally reaching the dignity of print in *Opticks* (1704). It is the first optical experiment there described by Newton, in two slightly different versions.[34] And always the lesson drawn from it is the same: "blew rays suffer a greater refraction than red ones." It is easy to see that this vital experiment could be suggested by casual handling of a prism, for example, looking through it at a woven fabric containing parallel blue and red threads.

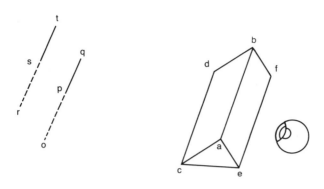

FIGURE 2.4 Dispersion observed with coloured threads

Whether these first simple experiments and the various notes in which they were embodied belong to Newton's last period in Cambridge before the plague, or to his Lincolnshire sojourn afterwards, is not a major issue. What is important is the message from these manuscripts that Newton struck out independently at a very early stage, retaining from Descartes little but the notion of the *globulus* or light particle, while redefining its attributes. Descartes had supposed – as most of Newton's contemporaries, whatever optical mechanism they favoured, continued to suppose – that a coloured beam of light is a qualitative modification of a white beam (for example, by its particles having been put into a spin by the process of refraction). Newton, ever an atomist, took the different view that the light particle once created (or rather, put into motion) does not change its physical properties any more than other atoms do. One of these causes our sensation of colour. The particular property perceived as colour is also in some way critical when a beam characterized by it is refracted, because blue refracts more easily than red. The property is also such that particles of different kinds (that is, causing different sensations of colour) can mingle to make new colours: thus blue and yellow make green, red and yellow orange. Newton understood already that refraction through a prism could make light that appeared to be of one colour resolve into two or more different colours.

Although the rules for the composition of white light were not as yet stated, Newton clearly recognized also that the sense of whiteness is one deriving from a mixture of all the colours (though this proved not to be exactly demonstrable with pigments). In other words, the atomicity of the particles guarantees the homogeneous nature of pure coloured light (the manifestation of one unadulterated form of the physical property), while white must necessarily be complex, hetero-genous, the totality of the full range of forms of the physical property. Newton would later draw the analogy between the pure colours of light and the pure musical tones of sound.

It is very remarkable that these fundamental ideas, so tersely expressed in the notebooks, remained with Newton throughout his life, and that the justification and demonstration of them was to be the chief business of all his optical publications. When he came to print, however, Newton was until late in his life hesitant to reveal his basic conception, that light is a rapidly moving stream of atomic particles – obviously smaller by far than the least atoms of ordinary matter, even of 'airs' [gases] – the basic attributes of each light atom (including that which we register as colour) being eternally fixed. He rightly regarded such a concept of light as debatable, and therefore, forbearing to affirm it, endeavoured to put forward views on the nature of light and colour that could be defended on experimental evidence alone. This move was not altogether a prudent one, but the epistemological reasons for it are obvious: Newton wished it to be understood that his vision of the natural world was independent of conjectures or assumptions about matters that could be neither proved nor disproved.

In Lucretian atomism, and to a lesser extent in Descartes's more sophisticated mechanical philosophy of nature, atoms were distin-guished by their shape, as being round or pointed, smooth or rough. Newton probably from the first dismissed such imaginings as ridicu-lous prevarications. There remained only two ways in which the atom (or group of atoms) might be distinguished from others: by its size (mass) or by its speed of motion. As we shall see later, there is a third way in which fundamental particles may be distinguished one from another but Newton in 1665–6 was not yet taking it into his considera-tion. Already he knew from Descartes, Galileo or other possible sources that the force of a moving body (such as a projectile) in impact is proportional both to its size and its speed. Newton believed – not uniquely – that the same would be true of fundamental particles, including light. Since he found a 'red' particle of light less easily displaced from its path by refraction than was a 'blue' particle, he judged that the 'red' was either swifter or more massive than the 'blue'. Either would give the 'red' greater momentum. He had no way of choosing between the possibilities and (again throughout his life) examined now one possibility, now the other.

Here, at the start, in the *Quaestiones*: "4.   Hence redness, yellowness &c are made in bodys by stoping the slowly moved rays without much hindering of the motion of the swifter rays, and blew, greene & purple by diminishing the motion of the swifter rays & not of the slower." (This was an early thought of Newton's about the reflected colour of bodies being caused by differential absorption.) Evidently he did not commit himself to such a choice of velocity as the critical attribute of the light particle, for only a few paragraphs later he wrote down the alternative: "8.   Though 2 rays be equally swift yet if one ray be lesse than the other that ray shall have so much lesse effect on the sensorium as it has lesse motion than the others &c." Newton goes on with an 'in principle' calculation of how the different sizes of light *globuli* might interact with the different sizes of matter particles upon the surfaces of bodies to cause the former, when reflected by the surfaces to the eye and brain, to give the impressions of different colours.[35] He could get no further with this topic. The notes continue with the physiological impressions of colour caused by pressing on the eyeball and other matters. Evidently Newton was now drawing upon Boyle's *Experiments and Considerations touching Colours* and has come to ground also covered in the other manuscript, Add. 3975. It is likely, therefore, that the *Quaestiones* material was begun before Newton had read Boyle's book, whereas the material in Add. 3975 was certainly written afterwards. One might even speculate that the *Quaestiones* notebook was not accessible to Newton at the time when he made the second set of notes.

So far there has been no word of an experiment in which a beam of white light is shone *through* a prism, forming a spectrum. This first appears in Add. 3975. As already mentioned, on folio 2 of this notebook Newton wrote five numbered paragraphs of extracts from Boyle; the sixth paragraph is the experiment of looking through the prism at the red-blue line; then follows:

> 7.   Taking a Prisme, (whose angle *fbd* was about 60^{gr}) into a darke roome into which the sun shone only at a little round hole *k* And laying it close to the hole *k* in such manner that the rays, being equally refracted at (*n* & *h*) their going in & out of it cast colours *rstv* on the opposite wall. The colours should have beene in a round circle were all the rays alike refracted, but their forme was oblong terminated at theire sides *r* & *s* with streight lines.

Having established that while the breadth of the spectrum corresponded to the sun's angular diameter of about 31 minutes of arc, its

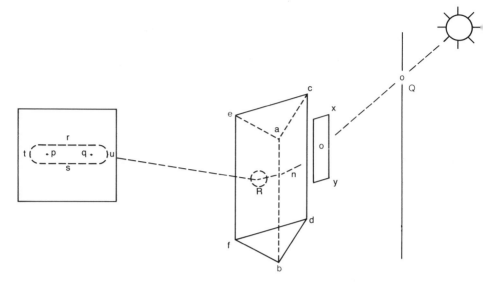

FIGURE 2.5   A spectrum is formed by a prism

length corresponded to a divergence between the extreme rays of about 1° 40', Newton made a second experiment to measure the difference between the refractions of the blue and the red rays more accurately. He made a diaphragm by boring a small hole through a board, placing it between the shutter and the prism so that the beam entering the prism was almost parallel-sided. The spectrum now formed was narrow but relatively long. Newton judged that on average the red and blue rays were inclined to each other at an angle of about half a degree, but the extremes of each at about a whole degree.[36]

Newton never invoked this experiment in print but the use of a diaphragm reappears in his first optical letter to the Royal Society (February 1672), in the so-called 'crucial experiment', a more sophisticated arrangement for achieving the same purpose. The notes support Newton's contention in that letter that he was greatly surprised to find the spectrum markedly oblong, rather than round; but the letter concealed the fact that the reason for its shape was at once evident to him from his earlier discovery of the unequal refraction of red and blue rays.

Here we should break off, since there are good reasons for supposing that these last two experiments, as well as the 'crucial experiment', were made during the second phase of Newton's optical work and therefore probably when he was back in Cambridge. After his rustication in Lincolnshire caused by the plague he had two long periods of work at Trinity College, from April to December 1667 and from February to August 1668. During either or both he could have made experiments with the prism. In the former of these years Newton bought (as his accounts show) tools, "glass bubbles" and a magnet; in the latter three prisms and putty for polishing optical surfaces. We have no positive evidence that Newton bought a second prism (required for his later experiments) before 1668, and he is very unlikely to have found one in Lincolnshire. The style of the later experiments, such as those just quoted, with their precise measurements, does not resemble that of the looser, earlier notes. It was in 1668 too that Newton constructed his first reflecting telescope as a further test of his ideas. When he returned to Cambridge in April 1667 he was certainly well embarked on the road to his new theory of light and colours, but he by no means as yet possessed it complete and fully formed in his mind. More experiments and fresh ideas were required before the theory was fully articulated and the evidence for it cogently assembled.

Before we leave optics for a while two matters of chronology relating to it should be considered. The first is the matter of Newton's "glass-works". In the great letter on his optical discoveries addressed to the Royal Society Newton wrote (in February 1672) that "in the beginning of the Year 1666 (at which time I applyed my self to the grinding of Optick glasses of other figures than *Spherical*,) I procured me a Triangular glasse-Prisme" and later wrote that after he had discovered the varying refraction of red and blue light "I left off my aforesaid Glass-works."[37] Now the glass-works are recorded plainly and at some length in the notebook (Cambridge University Library MS Add. 4000) which we saw earlier contained mathematical notes and dates extending from 1663/4 to 1665. It was certainly with Newton in Lincolnshire. The origin of the glass-works was in Descartes's *La Dioptrique*. His geometrical optics proved (as is obvious enough) that a spherical surface cannot refract incident light to a point. The focus of a spherical lens is a small circle, not a dot. Accordingly, no image formed by such a lens can be perfectly sharp; it is confused by 'spherical aberration'. Descartes appreciated that if a lens could be ground to the curvature of an hyperbola a perfect, aberration-free image would result. He proposed machines for grinding lenses to such a curvature, and Newton followed his example. His designs seem a little more realistic than Descartes's, but it is doubtful whether either man ever attempted to make an aspherical lens in practice. Newton may well have prepared these paper projects in Lincolnshire during 1665–6. There is no implausibility in his having done so after his own first, simple, optical experiments and the discovery of the red-blue differential in refraction. For this discovery by itself could not assure him of the greater relative importance of chromatic aberration, as compared with spherical aberration. The latter is a real defect of lenses and correction of it is significant. Only after Newton had satisfied himself that colour defects spoilt the image more than the spherical defects do was it reasonable for him to dismiss his Cartesian glass-works as futile. He could not reach this point before making the later series of experiments, therefore not before 1667 at the earliest. Once he had reached it the design of the reflecting telescope lay straight before him.

The other dubious matter of chronology is the date of Newton's reading of Robert Hooke's book dealing (in part) with light and colour, *Micrographia*, published late in 1664 (though dated 1665). The importance of this book is that it probably introduced Newton to a range of wholly new phenomena of colours and certainly set before him a series of theoretical ideas very different from his own. The new colours were those formed when light shines through, or is reflected from, a thin transparent plate, commonly seen, unfortunately, when oil is spilled upon water. Once this effect was recognized it became obvious why such glistening, shimmering colours as those of feathers are seen. The type phenomenon, because of his thorough study of it, became known as 'Newton's rings', unfairly to Hooke, who had first seen such repeated rings of colour in thin fragments of mica.

We can read Newton's notes upon *Micrographia*, making fourteen printed pages;[38] he studied the book with care. Since its principal subject was microscopy, Newton had no reason to doubt Hooke's reports. Though he sometimes entered minor criticisms, he seems to have regarded Hooke's work – not limited to optics and microscopy – as essentially sound, except as regards light. He copied out Hooke's table of experimental results demonstrating the truth of Boyle's Law, though he could have found analogous measurements in Boyle's own second edition of *Physico-Mechanical Experiments* (1662).[39] This may explain why, much later, he referred to this relation ($pv = k$) as "proved by the experiments of Hooke and others".[40]

Newton took three pages of notes on Hooke's account of insects' compound eyes and other microbiological observations (pages 100–217 of *Micrographia*, roughly) but omitting the last few pages of the book, in which Hooke suggested that the Moon, like the Earth, possesses a 'principle of gravitation'. Most of Newton's annotations were from the first hundred pages, in which Hooke considers a number of physical questions, including capillarity, in terms of the mechanical philosophy. Newton was interested in Hooke's hypothesis that the sociableness of things (as when mercury amalgamates with metals) or their unsociableness (oil and water) is the cause of many natural effects. Newton thus recorded Hooke's typically mechanical explanation:

[p.] 15. The reason why some bodys doe easily mix together others not some are congregated other segregated by motion, is the agreement or disagreement in their motions (caused by theire various bulkes, densitys or figures) to comply one with another or beate one another of[f], like concords or discords in musick.[41]

We shall see later how Newton utterly transformed this idea. Hooke's experience of the colours in mica is noted thus:

[p.] 48. Thin flake[s] of Muscovy glasse, aire, water, metalline scumme doe exhibit divers colour according to their thiknesse, if the midst be thinest there will bee a broad spot of one colour & coloured rings about it outward in this order. (white perhaps in the midst) blew, purple, scarlet, yellow, greene, blew, purple, scarlet, yellow, greene, blew, &c untill somtimes 8 or 9 such circuits. & the outmost limb of the flaws of Muscovy Glasse appears white because of its thinness.[42]

Many other observations of colours by Hooke were recorded by Newton as he had recorded those of Boyle before. One of the more interesting is the following, because later Newton was to make the same point himself, but his explanation of why bodies appear coloured was to be quite different from Hooke's:

[p.] 68. Noe colour can bee made without some refraction (& indeed the least visible particles of all bodys if strictly examined by a microscope are transparent, & apt for refraction some perhaps being flawed by grinding appeare opace) & were it not for such flaws & pores to let in the reflecting aire [or rather denser aether] noe body would bee opace [opaque] for wee see there may bee made of any metall christalls or vitriolls, which are transparent & variously coloured (as vitriol of Gold is yellow of silver blew &c) which variety of colours must proceede from the metalline and not intermixed saline particles.[43]

However, Newton rejected out of hand Hooke's own favourite hypothesis of light, to which he was to cling tenaciously. Against

Descartes and Newton, Hooke believed light to be a vibration or stream of pulses in an omnipresent aether, penetrating all bodies. White light, directly radiating from its hot source, consisted of a perfectly regular, uniform sequence of pulses (like a pure musical tone). When the sequence was disturbed, in a variety of ways described by Hooke, the non-uniform pulses created the appearance of colours. First, in his notes, Newton defended Descartes against Hooke's 'refutation' on page 60 of *Micrographia*:

> Though Descartes may bee mistaken so is Mr Hook in confuting his 10 Sec. 38 Cap. Meteorum[.] He says well that this Phaenominon of Muscovy Glasse cannot bee explained theirby, or that the turbinated motion [spin] of the Globuli signifys nothing unlesse they did not only endeavour but also [actually] move to the eye. &c[44]

Hooke had explained the successive rings of colour he observed in the mica flake by the supposition that the regular uniformity of the sequence of pulses in the incident light was disturbed, by the mixture in the reflected light of pulses reflected from the upper surface of the flake with others reflected from the lower surface. The latter he called a weak pulse, because twice refracted at the upper surface. Having summarized this, Newton objected: "Why then may not light deflect from streight lines as well as sounds do &c? How doth the formost weake pulse keepe pace with the following stronger & can it bee then sufficiently weaker."[45] To this objection against any theory of the aetherial transmission of light as waves or pulses Newton adhered throughout his long life, constantly repeating it. In his view wave motion must always spread round obstacles or corners, as sound seems to do, and was therefore inconsistent with the sharp boundary between light and dark characteristic of light.

Newton's knowledge of *Micrographia* and his reactions to Hooke's ideas expressed in it are of some importance, not only because the book contributed to Newton's scientific education (and especially his education in optics), nor because Hooke was to be Newton's most powerful critic and rival in years to come, but because reading Hooke seems to have confirmed Newton in the rectitude of his own independent path. No earlier author seems to have been so firmly rebuffed in Newton's private notes. Newton was strengthened in his attachment to the atomist/vacuist branch of the mechanical philosophy by Hooke's strong leanings towards the aetherist/plenist branch, almost more Cartesian than Descartes. Further, *Micrographia*, a book of wide-ranging, qualitative, physical speculations, did nothing to dissuade Newton from the mathematical ideas of Descartes, Galileo and indeed Isaac Barrow. It is therefore a pity that we do not know when Newton read the book. Nothing published by Hooke before 1674 is in Newton's own library. It is tempting to guess that he borrowed and read *Micrographia* early in 1665, before the plague. Style and writing support this suggestion. If it were so, then we have further evidence of Newton's commitment to the corpuscular theory of light from a very early stage in his scientific investigations. It is evident that Newton assimilated Hooke's point that because the rings in mica repeat periodically, there must be some periodicity in light itself; but at this stage the issue did not seem to be of major importance and Newton was not as yet approaching a dualistic account of optical phenomena.

Finally, we must not forget that while pursuing physical ideas of light and colour Newton was also increasing his mastery of geometrical optics on the foundations of Descartes's *La Dioptrique*. The first relevant note is dated September 1664.[46] Newton made a considerable study of the refraction of light by curved surfaces of various kinds, and similarly of reflection – a study of some mathematical intricacy. Some of it was linked with his "Glasse-works". One scrap deals with refraction through a 'lens' of water, confined between two convex glass lenses; this is one hint that as part of his theoretical glass-works Newton reviewed the possibility of a compound, colour-free lens, a hopeful idea of which in his maturity he despaired.[47]

"And the same year [1666] I began to think of gravity extending to the orb of the Moon . . ."

The most famous of all scientific anecdotes tells how Newton (a second Adam) found the law of gravity by watching the fall of an apple. The story was already known to Voltaire (about 1726 to 1729, when he was living in exile in London) but it is most directly and circumstantially related by Newton's friend and fellow-countryman, William Stukeley:

After dinner, the weather being warm, we went into the garden [of Newton's last residence, in Kensington] and drank thea, under the shade of some appletrees, only he and myself. Amidst other discourse, he told me, he was just in the same situation, as when formerly, the notion of gravitation came into his mind. It was occasion'd by the fall of an apple, as he sat in a contemplative mood. Why should that apple always descend perpendicularly to the ground, thought he to him self. Why should it not go sideways or upwards, but constantly to the earths centre? Assuredly, the reason is, that the earth draws it. There must be a drawing power in matter: . . . If matter thus draws matter, it must be in proportion of its quantity. Therefore the apple draws the earth, as well as the earth draws the apple. [And thus] there is a power, like that we here call gravity, which extends its self thro' the universe.[48]

The orchard behind Newton's house at Woolsthorpe still exists, and by a dubious line of descent pomologists have established that the apple grown there was of a kind called 'Flower of Kent', a greenish, acid fruit. But it may all have happened in Babington's orchard at Boothby Pagnell! However, it is one thing to imagine a 'drawing power' common to apple, Moon and Earth, and another to base a mathematical analysis of the planetary motions upon such a notion. Newton was not the first to entertain it, and he later paid a little tribute to Giovanni Alfonso Borelli as a precursor, though he was always deaf to Robert Hooke's claim to have advanced (from *Micrographia*, 1665, onwards) a clear notion of a *unique* power of gravity acting in the heavens in combination with the rectilinear inertia of moving bodies already defined by Descartes. (Borelli had postulated two active forces: a centrifugal force besides gravity's inward 'drawing power'.) To become more than a vague speculation the idea of such a power, or principle of gravitation, or centripetal force (as Newton was to call it) must be fitted into a set of ideas about the motions of bodies in curves, as yet little understood, on which Newton was to make a start (of a long-deferred promise) in his "Waste Book" (January, 1665). Galileo had laid the foundations of the science of mechanics when he discovered that the instantaneous speed of a falling body increases uniformly with the time of fall ($v = gt$) and so the distance fallen from rest is always proportional to the square of the elapsed time. (The even more subtle and far-reaching investigations of Galileo's English contemporary, Thomas Harriot, into the science of mechanics were to remain for ever in the realm of the might have been, since they were wholly unknown until the present century.) Galileo's influence upon his successors was profound. From his basic principles of uniform and natural accelerated motion he had demonstrated that the curved path of a projectile tending to fall back to Earth is parabolic. He also obtained, but could not rigorously prove, interesting results about the motion of pendulums, which he was the first to use as a scientific tool and the basis of a clock. He held that the time of a complete swing depended only upon its length, not on its amplitude, and comparing pendulums of different lengths the time increased as the square root of the length. Galileo also tackled (with less success) the issue of great concern to post-Cartesian philosophers: the laws of impact. Since, in Descartes's philosophy, all activity in the universe is ultimately reducible to the impact of particle upon particle, and 'motion' (momentum) must always be conserved (or the universe would run down, all motions ceasing) it seemed that the key to progress in mathematical physics lay in understanding impact. Descartes himself had made a false start here. Successful analyses covering all cases of impact were published only in 1669.[49] Newton also (about, perhaps, the years 1666–7) took up the same investigation with considerable sophistication and mastered it, but the future of mathematical physics did not lie in that direction.

The creative line was to be the study of the forces of rotation. Newton was by no means the first to approach this problem. Galileo had touched on it in his *Dialogue on the Two Chief Systems of the World* (1632) – probably read in English by Newton in 1664 – without making any progress. Descartes clarified ideas by explaining the regular, circular orbit of a planet as the product of a balance between its *conatus a centro*, or tendency to recede outwards along the tangent (since it is swirled round by the aether of the solar vortex), and an inwards aetherial pressure arising from the compression of the aether caused by its rotation within fixed bounds. The philosopher who solved the problem of estimating the force with which a revolving body seeks to escape outwards from its centre of revolution was the Dutch mathematician and physicist Christiaan Huygens (1629–95) whose career entwined with Newton's at many points. They were both Fellows of the Royal Society, Huygens being the first foreigner elected after the grant of its charter in 1663. Both were practical men, experimenters as well as mathematicians, though Huygens was more of an observational astronomer than Newton (he discovered Titan, the first known satellite of Saturn, and unravelled the mystery of that planet's rings). Both proposed major theories of light but Huygens never concerned himself with the problem of colour. Above all, in the history of mechanics Huygens comes next in line after Galileo and Descartes, though his work was in several respects less fundamental for the future than that of Newton.

Unlike Johannes Kepler – the first astronomer to think about the physical interpretation of the celestial motions – or Galileo or Descartes, Huygens did not set his interest in the force of rotation in a cosmological context. For him it presented a problem in pure mechanics, related rather to pendulum motions than to planetary revolutions. Indeed, when Huygens eventually (1673) published by investigations in pure mechanics he did so under the title of *Horologium oscillatorium, sive de motu pendulorum* ('The pendulum clock, or the motion of pendulums'). Dynamics figures only in the appendix *De vi centrifuga* ('On centrifugal force') containing only the enunciations (without proof) of thirteen propositions taken from a tract written by Huygens in 1659, but never published.

Among the earliest of Newton's notes on mechanics, stemming directly, it seems, from Galileo and Descartes (whose contributions 'constituted the chief base upon which Newton relied in order to construct the magnificent structure of his dynamics')[50] were investigations of rotational force. These notes are scattered in various manuscripts, whose dates must sometimes be estimated: one is dated 20 January 1665, but at least the sentences "On Violent Motion" in the *Quaestiones quaedam philosophiae* (see above p. 23) may probably be put back to 1664.[51] Here Newton argued that bodies are kept in motion (after being first impelled by some external agent) by an internal force, much like the medieval philosophers' 'impetus'. Later, following Kepler and Descartes (whose notions were far from identical, however) Newton called this force 'inertia' (literally, 'inactivity'). As he phrased it in the Third Definition of the *Principia*:

> The innate force (*via insita*) of matter is a power of resistance by which any body whatsoever, in so far as it is able, continues in its state either of rest or of uniform, rectilinear motion. This [force] is always proportional to the mass of the body, and in no way differs from the inertia of mass, save in our way of conceiving it. Inertia of matter renders it difficult to disturb any body from its state of rest or of [uniform] motion. Whence also the innate force may be called by the most significant name of *force of inertia*.

Not surprisingly, the paradoxical notion of a 'force of inactivity', a force that is not a dynamic force, has occasioned great difficulty to commentators then and since, particularly when linked with the idea of the relativity of local motion (powerfully exploited by Huygens) as it must be, since in the discussion of local events, as distinct from cosmological events, for the sake of symmetry the description must be independent of the observer's subjective sense of being in motion or at rest. In other words (as Huygens rightly pointed out) the rest inertia of one observer is the motion inertia of another who (it may be unaware) is himself in motion with respect to the first observer.[52]

Knowing nothing as yet of Huygens, Newton was struggling with these subtle matters in the years 1665–6, certainly already before the Great Plague. In his notes he took over from Descartes a primary "Axiome of Motion" that "Every thing doth naturally persevere in that state in which it is unlesse it bee interrupted by some externall cause" but this statement of a 'principle of sufficient reason' is at once extended into a formulation which is correct provided we understand by "determination" rectilinear direction: "hence a body once moved will always keepe the same celerity, quantity & determination of its motion." Inevitably Newton found himself in similar and conjoined difficulties with respect to the word *force* (*vis*). The same word is used by him (among other things) to identify both the agent that initially causes a body to move (or more generally to change its state in any way) and the resistance proportional to mass offered by the body to any such change, which Newton also calls "the force of a body's motion" (momentum), as in the following passage: "The force which the body hath to preserve it selfe in its state shall bee equall to the force which [pu]t it into that state; not greater . . . nor lesse . . ."[53] None of these formulations of mechanical principles so far considered is quantitative. But at some point Newton hit upon a key idea (known earlier to Huygens, and independently also to Christopher Wren), the idea that in considering the motions within a connected system of bodies the central point to determine is the movement of its common centre of gravity. For example, the mutual impact of bodies is best treated by reference of their motions to this common centre, whose motion (if any) is unaffected by the collisions. This was to be fully brought out in the *Principia*, where indeed it proves to be of great cosmological significance; the fact that the centre of gravity of the solar system lies within the body of the Sun provides the dynamic verification of the Copernican hypothesis.

We do not know exactly how Newton came to approach his first solution of the problem of rotational force but, as it appears on the first page of his "Waste Book" (CUL MS Add. 4004) the date must have been about the turn of the years 1664–5.[54] What he achieved there in a few lines was fundamental to all his far more rigorous future work in mechanics. There is a touch of supreme genius in the combined ingenuity and simplicity of Newton's demonstration, which depends upon the fact (utilized long before by Euclid in *Elements* XII, 2 and by Archimedes in his quadrature of the circle) that the circle can be considered as the limit of an inscribed polygon, each

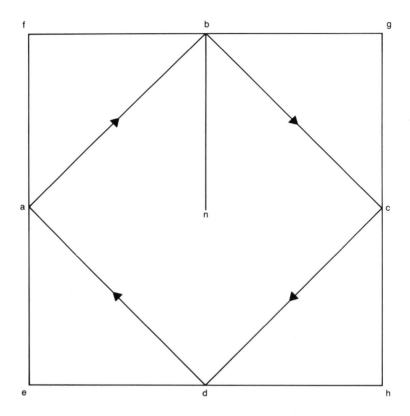

FIGURE 2.6   The force of rotation calculated

of whose sides becomes infinitely small; in this way a property true for the polygon is true for the circle also. Imagine, in ideal conditions, a ball bouncing indefinitely round the four sides of the square 'box' *efgh*, tracing the path *abcd*, a smaller square. Supposing the sides of this 'box' to be perfectly elastic the ball will bounce at each impact without loss of speed. By ingenious, simple reasoning Newton was able to calculate the force of each impact, needed to maintain the body in its square trajectory, as a ratio of the 'force of the body's motion'. The ratio is as *ab* (one side of the trajectory) to *af* (= *bn*, half one side of the 'box'). Therefore the total force of four impacts, or one circuit, will be to the 'force of the body's motion' as 4*ab* : *bn*, By analogy, if both 'box' and square trajectory were replaced by corresponding *n*-sided polygons, the ratio would become *n.ab* : *bn*, and when the polygon becomes a circle, 2π*bn* : *bn*, that is 2π : 1.

In the 'box', the inward-acting (centripetal) bounce on each of its sides forces the ball to continue indefinitely in its square orbit, and so (by analogy) with the polygon of more than four sides. When the polygon becomes a circle, there is no more bouncing on the sides, for elastic impact is replaced by a continuous centripetal pressure. Such a transition from a successive series of instantaneous events to a smooth, continuous process is a device used by Newton more than once in his thinking in mechanics, as we shall see; it is a device that enables him (in principle) to move from the unmathematical physics of impact to the geometrical physics of forces. But to return to the document: Newton next proves that the constant inwards pressure, or centripetal force, retaining the body in its circular orbit, is equivalent, in the course of one whole revolution, to 2π times the momentum of the body in its path: "the force from n in one revolution is to the force of the bodys motion as periphery to radius."[55] We may therefore express it as $2\pi mv$, where m is the body's mass and v its velocity around the orbit. To quantify the force acting at any instant, we need only divide this total, cumulative force by the time of a single revolution,

$$t = \frac{2\pi r}{v} \text{ obtaining } \frac{mv^2}{r}$$

The same expression appears in modern textbooks.

In other words, Newton now knew that the rotational force tending to disintegrate a rapidly spinning wheel increases in proportion to its size, and to the square of the number of its revolutions in unit time. It is obvious too that he understood even in this primitive calculation (as Descartes had before him) that to obtain circular motion the forces towards the centre and away from it must be equal and opposite.

A series of barely explained jottings of words and numbers upon the back of an old vellum lease granted by his mother surely also belongs to the Lincolnshire period.[56] The first of them seems to precede Newton's solution of the problem of rotational forces; others, obviously later, invoke it. The line pursued in the jottings was inspired by Galileo's raising in *The Two Chief Systems of the World* (1632) the question of the ratio between the outward force that must result from a diurnal rotation of the Earth, and the retaining pull of gravity; anti-Copernicans had surmised with a wild optimism that the former must greatly exceed the latter. The opposite view was indecisively defended by Galileo. After a number of crude attempts to compute this ratio, to discover which it is obviously necessary to know both the size of the globe and the force of gravity (or rather, the distance fallen under gravity by a heavy body during the first second) Newton decided to re-determine the latter but did not question the former (taken from Galileo as a diameter of 7000 miles, each of 3000 *braccie*). Newton knew that the force of gravity determines the period of a pendulum, but being as yet ignorant of the simple relation

$$t = \pi \sqrt{\frac{l}{g}}$$

he worked by a roundabout route, again (apparently) deriving from his own remarkable understanding of the basic symmetries of motion.[57] After several attempts Newton was satisfied with the value $g/2 = 196$ inches/sec^2, or a little over 16 feet, which is about right. Relating this to the Earth's rotation made its force about 1/300th of the force of gravity, confirming Galileo's assertion. A parallel computation, involving the unstated assumption that the Sun's distance is 5000 Earth-radii – making the Sun about eighty times further from us than the Moon – showed that the Earth's force of recession from the Sun in its orbit is about 1/7500th of the force of gravity upon Earth.

The jottings make no mention of the Moon, however, nor of any issue beyond that raised by Galileo. And everything basic in them had been anticipated by Huygens in 1659 (unknown to Newton, of course). The next document in the series, still following the same principles and linked to the vellum sheet by the same method of stating rotational force, by the same relation to Galileo, and the same parameters, goes much further and verges upon celestial mechanics.[58] Formally written out in Latin, it is obviously a more finished piece than its antecedents; in its extant form it may well have been composed some little time after Newton's return to Cambridge in 1667. In this sketch Newton uses Descartes's word *conatus* ('endeavour') for force rather than *vis*.[59] Moreover, he now calculates not the inward pressure required to keep a revolving body in a circular orbit, but the equal and opposite *conatus* with which it endeavours to escape from the orbit, that is, the acceleration required to bring the body in from the tangent to its circular parth. (This had been Huygens's method years before.) The numerical result is, of course, the same as in the previous calculation.[60] More interestingly, Newton now took the step, so obvious to us but which Huygens had avoided, of applying his new theorems to the Moon and the planetary motions, for which the necessary data were available (if inaccurately). First, Newton recalculated the ratio of the force of recession caused by the Earth's diurnal rotation to the opposite force of gravity (still using Galileo's underestimate of the Earth's diameter); in different units he rounded this out to 1 : 350. Then, comparing the Moon's revolution about the Earth with the rotation of the Earth, he found that the force of the Earth's recession at its surface from its centre is $12\frac{1}{2}$ times greater than that of the Moon from the Earth; hence the force of gravity "is 4000 times greater than the tendency of the Moon to recede from the Earth, and more". Indeed it is: the ratio is

$$\frac{1}{12\frac{1}{2} \times 350} \text{ or } \frac{1}{4325}$$

Applying a similar argument to the planets without going into details, Newton affirmed (after Kepler): "as the cubes of their distances from the Sun are reciprocally as the squares of the numbers of their revolutions in a given time, the *conatus* to recede from the Sun will be reciprocally as the squares of their distances from the Sun."[61]

This document presents many enigmas. It might seem to presume a future reader, but who could he have been? As we shall see, in the next few years Newton prepared other papers for a reader, whom commonly they never found. And why, at this promising point, did Newton put aside this line of enquiry? He never forgot what he had accomplished – he forgot nothing – and referred to it twice in later years. In 1673, thanking Christiaan Huygens for the gift of a copy of *Horologium Oscillatorium* (via Henry Oldenburg) Newton remarked upon the relevance of the theory of rotational forces to natural philosophy and astronomy, instancing the explanation of the apparent non-rotation of the Moon upon its axis and quoting from this document the ratio between the force of the Earth from the Sun and the force of the Moon from the Earth.[62] Much later, he used this letter as evidence – extended by the document on which it was based – as proof that he had been aware of the "duplicate proportion" or inverse-square law of gravitation at least twenty years before, that is, in 1666.[63] Later still he permitted his protégé David Gregory to read this document. Gregory found in it 'all the foundations of his philosophy' and, obviously on Newton's word, recorded its date as 'before 1669'.[64]

We may agree that the roots of Newton's celestial mechanics go back to his attack on rotational force in 1665. But the author of the document knew more than Newton knew in 1665. He had probably read Vincent Wing's *Astronomia Britannica* (1669) and become more expert in astronomy. Surely by this time (if not long before) he had read Robert Hooke's *Micrographia* (1665) and Borelli's *Theoricae mediceorum planetarum* (1666), both books developing the original Cartesian idea of the orbit as the resultant of a balance of forces. All these books were in Newton's library. Such evidence suggests that 1669 was the earliest year in which the document could have been written. Newton *never* wrote in Latin before his return to Cambridge in 1667; perhaps his embarking upon an academic career made him feel the need to express himself in the formal language of scholarship. His advance in this later document beyond the vellum sheet, and the authors just mentioned, is very marked. If we correct his erroneous reliance upon Galileo by giving the Earth a radius of 4000 miles in his calculation, the Earth's force of rotation becomes gravity/288, and that of the Moon in its orbit gravity/$(288 \times 12\frac{1}{2})$, that is, gravity/3600 or as the inverse square of the Moon's distance from the Earth at 60 Earth-radii.

What the document does not reveal, and today we would hardly expect it to do so, is a theory of planetary gravitation in accordance with the inverse square law. Newton, here combining an inheritance from both Descartes and Galileo, recognized a simple celestial law of centrifugal *conatus*, always proportional to the inverse square of the planet's distance from the Sun. This is a dynamic counterpart to Kepler's kinematic Third Law of planetary motion. (Unusually for this time, Newton fully appreciated the importance of this law, which he had encountered in Thomas Streete's *Astronomia Carolina* (1661) in 1661 or 1662, inscribing it in a notebook.)[65] Newton would only have been able to transform this dynamic law of *conatus* into a physical theory of universal gravitation if he had already imagined that the restraining force holding the universe together against its disruptive rotations was the same entity as that which we call 'gravity', or the cause of heaviness, here on Earth. There is no reason to suppose that he had yet got so far, and various considerations suggest that as yet (and until 1679–80, when Robert Hooke induced him to take another view) Newton believed the Cartesian vortex to be a necessary feature in the explanation of planetary motion. This becomes clearer when we consider the difficulty of accounting for the ellipse dynamically. In the circular orbit it is easy to see that the inward and outward forces are equal and opposite and unchanging at all points. In the elliptical orbit, however, Newton still believed (like his predecessors) that the planet edges towards the Sun from aphelion to perihelion because the inwards 'pressure' is greater than the outwards 'pressure', while in the other half of the orbit it edges out again because the outwards 'pressure' becomes greater after perihelion with the planet's increasing velocity.[66] He was still far from the dynamical sophistication of realizing that the inward and outward 'pressures' are the same at all points of the elliptical orbit too. Further, even after Newton had accomplished the great achievement of linking the centrifugal *conatus* of the planets by simple mathematical law, he was still far from the conceptual boldness requisite for postulating 'heaviness' as a universal and infinitely extensive property of every particle of matter. Such an hypothesis would have seemed to defy the basic principles of the mechanical hypothesis of Nature which Descartes and others seemed to have established so firmly.

Several recent writers have exploded the idea that Newton made a 'Moon test' of the theory of gravitation as early as 1665 or 1666. While his own claim to have done so is therefore falsified in a strict chronological sense, the present document confirms that he made such a precise test no later than 1670 – that is, sixteen or seventeen if not twenty years before writing the *Principia* and surely still, even when thus reduced, a remarkable antecedent. It also shows that the test failed to prove that the Moon's centrifugal acceleration is equal and opposite to the countervailing force of the Earth's gravity at that distance: instead of gravity/$(60)^2$ Newton could come no closer than gravity/$(65\frac{1}{2})^2$ at the best. The source of the error has long been obvious.[67] But Newton did not seek for it by checking his numbers and its importance has been exaggerated, for though he was later to claim that the test succeeded "pretty nearly", he well knew that even a more stringent confirmation of the Moon's involvement within the Earth's gravitational vortex could not prove that a similar vortex surrounded the Sun. Further, while establishing that the centrifugal accelerations of the planets are related by the $1/r^2$ rule,[68] the document conspicuously fails to connect this relationship with a similar increase in a solar inwards pull, however denominated. Newton had discovered an interesting mathematical correlation within the solar vortex, not a great new physical principle by which the stability of the planetary system might be accounted for.

It seemed worthwhile to review these early steps of Newton towards a new science of mechanics in some detail, first because they give an unrivalled glimpse of this inventive powers, second because they show how simple, when reduced to bare essentials, were the elementary forms of his dynamical ideas, and third because we can judge how limited they still were, at the point when Newton's life took a new course. His appointment to the Lucasian professorship at Cambridge and his subsequent decision to follow Barrow in taking optics as the subject of his lectures, were not in themselves events that offer sufficient reason for his neglect of this promising start in mechanics during so many subsequent years. Other interests pressed heavily upon this time. But Newton's long preoccupation with optics was certainly a factor in this neglect. After the inditing of these early records, we see scarcely anything further of Newton in mechanics till 1679;[69] thenceforward his progress to the accomplishment of the first *Principia* was enormously swift. Taking all this into account, it has become clear in recent times that the old idea, sedulously cultivated by Newton himself, of the 'discovery' in 1666 of the law of universal gravitation as a clear and complete explanatory concept is a fiction. What Newton discovered in 1666 was the outline of the inner dynamical relations of bodies in the solar system, an outline which would in the future integrate with and give mathematical structure to a law of gravitation when that concept was clearly present in Newton's mind. On present evidence this clarity was not achieved before the mid-1680s. It may seem paradoxical to declare that Newton had discovered the outline of dynamical structure, in mathematical form, which would sustain and explain the concept of a universal force, invariably associated with matter. One might have expected that the order in time would be the reverse of this: that the dynamical structure would have been developed by Newton *after* he had seized upon the idea of universal gravitation. Such indeed was the path followed unsuccessfully by Robert Hooke. To see that this was not Newton's way gives us a critical insight into the nature of his mathematical and scientific intelligence. He did not grope for the means to accomplish grandiose ends. Rather, he had created the means before perceiving the end to which it might be directed. When he understood the cosmic purpose to which the mathematical process whose outline he had already created might be applied, the means was at his command – though that too required much further development before its rich fruit appeared. Newton's debt, a great one, to Descartes and Hooke was that they made him confront this purpose.

3

Widening Horizons, 1667–1669

The Master of Arts

After looking so far forward to the origins of the *Principia*, it comes almost as a shock to realize that young Isaac Newton was still barely twenty-four years old, not yet a Master of Arts, when he returned to Cambridge in 1667. It would be strange indeed if he were not now far more conscious of his own potentiality in the world of learning than he had been before the plague. Five years into the future he would be arguing for his own discoveries on equal terms with the acknowledged leaders of the scientific movement in Europe. As yet, however, he stood on the lowest rung of the ladder of academic promise and his name was unknown. Newton had a chance of a minor fellowship at Trinity College at the next election, in the coming October, and no doubt hoped for something from the support of his family connection, Humphrey Babington. Candidates for the fellowship had to submit, at least in theory, to four days of oral examination by the Seniors in the college chapel. By whatever means, and whoever was convinced of his merits, Newton was indeed among the chosen. He was assigned the 'Spiritual Chamber' to reside in, but probably remained where he was with his friend (and amanuensis) John Wickins, renting out the room allotted to him. Trinity College now paid him 'wages' of £2 per annum, gave him allowances for livery and commons (that is, clothing and food) and allowed him his share ('dividend') of the college revenues. Trinity also assigned him his first pupil, a Fellow-commoner named St Leger Scroope, who made no mark in history. Newton never referred to him, and permitted him to leave the university (degree-less) without presenting the customary piece of plate to the college.[1]

Newton seems to have been disinclined to make himself financially independent of his mother; rather, in anticipation of his future inheritance of her estate he seems to have increased his (modest) demands upon her. He needed and obtained new clothes, as much as sixteen yards of "stuff" for a new suit, new shoes, and new academic dress, first as BA, then as MA, requiring eight and a half yards of "woosted Prunella", besides the lining for the gown, and presumably appropriate caps and hoods. There were fees payable for admission to these degrees, which demanded celebration with friends. Newton never in these early years avoided socially proper jollification, however contrary to the general bent of his nature. The cost of parties was relatively large in terms of annual dividends of £10. Some slight amelioration of his finances and greater security in his career came with advancement to a major fellowship in July 1668, following his inception as Master of Arts. Now he could dine at the Fellows' table. With the wind fair, Newton and Wickins embarked upon the beautification of their chamber: plasterers, painters and glaziers were called in and new furniture purchased. Such items as new bedding, a tablecloth and napkins also appear in Newton's accounts; presumably each young man had a separate bedchamber. Somewhere among these amenities – if not in some college outbuilding – room had to be found for the lathe Newton had bought, together with other tools, to which chemical apparatus was soon to be added. Though his BA gown was smartened up to suit Newton's higher dignity, he nevertheless bought for a new one eighteen yards of "tammy" (not to mention lining, hood and hat). All this finery cost almost £6.

Newton also added to his tailor's bill with new suits for himself. From this time with its varied expenses we may date the start of Newton's new life and his transformation from the former sizar to the man of rank and position. Like many other young men before and since, he shed the last traces of being a quiet country boy and perhaps his Lincolnshire speech too. With his fine new clothes and elegant surroundings he was ready for an evening at the tavern with his friends, or a few ends upon the bowling-green (this, too, not without cost to his pocket). He had his own 'gyp' (servant), Caverly, a laundress, and "Goodwif Powell", perhaps his bedmaker, to all of whom tips were given. As Whiteside has pointed out (in the introduction to *Mathematical Papers*, volume II) Newton was not simply a dour young Puritan, tender of conscience, nor was he always at his books. The obsessive absorption in work that so impressed Humphrey Newton later was by no means so deep when Newton was twenty-five. He was now unconsciously preparing himself for the social role he would play in the world later as a high officer of the State and the most celebrated philosopher in Europe.

One of the well-known tales about Newton is attributed by William Stukeley to the summer after his return, when the Dutch war-fleet (to Pepys's shame and despair) penetrated the Thames and the Medway, burning the dockyards and towing away the *Royal Charles* as a prize.[2] Newton is said to have inferred in conversation, from the growing loudness of the cannonade, that the Dutch were beating the English. More probable, however, is the alternative dating of this story to the summer of 1672 and the great naval battle in Southwold Bay, heard across East Anglia;[3] Newton's reasoning powers are more likely to have been remembered then, too. No one in his college seems to have left any record of Newton before his fame burst upon the world in that year, except his friend Wickins in his later memoirs. His chief personal recollection of Newton was of his hair's greying at the age of thirty!

That Newton had other friends, with whom he shared scientific interests, is obvious not only from his bizarre and pompous letter to Francis Aston (18 May 1669) but from the third letter in his extant correspondence. This is the one dated 23 February 1669,[4] in which Newton described his first reflecting telescope, constructed (it seems) near the close of the previous year. The unknown addressee might indeed be a patron rather than a friend, but its opening sentence "I promised in a letter to Mr Ent to give you an accompt of my Successe in a small attempt I had then in hand" invoking another doubtfully identifiable person hints, like the letter to Aston, at a world of the young Newton's relationships for ever concealed from us. It would be a mistake to confine this world to Babington, Barrow and Wickins. The widening of his horizons is also proved by his long, first visit to London in 1668. Including a visit to his mother at Woolsthorpe, he was absent from Trinity from 5 August to the end of September.[5] It is inconceivable to me that Newton went to London for a stay of three or four weeks (it cost him £10) without knowing a soul there, or at the least carrying letters of introduction; and (being Newton) he surely had some more cogent business there than admiring the lions in the Tower and observing the destruction of the City by the Great Fire. We may guess at a variety of reasons for Newton's journey; he wished to meet someone (for example, Robert Boyle), never in Cambridge; he wished to buy books, tools or apparatus; he wished to learn from opticians about the grinding and polishing of lenses and mirrors. At any rate, in view of the project he clearly had in mind, it seems likely that he would have sought out the best instrument-makers in London. (He might even, like James Gregory four years before, have thought of commissioning one of these to make the new instrument to his design.)

Work in Optics, 1667–68

Newton is unlikely to have perceived the (no doubt exaggerated) advantage of the reflector over the refractor before his return to Cambridge, that is, before making his renewed, more sophisticated analysis of refraction and its colours in the summer of 1667 or 1668. We can only guess that the visit to London, coming between Newton's developed comprehension of the nature of white light and the making of the first reflecting telescope, was a middle element in a sequence of events. That telescope, of which no vestige remains, was

> but Six Inches in Length, it beares something more than [one] Inch apperture, and a Planoconvex glasse whose depth is 1/6th or 1/7th part of an Inch, soe that it Magnifies about 40 times in Diameter which is more than any 6 foote Tube can doe I beleeve with distinctnesse . . . I have seene with it Jupiter distinctly round and his Satellites, and Venus horned.[6]

(How had Newton learned to make lenses as small as the expert optician could produce, and where had he looked through a six-foot-long telescope?) We have to infer – for it is not positively stated – that this little instrument was a reflector, like its successor three years later which came to the Royal Society, and that it was of the 'Newtonian' design, that is, with a small flat mirror deflecting the light coming from a curved speculum at the end, out through the side of the tube into the little plano-convex object glass inserted near the open front. The story of this original instrument is full of puzzles: its total absence from every document but this letter; Newton's skill in making it; his imparting knowledge of it to others, but in such a way that the secret did not leak; his setting the little telescope aside, then repeating his success in 1671. Did anyone but Newton look through it? Did Barrow know of it?

Thanks to the notebooks, we can offset our ignorance about these practical advances by some glimpses of Newton's progress in experiment and theory. If we are right in supposing that Newton's multiple-prism experiments (recorded in MS Add. 3975) are unlikely to have been performed before his return to Cambridge, a notion supported by the evident gap between the early notes on Boyle's *History of Colours* (in the *Quaestiones quaedam philosophiae*) and the later notes on the same book (in MS 3975, in the so-called 'Essay on Colours') which lead straight into these more complex experiments, then we can infer that the trial with the little refractor took place late in 1668. For, until that time, Newton had not been assured that the chromatic defects of lenses were much more damaging than the geometrical ones, whose amelioration had been the object of his former "glass-works". As Maurizio Mamiani has pointed out, it may well have been his second reading of Boyle's book that apprised Newton of the importance of the 'classical' experiment with a prism:[7] that in which, within a very dark room, a narrow beam of light is shone through the prism, to form (as he claimed) a large, well-defined spectrum upon a screen placed several feet from it. Descartes had never approximated to such an experiment, nor had Hooke. Marcus Marci had performed it in Prague and made observations upon it that anticipate Newton's, but Newton had never heard of his book.[8] Boyle had at least seen a spectrum.

As we have seen, in the new optical notes begun in Cambridge in MS 3975 Newton set down tersely many of the experiments, measurements and ideas that he was to elaborate in his optical lectures only a few years later, these lectures in turn serving as the base of Book I of *Opticks*, to be begun in 1687. They include experiments on refraction and experiments on interference (as we call it) or, as Hooke and Newton said, the colours of thin plates, in one form also known as 'Newton's rings'. In the seventh paragraph of these notes Newton came to the quantification of the differential refraction of blue and red light, an effect familiar to him for a couple of years at least, a quantification that he effected by separating the colours out of a beam of white sunlight by means of a prism (see figure 2.5). A few years later in his first letter to the Royal Society (6 February 1672)[9] Newton echoed this passage, declaring that it was the unanticipated elongation of the spectrum that had set him on the road to a new theory of light and colour. In the notebook experiment Newton took precautions (not mentioned in the letter) to ensure the near-parallelism of the Sun's rays – whose divergence after refraction he was to measure – before they entered the prism. Later in the same notebook, how much later in time we cannot tell, Newton recorded the first, primitive version of the "crucial experiment" that was to figure so largely in his famous letter to the Royal Society. Here is the first occasion of the use of more than one prism: "holding another Prisme about 5 or 6 yards from the former to refract the rays againe I found First that the blew rays did suffer a greater Refraction by the Second Prisme than the Red ones."[10] The "crucial experiment" of the letter incorporated refinements aiming at greater precision and, of course, at the additional major point that a pure homogeneous ray produced by the first prism is left unchanged by a second refraction. The unbreakable link between refrangibility and colour was to Newton's mind established beyond cavil.

Newton found other uses for his numerous prisms. Two tied together with thread enclosed a thin plate of air, which made visible colours. This experiment was then improved by pressing a convex lens of large radius upon a flat plate of glass – perhaps such a large telescope lens had been bought in London in the summer of 1668 – "this I observed by a sphericall object glasse of a Prospective tyed fast to a plaine glasse, so as to make the said [central] spot with the circles of colours appeare." Where Hooke had been unable to go beyond a qualitative account of the colours of thin plates and speculate upon their formation, Newton measured the thickness of the plates or films, supposing the glasses to have been pressed into contact at the centre of the circles. From careful measures of the diameters of the circles, knowing the radius of the spherical lens, this thickness could readily be established: "the diameter of 5 of the circles was one parte, whreof 400 was the radius dc of the [convex] glasses curvity. The said radius being 25 inches. Soe that the thicknesse of the air for one circle was 1/64000 inch, or 0.000015625 [which is the space of a pulse of the vibrating medium]."[11] Later still Newton went back to add: "By measuring it since more exactly I find 1/83000 = to the said thicknesse."[12]

Suddenly, without any explanation, physical speculation has appeared. Before taking that into account, we must note another group of experiments made with the pair of conjoined prisms of hardly less significance for Newton's treatment of optics. In these experiments he observed the fate of a beam of light shone through the side df of one prism edf, and either passing through its base ef into the second prism ebf, or being internally reflected from the base out through the side ed, or thirdly being partially transmitted and

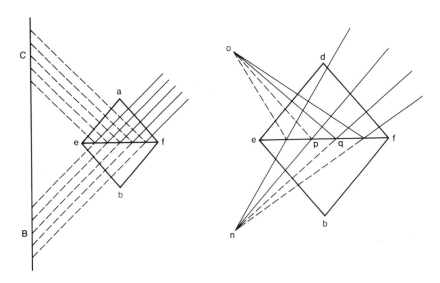

FIGURE 3.1   The partial reflection and partial transmission of light observed in two prisms tied together

partially reflected when the beam met *ef* at the critical angle. Such experiments provided the distinction fundamental to Newton's optical theory between easy/difficult transmission and difficult/easy reflection. At a certain angle, Newton found, the more copious white light transmitted to B become reddish in colour, while the reflected light at C (also white) became bluish. Newton concluded from this observation that just as blue rays refract more easily than red, so they reflect more easily also. In other words, the separation of the colours can be as well effected by reflection as by refraction. This point was to become, in later years, a major argument against the notion that light is not merely analysed but modified by refraction.

To finish with MS 3975, we have only to consider the "vibrating medium" that abruptly appeared in connection with the regularly increasing thicknesses of the air-layers forming Newton's rings. A late entry[13] – perhaps not long preceding the lectures – reads in part:

If rays be incident out of glasse upon a filme of air terminated twixt two glasses, the thicknesse of a vibration is 1/81,000 or 1/80,000 part of an inch. If water was put twixt the glasses the thickness of a vibration was 1/100,000 inch, or 3/4 of its former dimensions, viz as the densitys [optical] of the interjected mediums . . . And the thicknesse belonging to each vibration is as the squares of [the] secants of [the] celeritys. And the lengths of the rays belonging to each vibration as their cubes.

Note how the mathematical treatment is increasing in range and depth. It is as characteristic of Newton as of modern science in general to carry the mathematization of data from experiments as far as it can be taken. Clearly Newton was constrained by the phenomena of the regularly repeating circles to follow Hooke in postulating a periodic property of some kind in the rays of light, which he here calls a "vibration" or a "pulse". He never speculated that this periodic effect actually *was* light, as Hooke did; rather it modified the moving corpuscular beam that was light itself. In the notebook it is also evident that he supposed the smallest circles of successive colours to represent the least numbers of vibrations in the interval (nominally, one), the vibration for each successive colour being longer than for the one before. When the same colour recurred in a fresh series of circles, larger (integral) numbers of vibrations were required to span the interval. Hence the appearance of colour was in some way associated with, or indeed conditioned by, determinate vibration lengths.

"Of coloured circles"

The use Newton made of his optical experiments and the manner of his co-ordinating them into a coherent theoretical exposition are first to be seen in the optical lectures of the early 1670s. But because theoretical ideas do not figure prominently in the notebooks it would be rash to infer that they were not present in Newton's mind in (say) 1668 or indeed 1666. Almost certainly his fundamental concepts – the atomicity of light, the physical fixity of colours, the composite nature of white light – were with him from his first experiments upon reflected colours. But it was not his way to jot down the speculative framework of his investigations. This is usually to be found only in coherent drafts or essays, not intended for publication but sometimes circulated among friends. A notable example is the mathematical tract *On Analysis*; in optics there is a paper (discovered by Richard S. Westfall) headed: "Of the coloured circles twixt two contiguous glasses".[14] Newton opened the paper with six propositions, which he intended to justify by the experiments following (numbered as far as five). The propositions state that the proportions of Newton's rings

and of the interposed films causing them are geometrically determined; that the size of the circles varies with the density of the
"interjected medium" separating them; and that using monochromatic light, "that which is most refrangible makes the least circles."[15]
He wrote: "& the thicknes of a pulse for the extreame red, to that for the
extreame purple is greater than 3 to 2 . . . Viz about 9 to 14 or 13 to
20." In these propositions it is taken for granted that light is
corpuscular, and that its particles constituting a ray move with a
"force" or "percussion" whose variations in magnitude are responsible for some of the effects studied.

The close connection of these propositions with MS 3975, which
they must follow, is very clear; the experiments too are of the same
type as those in the manuscript, very careful measurements in
different experimental circumstances and computations of the
thicknesses of the films. The data are all independent of those in the
manuscript notebook because Newton was now using a lens of
double the radius, that is, of 50 feet, making the circles larger and
therefore easier to measure. The dexterity required to maintain with
one hand the correct steady pressure of the lens on the glass plate,
manipulating a prism with the other hand, while all the time the Sun
moved, are suggested by Newton himself:

After many vaine attempts to measure the circles made by coloured
rays alone I at last thought of casting colours upon a white paper, either
by a speculum or immediatly by a Prisme which turned about its axis
might make all the colours succeed on the same point of the paper; or
by casting integrated light[16] on a paper which might be varyed into all
the colours successively by stopping the rest. Then I laid the lentes so
that they wanted 3 or 4 rings depth of touching. And making the
colours be mooved to & fro whilst with a steddy ey I beheld the circles
observing always the spot in the middle how it increased or appeared
anew in going from the purple to the red light & how it decreased or
disappeard in returneing from the red to the purple & so many times as
there was made returnes or successions of the light & darke spot in
passing from one extreame to another soe many I reccond soe many
halfe circles difference twixt the red & purple. then the bright red being
held steddily in sight I waryly laid my hands on the upper glasse &
pressed it downward till I see the same spot in the midst which would
appeare in the deepest redd. After that gradually pressing more & more
I reconed how many light & darke spots succeded til I see the last darke
spot appeare which argued the glasses contact. And soe many halfe
pulses I reconed made by the extreame red light in its passage twixt the
glasses, then adding their difference found before, I had their proportion.[17]

Probably, for so early a date, such a detailed account of a delicate experimental manipulation is quite unique.

De gravitatione, c.1668–70

All these investigations going back to 1664 were carefully preserved by Newton, so that when he came to write his optical lectures, and still more when he came to *Opticks* itself twenty years later, he possessed rich files full of material. With the *Principia* the case was different. True, Newton retained his computations of central forces from long before and his reckoning that the centrifugal acceleration of each planet is as the inverse square of its distance from the Sun; he had written annotations upon Vincent Wing's *Astronomia Britannica* (1669); he had mastered motion in a cycloid and so forth, but he had sketched his general ideas in mechanics only in two essays. The second of these (*De motu corporum in gyrum*, or 'On the motion of bodies in orbits') was composed as late as 1684, that is, only just before the *Principia* itself, of which it is a preliminary sketch. The earlier essay, the only one that can be attributed to Newton's research fellowship years which seems to bear directly upon his greatest work, is known from its opening phrase as "De gravitatione et aequipondio fluidorum" ('On the weightiness and equilibrium of fluids', Cambridge University Library MS Add. 4003).[18] It is a strange document. When I first read it I could hardly credit that Newton was its author. Hydrostatics and aerostatics were indeed topics much discussed in the 1650s and 1660s, especially in connection with the newly obtained experimental 'vacuum'. Robert Boyle in England had made the 'spring and weight of the air' a famous issue. And in the *Principia* Newton himself would indeed lay the foundations of the science of rational fluid mechanics. While the small final part of MS 4003 concerned with hydrostatics may be viewed as an antecedent of Book II of the *Principia*, the principal interest of this essay lies in its exposure of Newton's metaphysical speculations (which relate to the motions of bodies) and his criticisms of Descartes in this area. One has the sense that he was stimulated to write – and what we have is surely not a first draft – by a careful reading of the volume of Descartes's *Opera Philosophica* (Amsterdam, 1656) which had presumably been in his possession for some years, and of the 1667 volume of Descartes's letters (in Latin) to which he alludes.[19] We have no notion of the purpose or occasion of the essay. Newton only began to compose his essays in Latin during the late 1660s, and it is inconceivable that this essay was the fruit of a *first* reading of Descartes's *Principia philosophiae*. Judging by its poor construction and evidence of immaturity, it may have been written as early as 1668 or in the early 1670s (as Whiteside estimates from the handwriting).

After initial astonishment the reader meets with much that is familiar in Newton. *De gravitatione* is deeply religious: Newton's greatest criticism of Descartes is that he has diminished God's necessarily continuous guardianship of the creation – an objection later urged against Leibniz's philosophy. As in the General Scholium which concluded the later editions of the *Principia*, young Newton stressed the real, ubiquitous presence of God in all space, which he (unlike Descartes) believed to be infinite, because God is infinite. Perhaps, he speculated, it is the constant (not only once in the past) volition of God that differentiates certain particles of space so that by us they are discerned as particles of matter – obviously in the composite form of tangible bodies. The constant presence of God may thus be matched by a constant creation of matter. Almost from the first, therefore, Newton reveals himself as an opponent of the Cartesian variant of the mechanical philosophy – the dominant variant of the period – which was to be carried forward into the eighteenth century by such 'neo-Cartesians' as Christiaan Huygens, Leibniz, Malebranche and their successors, not least Johann I Bernoulli. Newton always denied that the philosopher can properly treat the creation as a 'universe machine' complete and self-sufficient in itself, perfectly stable and unchanging for all eternity. Such a notion, Newton hints strongly, is conducive to atheism, for the philosophers who maintain it confuse God and matter, "And hence it is not surprising that Atheists arise ascribing that to corporeal substances which solely belongs to the divine."[20] To banish God from Nature is in effect to deny his existence; many years later Newton returned in both the *Principia* and *Opticks* to the contention that a true natural philosophy must prove God's perpetual maintenance of Nature, forming and reforming the parts of the universe.[21]

Descartes's mistakes (in Newton's view) were not limited to this highest level; his account of the fundamentals upon which a study of the universe must be based, the ideas of place, motion and physical causation were all defective. In all things Descartes was a relativist, Newton an absolutist. Descartes had argued, for example, that it is meaningless to define motion except as a displacement of one body among its neighbours. Newton believed (as again the *Principia*, as well as *De gravitatione*, would demonstrate) that space is a motionless absolute, and that motions should be defined in relation to the co-ordinates of this unvarying continuum. (In practice, Newton was prepared to take it as axiomatic that the Sun and fixed stars are permanent features of this continuum and so constitute the best reference points we have; he could have learned from astronomers that some stars at least are inconstant.) There was also the flow of real time, which Newton believed to be uniform and unaffected by any event within time; but he had to recognize that machines provide our most accurate means of measuring short intervals of time. However, though absolute time and space are metaphysical entities for ever beyond human grasp, there are important natural-philosophical consequences following from these distinctions between Cartesian and Newtonian metaphysics. The most obvious of these – the privileged status of absolute rotary motion as manifest by centrifugal force – had not yet occurred to him in 1668. Once he had recognized it, he could use it in the *Principia* to justify Copernicus, that is, he would demonstrate by dynamics that the Earth revolves about the Sun, and not vice versa.[22] But it is equally important that Newton's metaphysics was consistent with a concept of physical forces (of forces of Nature), indeed, required such a concept. The Cartesian universe knew no force: a motion was always caused by an antecedent motion. In this respect again *De gravitatione* foreshadows the *Principia*; its Definition 5 affirms trenchantly:

> Force is the causal principle of motion and rest. And it is either an external one that generates or destroys or otherwise changes impressed motion in some body; or it is an internal principle by which existing motion or rest is conserved in a body, and by which any being endeavours to continue in its state and opposes resistance.

Newton here avoids any ontological distinction between the force associated with motion (mv) and the force which is a "principle" (later called an "active Principle"). Gravity is brought into Definition 10 as an example of a "principle" of this kind: "Gravity is a force in a body impelling it to descend. Here, however, is meant by descent not only a motion towards the centre of the Earth but also towards any point or region, or even from any point."[23]

The next sentence reminds us how far he had to go in developing his own dynamics: "In this way if the *conatus* of the aether whirling about the Sun to recede from its centre be taken for gravity, the aether in receding from the Sun could be said to descend." But at any rate it is clear that gravity is much more than the cause of heavy bodies falling upon Earth.

Newton continues, following his essentially Cartesian definitions of the words force, *conatus*, impetus, inertia, pressure and gravity, by a revival of the technical vocabulary of medieval philosophers:

> the quantity of these powers, namely motion, force, *conatus*, impetus, inertia, pressure and gravity may be reckoned in a double way: that is, according to either intension or extension.
>
> *Definition 11.* The intension of any of the above-mentioned powers is the degree of its quality.
>
> *Definition 12.* Its extension is the amount of space or time in which it operates.
>
> *Definition 13.* Its absolute quantity is the product of its intension and extension. So, if the quantity of intension is 2, and the quantity of extension 3, multiply the two together and you will have the absolute quantity 6.[24]

There is a pre-Galilean not to say pre-Cartesian flavour about this passage which renders it implausible that Newton had as yet read Galileo's *Discorsi e Dimostrazione intorno a due Nuove Scienze*. This analysis is very far indeed from the gem-like clarity of the first pages of the *Principia*; even more confused is the last example illustrating the analysis:

> lastly, the intension of gravity is proportional to the specific gravity of the body; its extension is proportional to the size of the heavy body, and absolutely speaking the quantity of gravity is the product of the specific gravity and mass of the gravitating body. And whoever fails to distinguish these clearly, necessarily falls into many errors concerning the mechanical sciences.
>
> In addition the quantity of these powers may sometimes be reckoned from the period of duration; for which reason there will be an absolute quantity which will be the product of intension, extension and duration. In this way if a body [of size] 2 is moved with a velocity 3 for a time 4 the whole motion will be $2 \times 3 \times 4$ or 12 [*sic*!].[25]

Altogether Newton required, for a treatise he was to abandon uncompleted after a page or two more, nineteen definitions, "accommodated [he explained] not to physical things but to mathematical reasoning, after the manner of the Geometers who do not accommodate their definitions to the irregularities of physical bodies."[26] Newton may have recalled Galileo's words on this point in the

Dialogo, whose spirit he repeats; by the geometers he meant such authors as Euclid and Archimedes, who had treated fluids as frictionless, mirrors as perfectly plane or spherical, and so forth. He was clearly already aware of the problems involved in mathematizing physics, a matter about which he would have much to say in later writings.

Newton on Space and Time

Much learned ink has been spent upon the origin and meaning of *De gravitatione*, for it contains the first appearance of Newton the philosopher as distinct from Newton the mathematician and Newton the experimenter. Westfall has opined that it contains 'a view of matter intimately related to alchemical views', while admitting that it is not 'an alchemical essay'.[27] It is true that if the writing of *De gravitatione* is placed late enough – say in 1670 – it could have been written at the time when Newton was beginning to immerse himself in chemical literature and practice. But this thread of connection is very tenuous and in no way leads to the fascinating anti-Cartesian discussions in the essay. Hence others, among them J. E. McGuire, have suggested a strong influence from the Platonist of Christ's College, Henry More.[28] His is indeed the obvious, local, anti-Cartesian example. Of the links between the two men there can be no doubt, though they may be less strong than some scholars like to suppose.[29] Newton owned many of More's books – some presented by their author – but setting aside the thorny questions of biblical prophecy he seems to refer to only one, *The Immortality of the Soul*. He never owned and probably never read More's letters addressed to Descartes (first printed in 1657) wherein More recorded at length his general admiration for Descartes's philosophy, while at the same time rebutting a multitude of his positions in logic, metaphysics, epistemology and psychology. Newton in *De gravitatione* certainly echoed some of More's objections – both rejected Descartes's concept of the necessary relativity of all motions, for example – but the general pattern of the criticisms is so different that Newton's are probably independent. Nor is there evidence of Newton's reading More's increasingly sharp public attacks upon Descartes from 1662 onwards, in which (like Newton later) he maintained that Cartesianism encourages atheism. In philosophical matters the minds of More and

Newton were far apart, as they were also in biblical scholarship, and the case for the possible influence of the older upon the younger man rests on no more than such loose parallelisms as I have mentioned. Newton never repeated More's early encomia of Descartes, nor did he adopt More's positive answer to Descartes: the mysterious Spirit of Nature.

In his discussions of space and time, both early and late, Newton seems to follow Isaac Barrow's treatment of these themes in his first course of lectures on mathematics at Cambridge, begun in March 1664. His recollections render it almost certain that Newton did attend this course.[30] Unlike Descartes, Barrow taught the infinity of absolute space, which is distinct from matter (extension), and the flow of absolute time, independent of the motions used to measure it. Like Newton, he was a realist: 'each and every geometrical shape that can be comprehended is really present in every particle of matter whatsoever: is present, I repeat, in actuality and in perfection though it does not appear to the senses.'[31] From which it must follow that mathematical physics is not only impeccable logically by its internal construction, but actually corresponds to the reality of Nature. Again like Newton later, Barrow imagined vacuous space between material worlds. He seems to accept a single geometric concept of space, embracing all worlds, which exists where worlds are not and is the potential locus of all material existence.

Barrow's whole discussion is rational rather than religious. The complementary influence of More may perhaps be argued from the more markedly theological character of some features of Newton's treatment of metaphysical questions. But as always Newton was very much his own man, and if (like More) he sometimes extrapolated to our ideas of metaphysics from our conceptions of the divine power and will, this is done after his own manner, not More's.

Newton becomes Known

We cannot now expect fresh light upon the relations of Isaac Barrow, Lucasian Professor and Fellow of Trinity, and Isaac Newton, junior fellow, in the years 1667–9. What is certain is that Barrow launched Newton into the scientific world before surrendering his own chair to him (in effect). It is obvious that Barrow can have put no spoke in the successful roll of Newton's academic wheel, and far more likely that he helped to urge it forward. By July 1669 Newton was Barrow's 'friend' who, he told John Collins in London, had a very excellent genius in mathematics: '[He] brought me the other day some papers, wherein he hath sett downe methods of calculating the dimensions of magnitudes like that of Mr Mercator concerning the hyperbola, but very generall; as also of resolving aequations; which I suppose will please you; and I shall send you them by the next.'[32] Collins was a man of indifferent mathematical skill but considerable familiarity with mathematical literature. An autodidact, he lived as a government clerk – an analogy with Charles Lamb more than a hundred years later springs to mind – was a Fellow of the Royal Society and advanced his favourite study by keeping up a large international correspondence with mathematicians. He was a human news-sheet.[33] Collins was proudly conscious of the not inconsiderable British talent in mathematics at this time, both too reticent at home and too little valued abroad. Newton's he saw as just the sort of original capacity he longed to stimulate and publicize. He at once asked and received permission to show Newton's paper (*De Analysi*) to the President of the Royal Society, the mathematician Lord Brouncker.

This step might have ensured Newton's immediate launch upon the international world of learning, an event which was delayed a further eighteen months. Newton insisted upon the return of his paper, though Collins kept a transcript of it. However, the significance of his mathematical researches was not entirely confined to Barrow and Collins. The latter, in September 1669, briefed the Secretary of the Royal Society, Henry Oldenburg, on Newton's advances in analysis. Among other things, Oldenburg brought his Belgian correspondent René François de Sluse up to date on Barrow's forthcoming publications, and added:

Moreover he hath communicated an universall Analyticall method imparted to him by Mr Isaac Newton his Collegiate for the Mensuration of the Areas of all such Curves and their Perimeters wherein the Ordinates have one common habitude to the Baseline, and this is no other then the Method particularly applyed by Mercator for the finding of the Area of the Hyperbola, rendered universall . . .
And having shewed thereby the quadrature of many Curves he then comes to the Circle, and by turning the $\sqrt{aa + bb}$ or $\sqrt{aa - bb}$ into an infinite Series sheweth that there may be diverse such Series applyed to the Circle, so that giving any two of these Data, the Radius, Sine, Arch, and Area of the Segment either of the rest may be found infinitly true upon Demaund (a thing much coveted by all former Writers) he hath likewise thereby incredibly facilitated the finding the roote of any Æquation and of meane Proportionalls and gives a Series

for finding the Length of an Elliptick Line, he likewise to shew that his method extends to mechanick Curves, and their Tangents, squares and Cycloid and its portions, and finds the Area of the quadratick Curve and its Perimeter.[34]

With Oldenburg's letter to Sluse the breath of fame touched Newton at last, but for various reasons nothing came of it. Newton's repute as a mathematician followed his fame as an optical experimenter. It was blighted for many years by lack of public evidence of what Newton had actually accomplished, as not a word was printed. The papers of 1669 and those that followed were too long suppressed.

Newton's reticence, however unwelcome to Collins, did not mean that he rebuffed Collins's overtures. On the contrary, he welcomed Collins's acquaintance, which brought him into touch (directly or indirectly) with other British mathematicians, such as Michael Dary and James Gregory. The two men first met in London about the end of November 1669, so Collins related to James Gregory about a year afterwards:

> I never saw Mr Isaac Newton (who is younger than yourselfe) but twice viz somewhat late upon a Saturday night at his Inne, I then proposed to him the adding of a Musicall Progression, the which he promised to consider and send up . . . And againe I saw him the next day having invited him to Dinner: in that little discourse we had about Mathematicks, I asked him what he would make the Subject of the first Lectures, he said Opticks proceeding where Mr Barrow left [off], and that himselfe was a practicall grinder of glasses, and had ground lenses for a pocket tube, but 6 Inches long, that magnified the Object 150 times whereby he did frequently observe the Satellites of Jupiter, . . . having no more acquaintance with him I did not thinke it becomming to urge him to communicate any thing.[35]

Nevertheless, Collins went on to tell Gregory something of Newton's achievement in *De Analysi*. He seems to have missed the point that Newton was probably describing a *reflecting* telescope. Collins sent books to Newton: John Wallis's *Mechanics*, Dary's *Miscellanies*. He also began to pose questions that involved labour. Newton had to write at some length in January 1670 to deal with Collins's questions about the summation of series. Collins also put before Newton a Dutch work on algebra by Gerard Kinckhuysen (a translation into English by Nicolaus Mercator he had procured), which was to involve Newton in a good deal of futile work over the next five or six years. It was Collins too who, via Barrow, sent to Newton some mathematical work of the astronomer G. D. Cassini – just arrived in Paris – on the geometry of planetary orbits.[36] If Newton wrote to Collins: "I must needs acknowledg you more then ordinarily obliging, & my selfe puzzled how I should quit Courtesys" he was soon to find this the least of his problems.[37] It was difficult, he found, to participate in the world of learning without losing time from one's own studies and without being diverted from one's chosen path. To this problem Newton was to find no solution until he overtly renounced mathematics and science altogether.

De Analysi, *1669*

The papers about which Collins began to write to his friends were Newton's mathematical tract, *De Analysi per Æquationes numero terminorum infinitas* ('On analysis by equations unlimited in the number of their terms').[38] The odds are that this tract, a reworking of old material for a wide (but not a public) audience, was written in response to Nicolaus Mercator's *Logarithmotechnia*, published in London in September 1668. Perhaps therefore Newton wrote in the spring or early summer of 1669. In view of its evident purpose – to show that he had been earlier and more general than Mercator – Newton would hardly have delayed long, after perfecting the tract, in bringing it to Barrow about June or July of that year.

In general, Newton's immediate continuation of his calculus researches after an extraordinary thirty months of intense mathematical development is judged 'relatively spasmodic and jejune'. *De Analysi* was 'not of striking originality' as compared with the October 1666 treatise on fluxions, while the remainder appears as 'mathematical *divertissements* on which his mind could relax in an idle hour away from his researches in theoretical and applied optics, his over-riding preoccupation at this time'.[39] It would be a mistake, though, to suppose that Newton's mathematical interests (now or ever) were confined to calculus, as Collins certainly learned and presumably Barrow also knew. Besides the advances in algebra which Newton incorporated into his long-continued but ultimately unpublished improvement of Kinckhuysen's text (amounting to some 200 pages in Whiteside's edition) he devoted considerable attention to geometry, not least to dividing the general cubic curve into its component species, work whose ultimate fruit would be *Enumeratio linearum tertii ordinis* ('An Enumeration of [curved] lines of the third order'), an acute and subtle piece of mathematical taxonomy that was at last published with *Opticks* in 1704.[40]

De Analysi, the tract which might so easily have made Newton known to the world as a young mathematician of promise, remained semi-secret even longer. A copy was privately sent by Collins to John Wallis in 1677, for his use in writing an account of recent British mathematics (in *Algebra*, 1685) but Wallis at this stage did not make a great deal of Newton's discoveries. The tract was again forgotten until William Jones printed it, with other early mathematical essays by Newton, in *Analysis per quantitatum series, fluxiones, ac differentias* in 1711. Jones came across it by virtue of making himself master of the deceased Collins's papers (see chapter 14).

The tract opens abruptly with a statement ("Rule 1") of the primary process of differentiation and its reverse, integration, expressed in Newton's own way: Newton affirms that if x be the abscissa (AB) and y the ordinate (BD) of any curve (AD), and if x and y are related by the expression $ax^{m/n} = y$, then the area ABD is given by

$$\frac{n}{m+n} ax^{\frac{m+n}{n}}.^{41}$$

For the demonstration of this rule the reader has to turn to the very end of the tract (the "Praeparatio") where Newton justifies it by a process involving infinitesimal quantities added to the abscissa and the ordinate which are subsequently made to vanish. Newton gave no name to the infinitesimal ("o"), which does not figure earlier in the tract. The purpose of Rule I was that, as Mercator had already shown in *Logarithmotechnia*, one may divide out y, when it is a unit-function $1/f(n)$, to be a simpler sum-series. Thus $1/(1 + x^2) = y$ becomes $y = 1 - x^2 + x^4 - x^6 \ldots$ the pattern is obvious. In turn 'integrated' term by term, this series gives

$$x - \tfrac{1}{3}x^3 + \tfrac{1}{5}x^5 + \tfrac{1}{7}x^7 \ldots ,$$

the series for $\tan^{-1}x$.[42]

The treatise then presents an exemplification of these principles in the quadrature of a number of curves, including the circle and the hyperbola. In the resolution of numerical equations Newton showed how the work could be abbreviated by approximation (the so-called Newton–Raphson method). In treating algebraic equations by another method of division, so that "the value of the area should approach nearer the truth the greater x is", the example being

$$y^3 + axy + x^2y - a^3 - 2x^3 = 0,$$

Newton expressed the integral of one term of the series in a new notation (though privately used by himself before), that is,

$$\boxed{\dfrac{aa}{64x}}$$

signifying for us

$$\int \frac{a^2}{64x} \, .dx$$

but here left unexplained by Newton, the reader being left to guess that the 'box' signified 'the quadrature of' the expression within.[43]

The mathematician may turn to D. T. Whiteside's careful edition for further details. For the present, I note only that in exemplifying the usefulness of his method of series in handling 'mechanical' curves, Newton instanced the rectification and quadrature of the cycloid. Another is the quadrature of the quadratrix; he concludes this section:

> In this manner the length of the quadratrix arc VD is determinable, though by a more difficult computation. Nor do I know anything of this kind to which this method does not extend itself . . . And whatever common analysis performs by equations made up of a finite number of terms (whenever it may be possible), this method may always perform by infinite equations: in consequence I have never hesitated to bestow on it also the name of analysis. For reasoning in the latter is no less certain than in the former nor are its equations less exact, even though we men of finite intelligence can neither designate all their terms nor so conceive of them as to ascertain exactly the quantities we desire from them: just as the surd roots of finite equations cannot be expressed numerically, nor by any analytical artifice be so expressed that the quantity of any one of them can be known distinctly and exactly from the rest.[44]

Newton rightly saw before him the prospect of an almost limitless expansion of the territory that the mathematician might conquer, by means of the combination of his reversible process for relating an equation to the quadrature of the corresponding curve (and so, as he put it, reducing problems "on the lengths of curves, the quantity and surface of solids and the centre of gravity" to an "enquiry into the quantity of a plane surface bounded by a curved line") with his methods whereby this quadrature might be expressed as an infinite series. He would himself present much more in his 1671 treatise "De methodis serierum et fluxionum" ('On the methods of series and fluxions'), planned for publication in order to widen the boundaries of analysis and advance the doctrine of curves (but, alas, never completed) and other mathematicians, with James Gregory notable among them, were attaining results similar to those offered in *De Analysi*. Nevertheless Isaac Newton, second-year Master of Arts, had proved himself to a select few one of the most accomplished mathematicians of Europe. Whether Barrow, Collins or any of the few who read *De Analysi* before its printing in 1711 were able to judge the potentialities of the new methods as clearly as Newton himself did may perhaps be doubted.

Leibniz and De Analysi

In 1669 Gottfried Wilhelm Leibniz was not yet a mathematician. Seven years later, in the course of his visit to London in 1676, he was to be the most significant reader *De Analysi* ever had. John Collins freely permitted Leibniz to comb through his own large collection of mathematical letters and papers, though he withheld access to Newton's pre-1669 manuscripts of which he possessed copies. Leibniz, sad at heart, was about to enter the service of the Elector of Hanover, having given up hope of pursuing a scientific career as a member of the Académie Royale des Sciences in Paris. Only in the previous year had the principles of the differential calculus become apparent to him. Not only had he clarified existing calculus techniques, but he had also created fundamentally new methods of reducing integrals. All this, in his own notation, he was to set before Newton (in outline) in a letter of 11 June 1677. Of course Leibniz by no means found anticipations of his own investigations in *De Analysi*. Any instances of infinitesimal arguments he passed over, for perhaps they seemed to him not very different from such arguments developed by others. He saw no sign of the 'calculus of fluxions and fluents' that had since 1669 taken shape in Newton's mind, and was far from suspecting how close Newton's mastery of calculus was to his own. Hence, in the private record of his reading of *De Analysi* 'Leibniz's excerpts are confined exclusively to series-expansions and the general remarks accompanying them; the sections relating to infintesimals in *De Analysi* remained disregarded.'[45] Rule 1 and its sequelae contained nothing that was of interest to Leibniz now.

Later, Newton was understandably (though mistakenly in fact) to claim that Leibniz had pillaged the tract for his own benefit. Leibniz, on the other hand, though in public and in private he praised Newton's mathematical abilities unstintingly and particularly his treatment of series, would never allow that *De Analysi* contained even rudiments of the truly new calculus, which depends upon *differences*, and was his alone.

Newton's Chemical Experiments

It is true that in the post-plague years optics was the principal concern of the young Master of Arts, not yet a professor. Within a year and a half of his return from Lincolnshire to Cambridge Newton was preparing his first lectures upon this science, although neither optics nor mathematics were the sole objects of his interest. Either in the late summer of 1668 when he was in London or soon thereafter Newton ordered his first parcel of chemicals. Among them were the metals antimony, silver, mercury and white lead, and such reagents as nitric acid, alcohol, vinegar, saltpetre, sublimate of mercury, alum and salt of tartar. He spent eight shillings on building a furnace and bought a *Theatrum Chemicum*.[46] This was not Newton's first acquaintance with chemistry: his reading of Boyle had ensured that. But as the bulk of the information about Newton's activity as a chemist and alchemist belongs to a later time, I shall deal with this topic by itself in chapter 7. Only one or two matters in this context touch on Newton's early life in Cambridge. Later, chemistry was at times an absorbing interest, and he was still labouring in his laboratory after writing the *Principia* and indeed up to the time of his departure from Cambridge in 1696.

Virtually everything in this pyrotechnic aspect of Newton's life has stimulated speculation which, by reiteration, becomes fact. Much nonsense has been written as though it could be substantiated by documents. If we begin with the question of his laboratory, it is because some light is thrown by it upon the manner of Newton's life in Trinity College. It is unquestionable that in the 1680s – the time to which the unique account of Newton's use of it belongs – Newton had such a laboratory, 'On the left end of the garden . . . near the east end of the Chapel where he at these set times employed himself in with a great deal of satisfaction and delight,' according to the rather incoherent Humphrey Newton, Isaac Newton's sizar amanuensis from 1685 to 1690.[47] The 'set times' of Newton's pyrotechnic labours, stated twice by Humphrey, were of six weeks' duration in the spring and at the fall of the leaf, seasonal proclivities that no one has explained, 'the Fire scarcely going out either night or day; he sitting up one Night as I did another, till he had finished his Chymical Experiments, in the Performance of which he was the most accurate, strict, exact; What his aim might be I was not able to penetrate into.'[48] Humphrey also says that the laboratory was near Newton's garden, by which he may mean in it, and that the garden was downstairs from his chamber. All this fits well with what has long been supposed; that Newton in the 1680s lived in a set of rooms in the range of buildings running from the Great Gate to the chapel. David Loggan's engraving of the college, conveniently enlarged by Lord Keynes,[49] shows the street side of this range well and the gardens that (in this part of the college) then separated the buildings from the street. The drawing from which this plate was taken was probably made in the late 1670s. In line with the massive double chimney is a jetted-out structure supported on pillars from which a stair leads down to the garden. We again owe to Humphrey the information that Newton mounted a (refracting!) telescope of at least five feet in the balcony at the head of this staircase. The date and purpose of this added structure – long since removed – are not recorded.

In Humphrey's time, then, in the 1680s, Newton probably lived on the first floor to the north of the Great Gate, in rooms having this staircase structure built outside, giving access to the garden below. In 1683 the college had to repair the wall 'between Mr. Newton's garden and St John's College', which could hardly be elsewhere.[50] Humphrey recalled that he was 'very curious in his garden . . . not enduring to see a weed in it'. Not that Newton, who stoked the furnace, also weeded and pruned; for him the garden was a place for meditative strolling: 'When he has some times taken a Turn or two, has made a sudden stand, turn'd himself about, run up the stairs like another [Archimedes] with an [Eureka], fall to write on his desk standing without giving himself the Leasure to draw a Chair to sit down on.'[51]

C. E. Raven, the biographer of the naturalist John Ray (who vacated) his fellowship at Trinity in Newton's second undergraduate year), thought that this garden of Newton's had formerly been Ray's botanic garden.[52] This might account for the regular arrangement of little square plots in the garden thought to be Newton's, in Loggan's print, if we assume the layout unaltered during nearly twenty years. Possibly, therefore, Newton's first-floor rooms had also been Ray's, and the extrusive structure built for him (surely not by him, for Ray, a blacksmith's son, had been like Newton a subsizar).

This little spot has been visualized by Dr Dobbs as a veritable research centre. Newton's laboratory, she thinks, 'had been built by the earlier experimenters of whom [Isaac] Barrow spoke in 1654', so that Newton 'inherited the use of the laboratory and probably its equipment along with John Ray's rooms and garden'. Indeed, she supposes its history to be still older, since it had been 'probably maintained for a while' in the 1640s by the former Fellow of Trinity, John Nidd (d. 1659), who (supposedly) allowed Barrow and Ray to work there.[53] Lord Keynes, Dr Dobbs and others assume that this laboratory was a two-storey building, and one might imagine it to have been in the well-ventilated lower part of the structure already mentioned, containing the stair. However, these and other writers agree that the laboratory was the other visible structure, a kind of lean-to shed, at the chapel end of the garden (as Humphrey wrote). This, however, is clearly shown as containing one storey.

I think this a good candidate for the laboratory, but the rest seems as fanciful as the hypothesis that Newton poisoned himself with arsenical fumes generated by his pyrotechnics. Newton could not conceivably have moved into this prized Fellow's set with its private garden when Ray left Trinity; it must have been occupied by some other Fellow between 1662 and Newton's occupancy.[54] We have no positive knowledge of where Newton originally shared a chamber with Wickins or indeed of where he lived in college at all before Humphrey Newton's time. Stukeley thought that Newton had 'chummed' with Wickins on the north side of the Great Court. R. S. Westfall, on the basis of documents which he confesses to be incomprehensible in detail, suggests that Newton and Wickins moved into their rooms by the Great Gate in 1673. Items of undated expenditure recorded by Newton (perhaps in the 1670s) may be the relics of such a migration:

> Paid for making the Oven mouthed Chimney in the Chamber
> Paid for the fire irons there
> Paid for a stone roll in the garden besides the frame
> Paid for making a new Cellar behind the Chapel.[55]

The same list includes purchases of furniture (six "Russia leather chairs" along with ten others, two "Spanish Tables with neats leather" suggestive of comfort); over £14 was paid to an upholsterer. If shaky, Westfall's date is perfectly plausible.[56] If it is accepted, the biographer has to explain where Newton performed his chemical experiments before he took over the garden laboratory (assuming that this existed before 1673). A possible answer is obvious: in his fireplace, as Joseph Priestley was to do a century later. The peculiar chimney constructed in the new chamber may also have been intended for this purpose. Any of Newton's rooms in college at this date would have been warmed by a fire in a basket-grate within a large, open fireplace. It would have been simple to build or install a furnace (or furnaces) in the corner(s), with a flue extending into the wide, open chimney. (The free draught in such a fireplace would have reduced the risk of inhaling metallic fumes.)

It is therefore very likely that the *Principia* was written (and copied by Humphrey) in the room above the garden near the Great Gate; but equally likely that the great experimental work on optics, with which Wickins occasionally assisted Newton, as well as the writing of the optical lectures, took place in some other chamber in the college. To these lectures we now turn.

4

The Professor of Mathematics, 1669–1673

Newton's Teaching in Cambridge

The greater part of Newton's original work in mathematics and science was first expressed in the form of university lectures. This is true of his researches in geometrical and experimental optics, of his discoveries in theoretical and celestial mechanics, and of his investigations in algebra and some other parts of mathematics. He never lectured upon calculus, no doubt believing that this topic was far beyond the reach of his student auditors; and he did not lecture either upon his chemical, alchemical, biblical and historical studies, not only for the reason that they lay far outside the scope of his professorship. The details of Newton's lecturing are formally known to us from the manuscripts of his optical lectures (in two versions), of the *Arithmetica universalis* and of the *De motu corporum* which he submitted to the University Library in accordance with the statutes of his professorship – or nearly so![1] These statutes, indeed, required the Lucasian Professor of Mathematics to deliver upwards of twenty lectures in each academic year, neatly written copies of ten of them to be deposited with the vice-chancellor in the following year.[2] But although these statutes had been drawn up only in 1663, when the professorship was founded by Henry Lucas, their provisions were at once disobeyed. While his predecessor, Isaac Barrow, had probably lectured in two terms of each year, Newton went to the Schools in one only, half an hour being reckoned a near enough approach to the statutory 'about one hour'. He deposited copies of lectures three times only: in 1674 (thirty-one optical lectures), in 1685 (ninety-five mathematical lectures) and in 1686 or later (twenty lectures on mechanics and the "System of the World"). In effect, the ten deposited lectures became the sole teaching requirement of the professor. We cannot assume that any were read in their extant forms, or on the dates assigned to them by Newton.

However, the professor was also required to give open access to students during two hours each week out of term (if he was in residence) and four hours each week in term-time in order to assist them with difficulties in their studies and to instruct them in the use of the globes and other mathematical instruments. There is some slight evidence of the exercise of this privilege in Newton's time, and it may be that Newton's 1672 edition of Varenius's *Geographia generalis* (his first book) has some relevance to this informal instruction. Geography, its content by no means closely corresponding to the modern notion of this subject, was a popular branch of 'useful' mathematics.

Although Newton took the trouble to attach some precise dates to his deposited lectures, and although the form of some of the optical lectures particularly – such as the recapitulation at the beginning of a new lecture of topics treated in the last – suggests a text prepared for oral delivery, it is no longer supposed that these texts were actually read aloud by Newton on the dates indicated. The discrepancies between the two sets of optical lectures alone make this impossible, for Newton did not repeat his series. How he conducted himself in the Schools must therefore remain to a large extent a mystery. Fifteen years after his election to the chair his sizar Humphrey Newton observed (according to his recollections in old age) that 'when he [Newton] read [that is, lectured] in the Schools . . . so few went to hear Him, & fewer that understood him, that oftimes he did in a manner, for want of Hearers, read to the Walls.'[3] Lectures, even mathematical lectures, on optics and algebra may possibly have attracted more auditors than those of Humphrey's time on the subject-matter of the *Principia*. But there is every indication that few men ever heard Newton's lectures. Humphrey Newton may have attended in his time out of loyalty (for he could not have understood much); among more distinguished auditors were certainly John Flamsteed, shortly to become the first Astronomer Royal (a lecture on algebra, about midsummer 1674) and possibly Burchard de Volder (1643–1709), also in 1674.[4] De Volder was to become a celebrated professor of natural philosophy and mathematics at Leiden, who certainly visited Newton in Cambridge. As he introduced to him a certain J. C. Zimmerman, and later thanked Newton for his kindness in instructing this man, it is possible that Zimmerman too attended Newton's lectures.[5] William Whiston, Newton's deputy then successor in the Lucasian professorship, claimed to have attended one or two of the lectures which he had failed to comprehend well. He became an undergraduate of Clare College in 1686, and so could have attended the last, or penultimate, series of lectures that Newton ever gave, for after he had set the essence of the first two books of the *Principia* before the university he lectured no more.[6] When Newton had become one of the most famous men in all Europe, few registered a claim to have sat at his feet. One of these recollected being 'amongst a select Company in his own private Chamber', as well as attending his lectures in the Schools.[7] As has often been pointed out, only the most sporadic preparation was available for the advanced teaching that the Lucasian professors offered to *senior* students; the wonder is that any came at all. The statutes would have permitted the professor to choose more elementary and less abstract topics than his own recent researches, and late in life Newton opined that in a reformed Cambridge: "The Mathematick Lecturer [should] read first some easy & usefull practical things, then Euclid, Sphericks, the Projections of the Sphere, the construction of Mapps, Trigonometry, Astronomy, Opticks, Musick, Algebra &c."[8] No one could have listened to Barrow's lectures without intellectual effort, but at least he had begun by discussing mathematics in a general or philosophical manner. Newton, broaching his own recent discoveries in physical optics in continuation of Barrow's last course, by no means adhered to the plan that he favoured later. He soon rose to heights where few could follow his exposition.

At this point in the history of the university tutors and college lecturers had all the strings of educational power in their hands, while professorships were chiefly valued as sinecures creditable to their holders. College tutors probably sent as few pupils to professors of mathematics as to professors of Arabic. We may therefore safely conclude that while Newton put thought and labour into writing his lectures, his auditors can have been of little trouble to him.

Whether Barrow had some wish to proselytize his own favourite study (at that time) when he was translated from the chair of Greek to the new Lucasian professorship of mathematics in 1663, who can tell? If so he soon changed his mind. His wish to resign from it after only six years does not speak of great satisfaction in the job and it is probable that by 1669 he had an eye on high office in the future, to which his professorship might be a bar. Though only thirty-nine, he may also have felt that he had shot his mathematical bolt and that it was time for him to concentrate upon that highest of seventeenth-century studies: theology. That he did, with great applause: he was made a Royal Chaplain, and (in 1673) Master of Trinity College. This was an appointment at which Newton rejoiced, as well he might,[9] but Barrow was to enjoy it for only four years, during which he had a serious dispute with Newton.

The beginning of the acquaintance of these two men is unknown, and it is not at all unreasonable that it should have been some years old before leaving an historical trace. Feingold states (perhaps over-confidently) that 'undoubtedly Newton had had free access to Barrow's library ever since his undergraduate days', a suggestion already made (less positively) by D. T. Whiteside, even though the first evidence of the use of Barrow's books by Newton is as late as 1670.[10] It is no more than a guess that some of the scarce mathematical books read by Newton in early days came from Barrow; some certainly did not. Whether Newton's seeming failure to share his mathematical discoveries with Barrow as they developed – and because of the plague he could not have done so before 1667 – was due to Newton's timidity or secretiveness or to the distance in seniority between them cannot now be decided. In Whiteside's opinion (1969) Barrow had little direct stimulative effect upon Newton; still less (in any capacity) should it be held that Newton took further what Barrow had already begun. Yet even Whiteside wrote of a 'friendship' between the two men, and opined that 'Barrow was to afford Newton much-needed encouragement of his calculus researches.'[11]

There remains the question of Newton's reticence about his discoveries in the field of light and colour. It is hard to believe that Barrow had any knowledge of them when he wrote (and printed in October 1669) his *XVIII Optical Lectures*, a book proof-read by Newton. It is counter to all chronological likelihood that Newton had not by that time a high degree of confidence in his own investigations, upon which he was to lecture only a few months later. One can imagine several reasons for Newton's silence, if he was silent. He must have been aware that his revolutionary ideas about light were likely to evoke opposition. Perhaps at an early stage (say in 1668) he was still too diffident and uncertain to try to convert Barrow (who would later describe Newton's optical investigations as 'one of the greatest performances of Ingenuity this age hath afforded'); afterwards, perhaps, it seemed too late to urge Barrow to alter his text on what was, for him, a side-issue.[12] At any rate, it is gratuitous to imagine some malevolence on Newton's part.

In contrast with this shadowy question of Newton's relationship with Isaac Barrow before 1669, it is certain that Barrow was the chief promoter of Newton's early career. He probably encouraged Newton to prepare *De Analysi* in order to assert his independence of, and indeed priority over, Nicolaus Mercator's use of series in his *Logarithmotechnia*. If this was the case, some prior knowledge on Barrow's part of Newton's mathematical researches is implied (and is indeed plausible). Barrow then introduced Newton to John Collins, his mathematical friend of many years, and so to the London world. Collins then made Newton's name known to several other mathematicians in Britain (not least James Gregory) and in the rest of Europe, through the correspondence of his friend, Henry Oldenburg, the Secretary of the Royal Society. Collins and Newton met in London in November 1669, and engaged in an extensive correspondence; Newton's activity was continually stimulated by Collins's questions and gifts of books.

Barrow welcomed Newton's experimental researches upon light and colour, and encouraged him to print his optical letters (in vain). It was he who brought Newton's second reflecting telescope before the Royal Society in December 1671, an event marking the beginning of his wider fame. Best of all, Barrow in effect chose Newton to be his own successor. Newton's negligible public achievements before this appointment were not exceptional: few men in Britain were conspicuously equipped to be professors of mathematics before receiving their office. Of Christopher Wren, John Wallis and Isaac Barrow it might be said that their abilities were privately but not publicly displayed before their academic appointments. Further, Barrow had been concerned already with the implementation of Sir Henry Lucas's intention to enhance the teaching of mathematics in Cambridge and had the two elderly electors to the new chair in his pocket. Since he was convinced of Newton's outstanding quality, though even he did not yet understand how wholly exceptional it was, the way was open for Newton to receive the academic dignity and security which he enjoyed for the next twenty-six years.[13]

The Lucasian Chair

The salary of the Lucasian professor was set at the then high figure of £100 per annum, secured by the purchase of land in Bedfordshire.[14] He was also given solid privileges in that though he might hold a college fellowship and receive its dividends, he was debarred from all college and university offices (though perhaps he might be the head of a house). He was specifically prohibited from taking any pupil but a Fellow-commoner. Of these – young men of noble birth who shared the social amenities of the Fellows – two only were ever registered as Newton's pupils, in 1680 and 1687; the latter was a connection of Newton's grand relative, Sir John Newton.[15] Further – and this in the context of the age was extraordinary – the Lucasian professor was specifically forbidden to accept any benefice in the Church involving either a cure of souls or residence outside Cambridge. Most college statutes, requiring Fellows to take holy orders under pain of deprivation, were thus implicitly negated.

By January 1675 Newton's seven-year tenure of his lay fellowship at Trinity College had nearly expired, as he told Henry Oldenburg when asking to be excused his subscription to the Royal Society (this the Society granted). Another undated draft letter suggests that in 1674 or 1675, to avoid the necessity to take holy orders for the extension of his fellowship, Newton endeavoured in vain to change it for a lay fellowship in law, then vacant. The holder, in accordance with the slackness of the time, was not expected to be active in that discipline. Isaac Barrow, then Master of Trinity, received Newton's scheme "kindly" but the Vice-Master and Seniors would have none of it. Later, Newton followed the example of his friend Francis Aston (who was in a similar situation) in seeking from the Crown a dispensation from the necessity to take holy orders. Barrow officially stated the college's opposition to Aston's application, which failed, but said nothing against Newton's seeking the dispensation generally for all holders of the Lucasian chair, on the grounds that ordination ran counter to the intention of its founder that his professor should give all his attention to mathematics. Indeed, Barrow may have supported the request, for it succeeded. On 9 February 1675 Newton departed for London to see this business through, not returning to Cambridge till 19 March. This was perhaps his first visit to the capital since 1669, and he took the opportunity to be formally admitted to his fellowship of the Royal Society, to which he had been elected three years before. In this way Newton was relieved of the necessity to choose between losing his fellowship and becoming a priest of the established Church, as Barrow and most of his Trinity colleagues were. Nevertheless, in later life, not least as a servant of the Crown, Newton must have passed the ordinary religious tests without scruple of conscience.[16]

Dons might be totally idle, neglecting their studies, their teaching and their parishioners, but they were very rarely allowed to be laymen. Enforcement of the normal rule, like enforcement of the rule that Fellows should not marry (though heads of houses might) was one respect in which statutes were obeyed. These were rules that deeply affected the college system. The unwritten principle of this system was that Fellows should move on; a college fellowship was not to be a career office, even though the Fellow became dean or tutor or college lecturer, without service to the established Church. Taking holy orders enabled a man to retain a fellowship and be idle, at the price of remaining poor (Barrow regarded fellowship stipends as meagre) and celibate. Taking orders enabled a man, when he wished, to take his turn at the award of a college living according to his seniority, and so to embark on marriage and enjoy the comfort of his own home, with the hope of further preferment within the Church. Hence taking orders in the man, like confirmation in the boy, was an essential stage in the pattern of a don's life. Newton cunningly side-stepped into the studious, unemotional, experimental life he had always pursued.

As a professor, Newton could have married (as his opposite number at Oxford, John Wallis, did) and might not have lost his fellowship. As he evidently had no wish to marry the case did not arise, nor (considering his financial position) was there ever any inducement for him to seek the relative affluence of a country parsonage and the chance of clerical preferment.

The fruitful visit to London was not Newton's only excursion from Cambridge during the early years of his professorship, though such excursions were few enough and largely to family and friends in the midlands. His visit to London (26 November to 8 December 1669) following his election to the Lucasian chair has already been noted (p. 81); apart from his meeting John Collins nothing else is known about it and certainly Newton made no contact with the Royal Society then. Busy with his new responsibilities he did not leave Cambridge again until mid-April 1671 when (one can only guess) he took three weeks' Easter holiday at home. Shortly after his return he was afflicted "by the sudden surprisall of a fit of sicknesse, which not longer after (God be thanked) I again recovered of", as he told John Collins in a letter of 20 July 1671, in excusing a long silence of ten months.[17] At the same time he thanked Collins for the gift of G. A. Borelli's *De motionibus naturalibus a gravitate pendentibus* (1670). Newton became a regular recipient of books sent to Collins by scholars abroad for distribution to British mathematicians. Nearly eleven months later, by now a man of repute and a Fellow of the Royal Society, Newton left Cambridge for a visit to Bedfordshire of a few days (no doubt upon business about the professorial estates), then, after a brief return to Cambridge, set off on a long trip through Lincolnshire (staying with his mother) and Northamptonshire, returning to Cambridge on 19 July 1672.[18] His next absence was about Easter-time in the following year (10 March to 1 April). Perhaps the most interesting thing about these flickers of light upon Newton's private life is that they prove he had one, that he enjoyed friendships and family connections outside the well-known circle of his Cambridge friends, including the two Trinity men Francis Aston and John Wickins.

Newton Chooses to Lecture on Optics

As usual, the fabric of Newton's life was work. He began his first course of lectures in January 1670, some two months after his formal appointment as professor. What material he may have prepared in readiness for that event can one hardly guess. Optics presented itself to him as the most obvious topic for a variety of reasons: its choice was a compliment to Barrow; it enabled Newton to exploit a great deal of original material that he had by him (though as far as we can tell, much of it still in the form of rough notes); and his new material was of relevance to optical practice. Given the unlikelihood of his lecturing on advanced analysis or calculus to relatively unprepared students, this extraordinary young man could hardly have chosen a more suitable subject. What Newton surrendered by concealing the new mathematical techniques upon which he might have expatiated he gained by the opportunity to bring his experimental successes into the light. These sufficiently distinguished his lectures from those of Barrow, dealing abstractly with geometrical optics rather than with light and colour.

We can only presume that Newton had attended Barrow's lectures,[19] but Barrow himself is a witness to Newton's close connection with their printed text: *Lectiones XVIII Cantabrigiae in scholis publicis habitae in quibus opticorum phaenomenon genuinae rationes investiguntur &*

exponuntur (1669: 'Eighteen lectures given in the public schools at Cambridge in which the genuine reasons for optical phenomena are investigated and explained'). In an 'Epistle to the Reader' he names Mr Isaac Newton as 'our colleague (a man of outstanding abilities and notable experimental skill) who revised the text, pointing out some needful corrections and also offering some suggestions which you will find, suitably acknowledged, amidst my own work'.[20] (In fact, Newton's name nowhere appears in the text; his contributions are assigned to 'a friend'.) From Barrow's sentence it is clear that he had knowledge of Newton's skill as an optical experimenter, as well as his command of geometry.

Unfortunately, we know neither when Barrow's optical lectures were written, nor when they were passed to Newton. Some scholars think that both these events occurred long before their publication.[21] But it is equally possible (as D. T. Whiteside suggests) that Barrow's text was made available to Newton no earlier than the summer of 1669, in the form of proof-sheets. It is vain to speculate about the influence of one man upon the other in this context.

The most important of Newton's contributions was 'A neat and expeditious way of designating by geometry the [position of the] image [formed by a lens] in any given case; as also the construction of a lens that will project an image upon a given point.'[22] Here Newton presented a simple alternative to Barrow's prolix solution to the problem. The same elegance is seen in Newton's second contribution to the lectures: a neat construction for obtaining the tangential image-point for rays incident upon a lens at a distance from its axis.[23] Whether Newton was also the friend who informed Barrow about R. F. de Sluse's investigations in optics is less certain.[24]

These insertions by Barrow convince us that Newton was Barrow's peer in geometrical optics. Nevertheless, he was content to let Barrow's prior results stand without duplication, several times referring his audience to Barrow's book for proofs of propositions. However, whereas Barrow had claimed, over-modestly perhaps, that his aim was to do no more than explain systematically the work of earlier writers upon optics, Newton justly claimed to be explaining totally new ideas and discoveries about light and colour, and especially the improvement of telescopes.

It is difficult to discuss Newton's original presentation, to a Cambridge audience, of the work that had been a principal concern for him during the preceding six years, because the two extant manuscripts of the lectures disagree with each other at many points. The later text of thirty-one lectures differs from the earlier, and presumably incomplete, set of eighteen not only in containing extra material but in its order of exposition. The dates attached to material common to the two sets do not agree either. The later text was certainly written *post facto*, after a number of the lectures had been read, but it would be hard to prove that the earlier, shorter version had not been at least begun before Newton began his course, and so may in part represent his actual words. Both texts contain traces of a speaking voice. The auditors are directly addressed as "you", and the breaks between the earlier lectures are evidently original. It is worth while to quote here the first words ever uttered by Newton as a professor, on a January day in 1670, on an occasion that was then, as it still is, one of importance and trepidation:

> The invention of Telescopes has so exercised the majority of geometers that they seem to have left to others [working] in optics nothing which is not commonplace and no room for fresh discovery. Moreover, as the lectures which you heard here [in the Schools] not so long ago were composed of so great a variety of optical topics and an abundance of novelties, with very precise demonstrations of them, if I should again undertake to treat this science my endeavours may appear vain and my labour useless. But because I find the geometers hitherto to have been mistaken with regard to a certain property of light relating to refractions, while they tacitly postulate in their demonstrations a certain physical hypothesis [that is] not well founded, I judge that it will not be unwelcome if I submit the principles of this science to a pretty severe examination, and append to what my reverend predecessor last related to you in this place my own thoughts upon these matters, and my discoveries arising from multiple experiments.[25]

The first paragraphs of the lectures are the same in both versions. In my account of them I shall essentially follow the later text, made familiar to eighteenth-century readers after its first printing (in English translation) in 1728. For them it was an addendum to *Opticks* (1704). The earlier version was kept entirely private by its author and only recognized recently.[26]

The invention of the telescope had made the improvement of optical glasses a matter of widespread scientific interest; perhaps – as the French astronomer Adrien Auzout hinted – if telescopes were sufficiently perfected it would be possible to see buildings and even large animals if such things existed on the Moon! "Those knowledgeable in dioptrics", as Newton puts it, the theoreticians, had hitherto imagined that if only one could grind and polish lenses to the optimum geometric curvatures, telescopes might be brought to any desired degree of perfection. Such had been the message of Descartes in his *Dioptrique*, having recognized the impossibility that spherical lenses could form a point image. Newton now understood that such a hope was vain, at least without adopting a new plan of research.

The Announcement of Chromatic Aberration

The main purport of Newton's lectures begins with his announcement that the geometer's tacit belief in the equal refraction of all rays of light under the same conditions, and their assumption that colour is an unimportant variation in light, were both mistaken. The spherical aberration demonstrated by Descartes was not the sole nor even the chief defect of ordinary lenses, and "even if lenses were to be shaped to the best figure for their purpose that can be imagined, they would be no more than twice as good as spherical lenses equally well polished." What is at stake, Newton insists, is a certain irregularity in refractions, previously known to none, which overthrows this simple geometric theory.[27] He chose the same starting-point in his first optical letter, addressed to Henry Oldenburg on 6 February 1672:

> I left off my Glass-works; for I saw, that the perfection of [refracting] Telescopes was hitherto limited, not so much for want of glasses truly figured according to the prescriptions of Optick Authors, (which all men have hitherto imagined,) as because that Light it self is a *Heterogeneous mixture of differently refrangible Rays*.[28]

In the lectures he wrote more simply: "Concerning light, therefore, I have discovered that its rays differ from one another with respect to the quantity of refraction."[29]

Now this is an assertion as to matter of fact, and a large portion of the subsequent lectures is devoted to proving that this is a genuine fact and not an artefact. Newton had learned from his first prism experiments that (as he put it here) "the most refracted rays produce purple colours and those least refracted produce red, while those that proceed along intermediate lines generate the intermediate colours, blue, green and yellow."[30] That this is so, and that it is so not because of any adventitious circumstance but because its unique refrangibility is the defining characteristic of each individual coloured ray, is fully demonstrated in the lectures by experiments of which only a few are clearly recorded in Newton's notebooks. This is the best evidence we have of the extent and richness of his optical experimentation during

1667 and 1668, which enabled him to make his initial perceptions and reasonings more precise and complete. The only document extant from this period bearing directly on the lectures is the essay *On Colours* already mentioned.[31] In the longer version of the optical lectures the proof of Newton's fundamental physical assertion is made even more overwhelming than in the first version, perhaps (as Shapiro suggests) in response to the criticisms made against the first optical letter during the spring and summer of 1672, after its publication by Oldenburg in the *Philosophical Transactions* in March of that year. But no doubt Newton himself independently felt the need to rewrite, expand, clarify, modify . . . as was his wont. This rejection by the scientific world of his first attempt to improve natural knowledge deeply wounded Newton's sensitive ego, perhaps deepening the ambivalence in his nature between the desire to win fame by making his discoveries known and to avoid painful emotions by remaining silent. Whether Newton would have thus revised his course of lectures if some part of his mind had not considered publication of them as an ultimate objective, must remain a moot point; but it is certain that at more than one moment in the 1670s he did contemplate publishing them.[32]

Optical Atoms

It is perhaps worthwhile to remark at this point that Newton, like other deeply innovative scientists, tended unconsciously to write as though no sane person could possibly adopt any other view of the matters in question than his own. A ray of light, for example, previously a convenient fiction of geometrical optics, has become for Newton a physical entity, though one whose exact definition he was always reluctant to refine. One can no more produce such a Newtonian ray experimentally than one can produce a Newtonian atom; it is a theoretical construct. The non-producible, idealized, monchromatic ray of light was, as it were, the atom of Newton's theoretical concept of light. But it was his preference to justify his account of the formation of colours by refraction (and otherwise) pragmatically, without openly acknowledging the role of such a strange concept in his theory. Like all scientists at a major point of transition, Newton had great difficulty in persuading his contemporaries that the experiments pragmatically alleged in support of this account revealed facts of Nature, rather than artefacts. (Similarly, Darwin's critics regarded modifications of the races of animals and plants resulting from human selection in breeding as artefacts, not relevant facts about the possible evolution of creatures.) Robert Hooke was to protest – surely in all honesty – that the experiments alleged by Newton in support of his theory could be equally well explained by his own different hypothesis of light. Priestley made just the same point against Lavoisier, in defence of the phlogiston theory. In truth, Newton had gradually accustomed himself to a new language of optical theorization, which his critical readers failed to understand.

Newton's revolutionary concept of light must have seemed to the auditors of his very first lecture at Cambridge strange indeed. To challenge the perfectly uniform simplicity of white light, and the casual, illusory status of colour was to upset all received wisdom. The pure elementary nature of whiteness was as obvious as that of air, or water, or fire. As with these elements of being, light was an unanalysable natural fact, God's creation, whereas colour was mutable and changing, an unstable factitious quality modified by a reflection, a bit of glass, or a dye. If the essential nature of colour seemed indeed obscure its varied manifestations could be seen to arise from some interaction or other between light – in itself, paradoxically, invisible – and matter, which is revealed by light and also reveals light to our eyes. A long conventional theory that the darkness of matter when mingled with light produces not only all shades of grey but bright colours too is found in many early sources, including Theodoric of Freiberg's pioneering analysis of the rainbow, c.1300, which was the earliest attempt to link its colours with refraction. Descartes in Les Météores (1637) was responsible for a related, mechanical theory: contact with matter (depending upon its angle) variously causes the particles of light pouring out from the source to assume a spin, rather like a tennis ball striking the ground, and we perceive the varieties of spin as varieties of colour. Barrow in his recent lectures was not called upon to enter into the physical nature of colour, nevertheless, speculating 'uncharacteristically' (as he said) he silently passed by Descartes, reverting to the older notion of colour as arising from the modification of white light by its passage through the aether, or its reflection from matter.[33] Each of these three notions (and others, including Hooke's pulse-frequency theory) assume that white light is strong and simple, coloured light a weaker, altered form. In contrast, Newton set up an 'atomic' concept of light: we may liken each component ray in a white beam to an atom because it is infinitely small and its physical characteristics are never changed. In a refracted beam these infinite rays diverge from each other, in an infinite number of colours and refractions, though our eye observes these as falling into seven principal bands of colour, as in the rainbow. Subtraction of any ray or rays from the normal white composition causes the remainder to appear coloured. As well as refraction, reflection and the absorption of light by matter can cause such a subtraction of the rays. Newton's essential idea, that of early atomic theory, is that the character of each ray is unalterable: there is never any qualitative change of light though a quantitative redistribution of its rays may be effected in these and other ways. Because Hooke in particular argued against Newton's new theory in 1672, in favour of colour as a physical modification of white light, Newton went into the fallacy of this hypothesis more fully in the longer set of lectures.[34]

If modification was an erroneous concept, so also was 'splitting' or 'dilatation', the idea that the shape of the prism causes the light-beam to be stretched more in one direction than the other. Against this, Newton offered a cogent experiment that was to appear prominently in *Opticks*.[35] He shone a narrow beam of light through two prisms placed crosswise to each other so that the light was refracted in two planes at right angles. Any effect on the beam caused by the shape of the prism would now tend to stretch the beam in two directions, so that the image should be a square. In the event, the spectrum appeared as before, only a little longer, at 45° to the axis of either prism. This outcome justified Newton's contention that the spectrum was in reality a strip of overlapping circles, images of the round hole through which the beam passed, each individually refracted more or less, each in a sense an 'atom' displaced in one direction by the first prism, displaced again at right angles by the second prism and so now forming a diagonal strip.

The Improvement of Lenses

With the fourth lecture Newton took up 'The Measure of Refractions', continuing the treatment of the refraction of a homogeneous ray up to the end of Part I of the longer set of lectures.[36] The work here is partly experimental but in the main it is quite difficult geometrical optics. One of Newton's purposes was to distinguish chromatic from spherical aberration in lenses (to use later terminology)[37] and to demonstrate that the effect of the former in spoiling a perfect image was far greater than that of the latter, which Descartes alone had recognized. He also wished to discover whether the dispersion (another non-

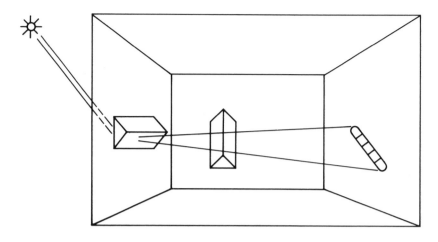

FIGURE 4.1   A beam of light is passed through crossed prisms; redrawn
from the first manuscript of the Optical Lectures

Newtonian term: the divergence of the coloured rays after refraction)
was proportional to the amount of refraction undergone by the beam,
as one would naturally expect; and again, whether the dispersion
was greater or less in proportion to the refractive power of the
transparent substance (its refractive index). Again, Newton expected
that when the refractive power was greater, the dispersive power
would follow suit.[38] He used these twin properties of transparent
materials in proposing that spherical aberration might be corrected by
a double glass lens enclosing water, the spherical surfaces being
suitably curved.[39] A passage in one of Newton's 1672 letters can be
read (but it bears more than one interpretation) as hinting that
chromatic aberration also might be corrected by the same kind of
compound lens: "it seem'd not impossible for contrary refractions so
to correct each others inequalities, as to make their difference regular;
and, if that could be conveniently effected, there would be no further
difficulty."[40] In *Opticks*, twenty years later, Newton was to describe
his compound water-lens in print for the first time, again for the
correction of spherical aberration, there remarking that such a lens
might serve to bring telescopes to "sufficient perfection, were it not
for the different refrangibility of different sorts of Rays. But by reason
of this [difference], I do not yet see any other means of improving
Telescopes by Refractions alone, than that of increasing their

— 104 —

lengths."[41] This counsel of despair entailed that such telescopes could possess, proportionately, but small apertures.

It seems certain that Newton never appreciated the fact that different kinds of glass possess different refractive indices, so that his water-lens was needless. The experiment and computation required to demonstrate the mediocrity of even a lens corrected for spherical error, if not also corrected for colour, took much space in the lectures, but Newton's solution to the whole problem, the reflecting telescope, was barely alluded to: "the ultimate perfection of [practical] optics (contrary to the received opinion) must be sought in dioptrics and catoptrics combined."[42]

Thus was one great problem of optics disposed of, by a typically Newtonian alliance of mathematics with experiment; an even more celebrated problem, the explanation of the rainbow and its colours, yielded to the same alliance.[43] The crucial medieval analogy between rainbow colours and refraction colours had been carried further by Descartes, who could submit the two types of joint reflection and refraction in the raindrop to calculation with the aid of the sine law.[44] Thereby he correctly computed the mean angular sizes of the primary and the secondary bow and (by a very ingenious mathematical process) derived their mean radii also; but he had nothing to say about the coloration. Newton went over Descartes's ground with greater exactitide, allowing that he had made "a most ingenious discovery about the refraction of the drops, and their limits", and was now able to explain that the width of the bow is due to the dispersion of refracted light, here given as 2° 6', the rainbow colours being produced by this dispersion as in the spectrum.[45] With this explanation the fuller text of the lectures ends.

The Physical Nature of Light Unexplained

It should be noted that nowhere in the lectures did Newton explain the physical nature of light, nor the nature of the light ray which is his ultimate unit of light. Only in Query 29 of *Opticks*, added to the 1706 Latin translation of that work, did he at last propound boldly: "Are not the Rays of Light very small Bodies emitted from shining Substances?" But no one doubts that Newton held this view from the first, while cautious never to make any of his arguments in optics depend overtly upon it. He would have found this corpuscular hypothesis of light in Descartes and in James Gregory's *Optica promota* (1663), the latter containing the best proposal for a reflecting telescope before Newton's own.[46] Openly, Newton confessed only to the two conditions that (1) light-rays be rectilinear and infinitely small in their cross-section; and (2) though they mix (in our vision) they do not blend. This second condition led him to conclude that (as he was to write later in Definition 1 of *Opticks*) the light-ray has "least Parts, and those as well Successive in the same Lines, as Contemporary in Several Lines."[47] In support of this concept he offered a conclusive (as it seemed to him) experiment: a beam of light refracted by the prism A shines upon a broad lens B and is brought to a focus at point C^2. If a white screen is placed at C^1 a spectrum is seen, and the same inverted is seen at C^3. At the focus C^2 only a white spot is visible on the screen because there all the dispersed rays are reunited into white light; but they are not *blended* there because they reappear at C^3 without further reaction. Moreover, blocking one or more rays at C^1 causes consequential subtractive coloration at C^2 and disappearance of these rays from the image at C^3. Rapid interruption of the rays at C^1 (as by a swiftly rotated toothed wheel) has no effect at C^2 or C^3; if the rays can be so 'chopped' mechanically without any effect upon them it must be because they are already 'chopped' in nature.

Whether or not this experiment can be as clearly contrived as Newton claimed, or his argument from it is really decisive, it does *not* prove that the infinitely small and discrete rays are streams of

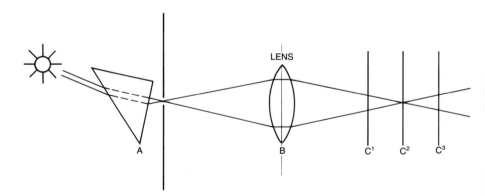

FIGURE 4.2 Prismatic colours are recombined into white at the focus of a lens, after Newton's diagram in the first manuscript of the Optical Lectures

material particles. Nor did Newton ever make such a weak claim. Though necessarily divided into 'packets' light might be a substance, or any force or quality of a substance. But Newton had, as he thought, very strong independent reasons for believing that light could not be a wave-motion, even though he understood very well that colours could be hypothetically accounted for as a function of the varying frequency of waves all having the same velocity.[48]

Evidently, though Newton in *Opticks* cautioned readers against studying imperfect versions of his researches, that is (chiefly) the optical lectures, rather than the printed book of 1704, Book I of *Opticks* is no more than a skilfully reworked version of the thirty-one optical lectures. After its opening pages, the printed text added neither fundamentally new materials to the lectures (still unprinted) nor new ideas.[49] Indeed, a good deal that was elaborately spelled out in geometry in the lectures is passed over rapidly in the book, which was intended for a wide readership. The outline, but by no means all the detail, of what was later to be Book II of *Opticks* was also present in Newton's mind in 1670, but instead of working this material up he decided (leaving some of his announced plan unfulfilled) to break away from optics in order "to turn to the more abstract parts of mathematics", upon which he made a start in October 1673. According to Newton's own arbitrary division of the manuscript, he had abandoned the lecture-room in the autumn of the previous year after delivering only his final three optical lectures!

In some respects the modern translation of the thirty-one lectures reads more like a modern monograph than does the *Opticks* book. This is partly, of course, because the book is couched in seventeenth-century language in its vocabulary and sentence structure, but partly also because the form of the lectures is extraordinarily modern. Newton gradually sets up a geometrical structure, introducing new concepts as he goes along, illustrating and justifying his evolution of the subject by means of experiments, whose quantitative results are in many cases all too precisely compared with the predictions of theory; this is mathematical physics of extreme power and accuracy. The elaborate computations which permit such comparisons to be made were largely excised from the more popular text of *Opticks*. This explains why, after the *Lectiones Opticae* had at last been printed, they were regarded as providing the more thoroughly scientific account of the earlier phase of Newton's investigations into that science.

Newton turns to Algebra

It is inconceivable to me that Newton took so much trouble with the text of his optical lectures without any idea of their future publication, following Barrow's double example. Ample later evidence suggests that he long retained the idea of publishing his experimental discoveries, in more than one form. But from the first the plan was confused with a quite different enterprise, also occupying Newton's time from December 1669 onwards, a project for a new Latin algebra. Newton's and Barrow's friend John Collins was oppressed by the deficiencies of British textbooks of algebra and, being unable to prepare such a book himself, still less to finance it, he asked Newton to edit a Latin translation of *Algebra, ofte Stel-konst* (Haarlem, 1661) by Gerard Kinckhuysen, one of the group of Dutch mathematicians who had carried on Descartes's work during the middle years of the seventeenth century. Whiteside calls his book 'an unlikely blend of basic elements and long-winded technicality'.[50]

This translation, the work of Nicolaus Mercator,[51] became the subject of "some notes" written by Newton in January 1670. Newton thought less well of the Dutch algebra than Collins did, opining that it contained "nothing new or notable which is not to bee found in other Authors of better esteeme". Rewriting the text in some places, enlarging it in others, Newton's "Observations" upon it were far from flimsy, taking up about forty pages in the *Mathematical Papers*. Some of this material was to be used by Newton later in his Lucasian lectures on algebra.[52] However, after receiving back the translation with Newton's "Observations", it was now Collins's turn to be dissatisfied with Kinckhuysen's treatment of binomial surds, and to beg Newton to amplify it, enclosing three other books for comparison.[53] This was further occupation for Newton's summer; he may well have felt that Collins was drawing him into a quagmire far from his own path. Collins's bait that 'your paines herein will be acceptable to some eminent Grandees of the R. Societie' would hardly have tempted him.[54] Delaying an answer till the end of September upon the excuse that he had had to wait for Barrow's return to Cambridge in order to look into his books, Newton made clear his view that Collins was seeking the impossible. Nevertheless, Collins's imposed task made Newton consider "writeing a compleate introduction to Algebra" because his own treatment of topics and Kinckhuysen's made an ill marriage. Kinckhuysen's tendency to solve problems by particular techniques rather than by applying a general method was in Newton's judgement "lesse propper to instruct a learner, as

Acrostick's & such kind of artificiall Poetry though never soe excellent would bee but improper examples to instruct one that aimes at Ovidian Poetry."[55] But "I have chosen rather to let it passe" and after his rare literary comparison Newton heard no more from Collins for a long time.

Although this affair reveals Newton's willingness to assist others' schemes, it would scarcely be worth a mention (save as a hint of some occupations of the new professor and a suggestion of background to his choice of algebra as his next topic for Lucasian lectures) were it not for the next development, which came as a consequence of Collins's failure to find a publisher for Kinckhuysen's improved text in Latin. While he had dallied, John Kersey, a mathematical practitioner, had put his own (English) *Elements of . . . Algebra* in the press.[56] After that, no bookseller would touch a competing text, especially as Barrow's geometrical lectures had proved a financial disaster. Collins now formed the rather unlikely notion that something from Newton's own pen would find 'better entertainement and more Speedy Sale'. Newton, who still had the Kinckhuysen text on his hands – indeed, he had "reviewed" it again during the long months of Collins's silence – welcomed the idea (July 1671), now announcing that he had also begun, partly upon Barrow's instigation, "to new methodise the discourse of infinite series, designing to illustrate it with such problems as may be more acceptable than the invention it selfe of working such series". He hoped to complete this work during the following winter, after which he might be able to annex to Kinckhuysen "something which I may call my owne [in algebra], & which may be acceptable to Artists as well as the other to Tyros."[57]

The Method of Series and Fluxions, 1671

None of this came about. The new methodized tract was of course *De Analysi*, greatly amplified by material from the 1666 tract on fluxions, the product being the treatise afterwards referred to by Newton as his 1671 *Tractatus de methodis serierum et fluxionum* ('Treatise on the methods of series and fluxions'). We do not know what title, if any, Newton gave to it since the first leaf is missing from the manuscript, perhaps worn away by the favoured few to whom it was shown. Never published in Newton's lifetime, it was first printed in an English translation by John Colson in 1736.[58] As for the Kinckhuysen translation, it remained for some years in Newton's hands, while no

Cambridge printer or bookseller would handle it. Somehow it was conveyed to John Wallis in Oxford, among whose papers it was discovered not many years ago. The £4 paid by Newton to the bookseller Moses Pitt for the copyright was money wasted.[59]

Therefore it happened that the only printed book to be credited to Newton before the *Principia* in 1687 was the first British printing of *Geographia generalis* by Bernard Varenius (Amsterdam, 1650). Newton's Cambridge text of 1672 (in Latin, of course) shows that he had corrected some slips (introducing others of his own) while leaving it unaltered. The book taught the Copernican doctrine and stated Snel's nearly exact value for the circumference of the Earth. It contains much mathematical science and is a sound introduction to the natural philosophy of the globe. It may be that Newton read it with classes or private pupils. He made a few more improvements in the 1682 edition, reprinted in 1712; then it had a further life in an English translation (by Douglas, 1733).[60]

Now to return to the optical lectures. Collins is our informant that in the heady days after the warm welcome given to his reflecting telescope in December 1671 Newton planned to publish '20 Dioptrick Lectures and some about infinite series, with his additions to Kinckhuysen's Introduction'. At the end of 1672 Collins offered to get this book published in London. Tales about Newton's activities got abroad, and it is interesting that twenty lectures were in question, for this points to the unfinished first text of the *Lectiones Opticae*, abandoned (it seems) about the end of 1671, rather than to the later version with thirty-one lectures, which perhaps Newton had not yet begun when Collins received the story from Barrow. However that may be, in later years Newton related that "friends" had urged him to prepare a book on optics for publication at this time.[61]

Though Newton had the fixity of purpose to complete, revise and fair-copy the longer text of thirty-one lectures after his engagement in optical controversies with Hooke and Huygens, the controversies impelled him to renounce the idea of publication. As he wrote to Collins:

> Your Kindnesse to me . . . in profering to promote the edition of my Lectures which Dr Barrow told you of, I reccon amongst the greatest, considering the multitude of buisinesse in which you are involved. But I have now determined otherwise of them, finding already by that little use I have made of the Presse, that I shall not enjoy my former serene liberty till I have done with it; which I hope will be so soon as I have made good what is already extant on my account. Yet I may possibly

complete the discourse of resolving Problemes by infinite series of which I wrote the better half the last christmas with intension that it should accompany my Lectures, but it proves larger than I expect & is not yet finished.[62]

Six weeks later another letter to Collins confirmed his decision about the lectures, while leaving the completion of *De methodis serierum* still uncertain. Henry Oldenburg, with whom and his *Philosophical Transactions* Newton had also been involved, received the same message: Newton "intended nothing further for the public".

As we shall see in the next chapter, these were not pronouncements to be read as literal and final. Although one may reasonably see in Newton's withdrawal into his college shell a first ominous sign of the paranoia which later exercised so baneful an influence upon his career, one must not exaggerate its effect. Newton protested but he published when he had a mind to do so. During the next few years he wrote a bookful of explanations of his experiments, most of which was printed in the *Philosophical Transactions*.[63] It was hardly unbalanced to be impatient with repeated objections after three and a half years of replying to them at length. Further, at the end of 1675 he sent to the Royal Society another large body of material on light and colours, in effect Book II of the future *Opticks*. And twice thereafter, despite renewed criticisms from Robert Hooke provoked by this second batch of material, Newton responded cordially to overtures from Hooke. With Robert Boyle, John Flamsteed and others he corresponded freely. His reticence did indeed extend to his mathematical correspondence with Leibniz.

The Fate of the 1671 Tract

As already noted, the writing of the treatise *De methodis serierum*, Newton's chief mathematical work in this period which forms a large part of volume III of the *Mathematical Papers*, was not resumed after the end of 1671. Its early history, apart from its use by Newton in composing his long calculus letters to Leibniz in 1676, is obscure. Collins is not known to have seen it. About 1691 Edmond Halley and Joseph Raphson tried without success to persuade Newton to revise the already superannuated work and prepare it for publication. At this time it was (by Raphson's report) already 'much worn by having been lent out' but to whom it was lent is not known. There is some evidence that inaccurate copies of the treatise passed from hand to hand in the early eighteenth century. Finally, in 1710 William Jones was allowed to prepare an 'authorized' copy, from which Colson's 1736 translation was made.[64]

This 1671 treatise was a revision of earlier studies, rather than a wholly new work; that is, it combines an enlarged *De Analysi*, dealing with the expansion of algebraic expressions into infinite series; a section on the inverse operations of deriving the fluxion of a quantity (the fluent) and the fluent from the fluxion, employing this pair of terms for the first time;[65] and ten problems revised from the October 1666 tract.

The formulation of fluxion and fluent is of course of the greatest importance. It is well worthwhile to quote here, as Whiteside does, Newton's own justificatory lines. "In this Compendium", meaning *De methodis serierum*, Newton wrote in an anonymous paper forty-five years later,

Mr. Newton represents the uniform Fluxion of Time, or of any other Exponent of Time by an Unit; the Moment of Time or of its Exponent by the Letter *o*; the Fluxions of other Quantities by any other Symbols; the Moments of those Quantities by the Rectangles under those Symbols and the Letter *o*; and the Area of a Curve by the Ordinate inclosed in a Square, the Area being put for a Fluent and the Ordinate for its Fluxion. When he is demonstrating any Proposition he uses the Letter *o* for a finite Moment of Time, or of its Exponent, or of any Quantity flowing uniformly, and performs the whole Calculation by the Geometry of the Ancients in finite Figures or Schemes without any Approximation: and so soon as the Calculation is at an end, and the Equation is reduced, he supposes that the Moment *o* decreases *in infinitum* and vanishes.[66]

(In parenthesis, note in this passage the typically Newtonian insistence that the method of fluxions proceeds in essence by the geometry of the ancients; it involves no approximations, no limit-increments.) In the same paper, a few pages later, Newton continues his defence:

And whereas it has been represented that the use of the Letter *o* is vulgar, and destroys the Advantages of the Differential Method: on the contrary, the Method of Fluxions, as used by Mr. *Newton*, has all the Advantages of the Differential, and some others. It is more elegant, because of his Calculus there is but one infinitely little Quantity represented by a Symbol, the Symbol *o*. We have no Ideas of infinitely little Quantities, and therefore Mr. *Newton* introduced Fluxions into his Method, that it might proceed by finite Quantities as much as possible. It is more Natural and Geometrical, because founded upon the *primae quantitatum nascentium rationes*, which have a Being in Geometry, whilst *Indivisibles*, upon which the Differential Method is founded, have no Being in Geometry or in Nature.[67]

(It should perhaps be emphasized at this point that in the original Leibnizian conception – before Jean Bernoulli modified it – the differential was a minute fixed quantity, a true infinitesimal.)

In Whiteside's opinion the publication of Barrow's last series of geometrical lectures in the summer of 1670, coupled perhaps with a rereading by Newton of his earlier series, had a creative effect upon Newton's introduction of this more definitive algorithm for his calculus. Indeed he seems to echo some of Barrow's phrases. Newton's method of expressing his method of fluxions was, it must be said, still clumsy, where fluents are designated by the letters v, x, y and z:

> the speeds with which they flow and are increased by their generating motion (which I might more readily call fluxions or simply speeds) I will designate by the letters l, m, n, and r. That is, for the speed of the quantity v I write l, and for the speeds of the other quantities x, y and z I put m, n, r respectively.

After this follows the algorithm for differentiating an expression, just as in the 1666 tract.[68]

De methodis serierum is a very substantial monograph, about 140 pages in Colson's English text and about 160 in Whiteside's annotated translation. As Newton himself probably realized in mature years it represented his last possible basis for a claim to have made a public declaration of his new mathematical method – if he had published it. As Whiteside says:

> It is tragic that he himself could never feel an urgent necessity to publish them [Newton's mathematical writings of the 1670s] – if he had done so, scientific history might well subsequently have taken a different course. [In] all these papers – above all the 1671 treatise – Newton's breadth of vision and his complementary mastery of technical detail are as impressive as ever, and they will remain always of fundamental importance to any understanding of his methods of fluxions and infinite series.[69]

A Paper on Dynamics: the Cycloid

Of Newton's mathematical work at this period – by no means limited to the optical lectures and the tract just discussed – one other short draft may be mentioned merely because it shows his ready capability in dynamics, an applied mathematical science in which the young Lucasian professor seemed to interest himself little. For until this brief paper was written, post-1670, only the single sheet of central force calculations (see p. 60) reveals any interest in mechanics on Newton's part since the plague period.

Whiteside has guessed that the present paper, linked by 'strong internal mathematical evidence' with the '1671 fluxional tract', was composed by Newton after reading James Gregory's little eight-page tract 'Tentamina quaedam geometrica de motu penduli et projectorum' (that is, 'Certain geometrical essays on the motion of the pendulum and of projectiles') in 1672. This may well be the case, but we have no evidence of Collins's ever sending this pamphlet to Newton, despite his allusion to it in a letter of 30 July 1672.[70] By 1672 every person in Europe interested in mathematics knew that Huygens had perfected the pendulum clock by making the arc cycloidal and isochronal, and that Huygens had a demonstration showing why this was necessary. Huygens had made known this technique for perfecting the oscillations of a short (half-second) pendulum in 1659.[71] In *Horologium Oscillatorium* (1673) he at last demonstrated in print (1) that a cycloidal path of motion is isochronous and (2) that because the evolute of a cycloid is the same cycloid, confining the pendulum thread between two cycloidal 'cheeks' of metal placed back-to-back will render the track of the bob cycloidal. It seems likely that Newton's paper was written before Oldenburg sent him a copy of this book, as its author's gift, in early June 1673.[72]

One may, perhaps, draw confirmation that this was so from Newton's somewhat unexcited reception of this important book, which was the first public key to the study of mechanical forces.[73] His acknowledgement of the gift copy opens a letter whose main content is Newton's further explanation of his (now published) new theory of light and colours, which had so far seemed to Huygens neither clear nor convincing.[74] Newton had "viewed" Huygen's treatise "with great satisfaction, finding it full of very subtile & useful speculations very worthy of the Author". After this rather doubtful encomium – for "speculations" seems almost the last word to apply to Huygens's very solid geometrical reasoning – Newton silently passed by Huygens's pendulum theorems to refer to his (undemonstrated) theorems on centrifugal force, and to point to a possible error.[75] One would certainly not guess from this letter that the question of pendulum motion and the geometrical definition of the exactly isochronal path had ever held any interest for Newton.

Christiaan Huygens certainly, and James Gregory presumptively, were familiar with Galileo's treatment of pendulum motion in *Discorsi e Dimostrazione matematiche* (Leiden, 1638), and took their start from Galileo's theorems.[76] Not so Newton; almost incredibly, it seems that this celebrated work was unknown to him. Instead, Newton's investigations go back independently to mechanical theorems noted on the very first page of his "Waste Book" (1664). He states in the present paper that "if two heavy bodies are poised to descend to C [the lowest point of a vertical circle], one straight down along the diameter BC, the other at a slope along the chord YC . . . both heavy bodies . . . will reach C simultaneously."[77] Starting from this same principle, Huygens in *Horologium Oscillatorium* generalized by a kinematical argument to the cycloidal arc as being the curve (as distinct from the various isochronous chords) along which a body will always descend to the centre in the same time, whatever the length of its descent. Newton reached the same result by a dynamical argument – in a manner that was to be paralleled, strangely enough, by Huygens shortly afterwards in utter independence of Newton.[78]

D. T. Whiteside brings out the significance of such a dynamical process of argument for Newton's future investigations in mechanics, when he returned to this field:

> this dynamical approach by equating instantaneous acceleration in the direction of motion to the component of the force of 'gravitas' which provokes it was radically novel, indeed unprecedented. From it was to evolve the powerful general theory of central force orbits which Newton later present in Proposition 41 of the first Book of his *Philosophiae Naturalis Principia Mathematica* (London, 1687).[79]

To this work indeed was to be transferred an improved version of the demonstration just considered, but this was some years in the future. Meanwhile, Newton's knowledge of technical astronomy and especially of the problems of planetary motion was to be considerably advanced by his study of such works as Nicolaus Mercator's *Institutionum astronomicarum libri II* (London, 1676).[80] Then, at the end of the decade, it was Robert Hooke who inspired a reluctant Newton to apply his dynamic principles to planetary orbits.

5

Publication and Polemic,
1672–1678

The Reflecting Telescope

The close of the year 1671 was marked by the public appearance of
Newton upon the national stage of the Royal Society of London. It is
strange that with both Newton and his arch-rival Leibniz this event
happened in the same unexpected and unlikely way: each of them
presented a gadget to the Royal Society. In Leibniz's case it was an
arithmetical machine, a toothed-wheel device of the type pioneered
by Blaise Pascal, in Newton's his miniature reflecting telescope.
Newton's introduction to the Society (December 1671) preceded
Leibniz's by just over a year. Both men were, for their pains, severely
mauled by Robert Hooke, with assertions that their inventions were
valueless and much inferior to his own schemes.[1] Both were deeply
wounded by this reception, though also conscious of the warmer
welcome extended to them by other Fellows of the Royal Society, not
least its Secretary, Henry Oldenburg.

   Newton and Leibniz certainly did not meet during this first visit of
Leibniz to London (January–March 1673), though before long Collins
would be sharing with Leibniz an outline of the mathematical

progress accomplished by James Gregory and Isaac Newton. Yet curiously there may be a link between Leibniz and Newton's reflecting telescope. In a letter that Oldenburg received probably about 20 October 1671, Leibniz (writing about the innovations made by a young German optician) mentioned 'Tubos Catadioptricos, quales mihi in mentem venerunt' ('reflecting-refracting tubes, such as occurred to my mind').[2] No explanation of this phrase has been found, but Newton had certainly invented a catadioptrical telescope. Oldenburg made a note to read Leibniz's letter to the Royal Society, as usual, but an imperfection in the minutes at this time prevents our confirming that he did so. It is therefore a conjecture that these few words of Leibniz came, by some friend who already knew of Newton's telescope, to the ears of Newton himself. If they did, they might have stimulated him to prove publicly what *he* could achieve in the catadioptrical way. There is no other guess at the reason why, in the autumn of 1671, he set about repeating, on a slightly larger scale, his former success in constructing the type of instrument of whose merits Leibniz was clearly apprised, though of course only in relation to the correction of spherical aberration. As soon as it was finished, Newton ensured that it was laid before the Royal Society. He had no need to satisfy himself and a few private friends that such a reflector would really work, for this he had done in 1668.

That concave mirrors like convex lenses enlarge objects had long been known, and the idea of employing mirrors in a complex lens-system is at least as old as the English mathematician of Elizabeth's reign, Leonard Digges. However, James Gregory's design of 1663, involving a large pierced primary speculum (through which the rays reflected back from the secondary mirror could pass to the eye) was the first precise proposal of the kind. Gregory's attempt to have his design realized by the London instrument-maker Richard Reeves failed, probably because of his inexperience in polishing a large primary mirror of metal. It may be that Newton deliberately avoided Gregory's axial design, just as he certainly criticized another axial proposal, that of Cassegrain, because of the awkwardness of polishing a pierced mirror; in correspondence he advanced many reasons for preferring his own design, in which the secondary mirror (or prism) requires only plane surfaces and the whole of the primary is available for reflection. At this time no process was known for silvering an optical surface of glass; several experimenters (including both Newton and Hooke) therefore tried curved glass mirrors silvered on the back, convex surface in the usual way, but these never answered to expectations. Hence, like Gregory before him, Newton resorted to speculum metal, making many experiments to find a hard, white bronze. It proved impossible to make an alloy that reflected as well as silvered glass does, that was as hard as glass and as resistant to tarnishing. But Newton's experience – confirmed by telescope–makers in later generations, not least James Short and William Herschel – was that astronomically useful metal specula could be cast and polished. To attain success with even a tiny mirror requires no mean skill and patience; it is a tremendous testimony to Newton's persistence and artisanal skill that he did succeed. To the end of his life Newton admired his own dexterity. If he had waited for others to make his tools for him, he declared, the thing would never have been done.[3]

Newton's new telescope was brought to London by Isaac Barrow and shown to several Fellows of the Royal Society, then in its Christmas recess. It also amused Charles II at Whitehall. Its effect for its short length – six inches – and small aperture amazed everyone save a few sceptics. Defence of Newton's priority seemed very important; within a few days of the telescope's first being seen in London Henry Oldenburg sent an account of it to Christiaan Huygens in Paris.[4] Steps were taken at once to develop Newton's small instrument into one of workable size, but these (like Gregory's earlier) failed – as did Huygens's attempts of the same sort – and it is unlikely that any scientifically useful reflector was made in the seventeenth century.

Its significance, then, is that it made Newton's name known, especially to such astronomers as Huygens, John Flamsteed or Richard Towneley, who had barely heard of Newton as a mathematician, if at all. And more important still, the success of his miniature instrument rendered Newton untypically communicative. In response to Oldenburg's news that he had been elected a Fellow of the Royal Society, Newton wrote (18 January 1672):

> I desire that in your next letter you would inform mee for what time the [Royal] Society continue their weekly meetings, because if they continue them for any time I am purposing them, to be considered & examined, an accompt of a Philosophicall discovery which induced mee to the making of the making of the said Telescope, & which I doubt not but will prove much more gratefull then the communication of that Instrument, being in my Judgment the oddest if not the most considerable detection which hath hitherto beene made in the operations of Nature.[5]

True to his word, Newton little more than two weeks later posted off to Oldenburg what was to be his first printed work, the first letter on his new theory of light and colours.

The New Theory of Light and Colours, 1672

This letter, in John Wickins's fair copy some ten pages long, is puzzling in more ways than one. Its distorted account of how Newton came to make his optical discoveries misled posterity down to recent times, and for the exposition of an experimental discovery it is itself astonishingly non-experimental. The one experiment upon which Newton laid great emphasis in his letter is largely fictitious, newly minted for its present purpose. I mean that while Newton certainly did make what he called his "crucial experiment" in one way or another, as his notebooks and lectures reveal, it was never perhaps exactly as he so confidently described it in the letter, and this experiment never carried, in the historical development of Newton's optical ideas, the uniquely decisive role there assigned to it.

As evident bases for his new theory, Newton presented in the letter two phenomena: (1) the spectrum cast by a prism refracting a beam of light is elongated, not circular, as the merely displaced image of the round hole admitting the beam would be; (2) if a large spectrum is carefully formed and by means of diaphragms a single coloured ray is separated out, a second refraction does not alter its colour. In any predominantly theoretical paper the balance between evidence and theory is always difficult to strike. Newton's somewhat cavalier attitude on this occasion to the massive amount of experimental evidence that he had collected in favour of his new theory was to cost him dear in subsequent controversy. His assertion that a full spectrum – in itself a phenomenon that few at this time had witnessed – is elongated rather than round was not much challenged, though some critics denied that the length/breadth ratio was as high as Newton claimed, or doubted that the coloured bands were quite as he had described them. The "crucial experiment" was much harder for readers to swallow. It has never been repeated without the exercise of great care, and often (to obtain success) experimenters have found it necessary to embellish Newton's procedure for producing a homogeneous ray. Casual repetitions, in less than perfectly black rooms, with apertures too large and so on seemed to falsify Newton's description. This was the case with the experimenter of the Académie Royale des

Sciences in Paris, Edmé Mariotte, in 1681; for him a seemingly pure ray, on second refraction, produced new colours. Such was Mariotte's authority in France that his proclamation of Newton's error banned Newton's optics from France for a generation.

Newton's rhetorical exposition of his interpretation of the new effects that he had discovered was much more effective. The physical essence of his new conception of light emerges more clearly than it does in the optical lectures. It reduces to two propositions: (1) every pure coloured ray possesses its own unique refrangibility; (2) white light is a heterogeneous mixture of such variously refrangible rays, in due proportions. Newton combines the two in the statement: "Light consists of rays differently refrangible."[6] Newton did not here attend to the point that the colours in the spectrum are not seen by the eye as continuously changing, but (as he defined them) in seven bands, so that red, for example, is a broad strip rather than an infinitely narrow band of unique refrangibility. Nor did he explore the distinction between prismatic colours and pigments, which some of his critics were to find a source of confusion.[7] The twin propositions introduce the concept of chromatic aberration (though not the term), the assertion that this aberration is "some hundreds of times" more damaging to the image than spherical aberration, and the recommendation of mirrors in preference to lenses for optical instruments. Newton then explained very tersely how his new concept of light accounted for the colours of both the spectrum and the rainbow, the odd ability of some substances to transmit light of one colour, but reflect that of another, and the variety of hues of natural bodies, these being "variously qualified to reflect one sort of light in greater plenty than another".[8]

Finally, Newton narrated a second experiment to confirm his idea of the mixture of primary colours, forming white or some hue. It is the one already mentioned, in which the divergent beam from a prism falls upon a lens (figure 4.2); thereby the separated coloured rays are united at the focus of the lens or appear again, inverted, on the distal side of the focus. For Newton, this experiment gave ocular testimony to the heterogeneity of whiteness, only produced by the presence of all the colours. However, his assertion that "the new Production of Red, or any intercepted colour will be found impossible" seemed to critics incompatible with the earlier correct statement that "The same colours *in Specie* with [the] Primary ones may also be produced by composition: For, a mixture of *Yellow* and *Blew* makes *Green*; of *Red* and *Yellow* makes *Orange*." Newton was of course aware that if alike *in specie* the two types of apparently identical colour were not alike *in essentia* since the primaries could not be resolved into components by the prism, as secondary shades formed by mixing could be. Nor could red and blue be formed by mixing. It would have been wiser, perhaps, in so short a communication not to embark upon these thorny questions which remained troublesome into the nineteenth century.[9]

In the course of the letter, Newton committed himself to two interesting affirmations, both expunged from the printed text in the *Philosophical Transactions*. In the first he seems to claim a mathematical certainty for his new theory:

> A naturalist would scearce expect to see the science of [colours] become mathematicall, & yet I dare affirm that there is as much certainty in it as in any other part of Opticks. For what I shall tell concerning them is not an Hypothesis but most rigid consequence, not conjectured by barely inferring 'tis thus because not otherwise or because it satisfies all phenomena (the Philosophers universall Topick,) but evinced by the mediation of experiments concluding directly & without any suspicion of doubt.[10]

This is a curious statement. That the science of colours became mathematical by Newton's discovery is true, to the extent that geometrical optics was broadened to take in the separate refraction of each distinct ray. Thus Newton's discovery added a new dimension to Snel's Law. But Newton had hardly yet constructed a mathematical science of colours in the same sense that Galileo constructed a mathematical science of natural motion. Again, what did he mean by his Baconian confidence in "experiments concluding directly and without any suspicion of doubt"? Did he mean that such a complex proposition as white light is heterogeneous, not homogeneous, had been as it were dictated to him by the experiments he had made? If he really believed that no sane human being could fail to come to the same interpretation of his experiments as himself, he was to be greatly surprised by the reception his letter received. For as yet Newton seems not at all to have asked himself the basic question: when an experimental arrangement produces a certain effect, Nature having been perhaps thrust out of its usual course (as Bacon said), how can we be sure that the effect reveals a normal aspect of Nature rather than some artefact caused by the experiment itself? Indeed, we have an example in the second of the experiments recited by Newton in his letter: the white patch produced at the focus of the lens is an artefact. It is not natural white light, for the beam at this point diverges and, after a little space, yields a spectrum, neither of them features of an ordinary beam of white light. Newton's critics would soon protest that his theoretical propositions were far from being obvious and patently true.

The second of Newton's affirmations, to which we now turn, was a simple error in rhetoric. Newton intended to adopt a pragmatic position: *this* stated characteristic of light follows directly from *this* evidence. He meant to avoid proposing any physical hypothesis defining the nature of light and colour (in the manner of Descartes and Hooke, for example); even more, he meant to avoid deducing his fundamental proposition – that white light is a mixture of homogeneous coloured rays – from some physical hypothesis of light. He adhered to this pragmatism while discussing refraction and its colours in the letter, but when he came to the colours of natural bodies, made apparent by their absorption of those constituent rays in white light which they fail to reflect (a process of subtraction at which Newton barely hinted in the letter) he was so taken with his own idea that he slipped unguardedly into dangerous scholastic language:

> These things being so, it can be no longer disputed, whether there be colours in the dark, nor whether they be the qualities of the objects we see, no nor perhaps, whether Light be a Body. For, since Colours are the *qualities* of Light, having its Rays for their intire and immediate subject, how can we think those Rays *qualities* also, unless one quality may be the subject of an sustain another; which in effect is to call it *substance*. We should not know Bodies for substances, were it not for their sensible qualities, and the Principal of those being now found due to something else, we have as good reason to be that to be a substance also.[11]

We may believe Newton's logic to be inappropriate, and that he was imprudent to give it currency; it was removed from the printed text.

To contemporaries, Newton seems gratuitously to have stated that colour is our perception of some quality or attribute of a substance, not the body reflecting or refracting light, but light itself. To say that light is a substance, or material, was in the seventeenth century to declare it corpuscular or atomic, and to enter a world where the 'redness' of one atom is like the 'sweetness' of another, as no doubt (in a more sophisticated way) Newton actually believed. But to argue from the world of mechanical hypotheses is precisely what Newton had set out *not* to do, because nothing could be proved or calculated of these imagined particles.

Reception of the New Theory

The Royal Society audience at the reading of Newton's 'very ingenious discourse' was warmly receptive to it. It was to be officially registered and with Newton's consent published, 'as well for the great convenience of having it well considered by philosophers, as for securing the considerable notions of the author against the pretensions of others'. In accordance with ordinary practice Robert Hooke, as Curator of Experiments and a writer upon optics, was asked (with others) to read the letter carefully and report upon it. Only Hooke did so in writing, at some length.[12] There can hardly be a more paradigmatic example of how a review of an original paper should not be written! Newton had, it is clear, imagined that the Society would, as a first step, repeat his experiments and make others to discover whether the factual foundations of the dual propositions about light that he had put forward were valid. Hooke did nothing of the kind. Airily admitting that Newton's experiments might be admitted as true, 'as having by many hundreds of trials found them so', he asserted that Newton had written nothing to convince him that Newton's 'hypothesis of solving the phenomena of colours' was correct. His own hypothesis of light as 'nothing but a pulse or motion, propagated through an homogeneous uniform and transparent medium', a simple motion giving white, more complex ones colours, would account for every phenomenon that Newton had adduced, and indeed other hypotheses might do the same. Hooke was not to be shifted from his belief in the existence of two primary colours, all others being manifestations of mixtures of the compound motions producing these two, and white the purest, simplest motion (much as in music). Hence he rejected Newton's detection of an infinity of shades in the spectrum, and in the white light forming it:

> But why there is a necessity, that all these motions, or whatever els it be that makes colours, should be originally in the simple rayes of light I doe not yet understand the necessity; noe more than that all those sounds must be in the air of the bellows which are afterwards heard to issue from the organ-pipes.[13]

Hence, while Newton regarded the physical rays constituting light as being as fixed in their properties as atoms are, and the ordinary operations upon light such as reflection and refraction as dividing or subtracting one or more of these constituents from the rest without changing their characteristic properties, Hooke regarded light as a mode of motion which could be infinitely varied by these operations and the variations again reversed. Colour for him was an indication of change, whereas for Newton it was an indication of constancy.

Hooke did not fail to take note of Newton's rhetorical mistake. Incredibly proclaiming 'light is a body' to be Newton's 'first proposition', Hooke graciously allowed that 'there will be no great difficulty to demonstrate all the rest of his curious theory' if this 'proposition' be granted. But he himself could not grant it (holding a different hypothesis) and was confident 'that all the coloured bodies in the world compounded together should not make a white body.' And so 'If Mr. Newton hath any argument, that he supposeth an absolute Demonstration of his theory, I should be very glad to be convinced by it: the Phaenomenon of light and colours being, in my opinion, as well worthy of contemplation as any thing els in the world.'[14]

At first sight one can hardly believe that Hooke is reacting (albeit hastily) to a paper that has become classical in later eyes. Not overlooking the fact that Newton's paper has real faults, Hooke seems to show little grasp of the structure and level of Newton's discussion. Deliberately ignoring experiment and calculation as pointless, Hooke has flown straight to the high level of speculative mechanics where, in effect, he reiterates 'my guess is as good as yours.' Newton's careful distinction between spherical and chromatic aberration is not the only major point to escape Hooke's notice completely, and so he continued to assert dogmatically that by improved refractions better telescopes might be constructed than by reflection.[15]

Perhaps it would be pointless to insist upon Hooke's lack of depth and perspicacity in his careless reading of Newton's letter, nor upon Christiaan Huygens's comments which were only slightly more perceptive of its intent, were it not for the effect of their criticisms upon Newton himself, and the fact that Hooke's paper in particular illustrates with such startling clarity the difference between the old, qualitative mould of the mechanical philosophy and the quantitative mould which Newton – and indeed Huygens before him – were bringing in. Of course the distinction is not wholly one of mathematics; Newton's letter contains none. But 'mathematics' expresses most neatly the divergence in cogency and rigour between the two types. Although there are some schematic diagrams illustrating Hooke's exposition of his theory of light in *Micrographia*, this theory is quite innumerate. It was never within Hooke's power, as it came to be within Newton's, to express a theory of pulses or waves in a mathematical form.

It was far otherwise with Huygens, the finest and most subtle mathematician in Europe (after Isaac Newton) in the 1670s. Huygens was Leibniz's master in mathematics while they were both in Paris, and also a master of the geometry of his own wave theory of light. Himself a practical lens-grinder and constructor of refracting telescopes, he welcomed the news of Newton's miniature reflector and tried to produce a larger metal mirror, without success. For practical reasons, therefore, Huygens (like everyone else) proposed to soldier on with refractors.[16]

Huygens made no comment upon Newton's new theory of light and colour until 21 June 1672; then, besides stating that he was not satisfied with Newton's opinion about the damage done by chromatic aberration to the image formed by a lens, he added that 'his new hypothesis of colours . . . seemed very probable to me, and the crucial experiment (if I understand it aright, for he has written a little obscurely) confirms it very well.'[17] Writing from Mrs Arundell's house at Stoke Park in Northamptonshire Newton explained his aberration calculation to Huygens on 8 July; however, Huygens was not to be convinced that Newton did not exaggerate the error, and pertinently asked why refracting telescopes perform so satisfactorily as they do – a point with which Newton had to deal in *Opticks*. By autumn 1672, in fact, Huygens found Newton too positive in his claims for the new theory in his communications to the *Philosophical Transactions*, for 'the thing could very well be otherwise.' It appeared to Huygens that Newton ought to be content if the 'new theory' were accepted as a very probable hypothesis, particularly as it had nothing to say about the physical nature of light in terms of the mechanical philosophy.[18] Early in the new year, having seen Newton's reply to Hooke in the *Transactions* (see below) and evidently feeling some sympathy with Hooke's attitude to the 'new theory', Huygens returned to the last point again. Are there (he asked) in truth more than two primary colours, yellow and blue, from which others arise by mixing, so that a mechanical hypothesis explaining these two would account for all? It would be far 'more easy to find an *Hypothesis* by Motion, that may explicate these two differences, than for so many diversities as there are of other colours'. And yellow and blue together in due proportion might make white.

By now Newton's temper was short. He had devoted a great deal of time during the previous nine months to explaining his optical discoveries and their application. It must have seemed to him that the whole world was obtuse. He did not reply to Huygens in haste – not till 3 April 1673 – but his reply then was downright and rough, bearing in mind that Huygens was an older man of great authority. A month before Newton had declined to answer at all, and had offered his resignation from the Royal Society, "since I see I neither profit them nor, (by reason of this distance) can partake of the Advantage of their Assemblies."[19] Long before, in a letter subsequently published in the *Philosophical Transactions*, Newton had written to Henry Oldenburg:

> I cannot think it effectuall for determining truth to examin the severall ways by which Phaenomena may be explained, unless where there can be a perfect enumeration of all those ways. You know the proper Method for inquiring after the properties of things is to deduce them from Experiments . . . The way therefore to examin it [the Theory] is by considering whether the experiments which I propound do prove those parts of the Theory to which they are applyed, or by prosecuting other experiments which the Theory may suggest for its examination.

He then proposed a set of questions about the theory which could be answered by experiment.[20] Now here was Huygens once more bringing up the will-o'-the-wisp of mechanical hypotheses, which had no relevance to the question of the validity or otherwise of the 'new theory'. "If you ask what colours cannot be derived out of Yellow & Blew, I answer none of those which I defined to be originall." Huygens simply had not understood Newton's use of the prism as an analytical tool, nor his mathematical labelling of the separate coloured rays. A sentence of Newton's echoes Hooke's simile of the organ: "No man wonders at the indefinite variety of waves of the Sea or of Sands of the Shore, but were they all of but two sizes it would be a very puzzling phaenomenon." Newton was not gentle with Huygens's failure to take his point; he, as he told Oldenburg, finding Newton over-concerned to defend his 'new theory', held his peace. Huygens would never yield himself to it, nor would he ever publish a word more about colours. Huygens's mind and Newton's could meet only in geometry; in physics Huygens was always a neo-Cartesian, a generation behind Newton.

Misunderstandings

The failure of both Hooke and Huygens to understand the nature of Newton's objectives in the letter, never mind the question of whether or not he had attained them, is by no means unique in the history of science. It is perhaps the rule rather than otherwise that the point or thrust of major innovations has been at first ignored or misconceived. In this case, despite their eminence in other parts of optics, neither of Newton's critics possessed his extensive familiarity with prism experiments and the phenomena of the spectrum; therefore, like others (including Fr Pardies) whose initial incomprehension of the printed letter was of a simple order arising from the letter's being published in English (at least in part) both Hooke and Huygens failed to attend to the experimental character of Newton's investigation. His first question, as posed in the letter – why is the spectrum long rather than circular? – was too simple for them. Just as critics of *The Origin of Species* accused Darwin of degrading humanity, though he wrote no word in that book of the origin of the human species, so Newton's critics accused him of belittling the settled purpose of seventeenth-century science – to explain Nature by reference to the fundamental motions of matter. The mechanical philosophy, set out in geometrical style by Galileo and in a more rhetorical guise by Descartes, was conceived as the great advance made by the Moderns over the Ancients, with their forms and qualities, sympathies and antipathies. Newton, outdoing Hooke and others like him in strict adherence to Francis Bacon's precepts – though what study, if any, Newton can have made of Bacon's writings directly is barely known to us[21] – seemed to deny the great principle that the way forward in natural philosophy lay in devising plausible mechanical models to account for what is observed in Nature. Or, to put it another way, critics found in Newton's letter a positivism (as we should say) or even a nihilistic attitude towards such speculative mechanism, which may indeed be found in Bacon too, for he was no admirer of the mechanical philosophy of his own day. It is less easy to understand why both Hooke and Huygens, from their predilection for aetherial wave or pulse hypotheses of light, seemed to regard the alternative corpuscular or emission theory with so much dismay. It is equally mechanical and equally Cartesian. Newton's later, acute contention that the content of his letter was perfectly consistent with a wave theory for those who preferred this view (each distinct ray being correlated with a particular frequency of the wave) came too late to have any contemporary influence.

The genuine difficulties encountered by readers of Newton's first letter on the new theory of light and colour, and the weaknesses latent in Newton's own evidence and arguments have been exposed by recent scholars such as Sabra and Lohne.[22] Rarely in the history of science is a new idea initially 'proved' to be true: Lavoisier did not 'prove' the existence of oxygen nor Darwin 'prove' the existence of evolution by natural selection. Newton, for all his great stature, was no exception. Modern critics of Newton's expositions of his theoretical ideas in his communications of the 1670s and in *Opticks* much later have argued particularly that he was mistaken in supposing white light to embrace seven coloured constituents – most clearly, but not solely, revealed by prismatic analysis – and mistaken also in his confidence that these constituents were actually inherent in white light. The former criticism is itself an error; Newton realized that an infinity of rays of diverse refrangibilities paint the spectrum, though he judged that *to the eye* these colours fall into seven bands, divided by his rather bizarre musical rule.[23] The second argument is an illusion, arising from the difficulty of convincing oneself that the products of an analytical process are genuine and not artefacts of the analysis. One electrolyses water into two gases, hydrogen and oxygen; a spark reunites the separated gases, re-forming water. Is it to be argued that the gases are artefacts of electrolysis, not chemical elements? If it is pursued, such an argument becomes a dispute about names, just as our 'electron' was for J. J. Thomson a 'corpuscle' when in scientific language the word 'electron' signified a different entity. We might, if it were not absurd, always call hydrogen 'α-water' and oxygen 'β-water'. The second experiment described in Newton's letter, on the recomposition of the spectral rays into white light by a lens, is closely analogous. Newton's claim that the coloured rays are real constituents of heterogeneous white is, in the last resort, justified by conceptual elegance, convenience and economy.

In *Opticks*, a book whose construction and language at a number of points was shaped by criticisms of the 1672 letter, Newton opened Part II of Book I precisely with a double refutation of the idea that colour is a result of the mixing of white light and shadow (darkness) and of the more general notion that colour is whiteness modified: "The Phaenomena of Colours in refracted or reflected Light are not caused by new Modifications of the Light variously impress'd, according to the various Terminations of the Light and Shadow." Four experiments are brought forward in proof of this assertion, of which one is that with the lens just mentioned. While much of the argument is concerned with this non-effect of the bounding shadow, Newton's utter rejection of any type of modification hypothesis is crystal clear, as in the proposition's last sentence:

> in general we find by other Experiments, that when the Rays which differ in Refrangibility are separated from one another, and any one

Sort of them is considered apart, the Colour of the Light which they compose cannot be changed by any Refraction or Reflection whatever, as it ought to be were Colours nothing else than Modifications of Light caused by Refractions, and Reflexions, and Shadows.[24]

Newton responds to Critics

The most constructive way in which to regard Newton's replies to correspondence about the new theory, replies which occupied his time from the late winter of 1672 to the end of 1675 (and beyond),[25] is to see them as amplifications and explanations of his original brief account. Newton's reluctance to prolong this task appears more than once; as early as June 1672 he implored Henry Oldenburg not to "print any thing more concerning the Theory of light before it hath been more fully weighed". The first of these replies, that to Hooke's 'Considerations' on the letter, is probably the most significant of all. Newton was not the only person to be "a little troubled" by Hooke's swift rebuttal of a letter that had won so much applause when it was first read at the Royal Society; the Society itself was embarrassed and in the end failed to order the printing of the 'Considerations', 'lest Mr. Newton should look upon it as a disrespect, in printing so sudden a refutation of a discourse of his'.[26]

Newton opens with a quiet, polite protest against Hooke's manner of criticism, then delivers a painful jab by calculating the spherical aberration of a concave mirror to be *less* than that of an equivalent lens; Hooke, not troubling to make the computation, had asserted the opposite. Then he again outdoes Hooke in explaining how, if light were a wave motion, the shortest waves producing violet and the longest red, the colours of thin plates and those of natural bodies, as well as many other phenomena, might be accounted for. Hooke had not correlated colour and wavelength. But, Newton reiterated, the message of his letter in no way depended upon such an hypothesis or any other. In the rest of his answer he defends in greater detail the statements made in the letter: a light-beam is neither split nor dilated by refraction; there are more than two original colours; white light is heterogeneous, and so forth.

After this reply of Newton's had been read to the Royal Society Hooke did privately repeat a few of Newton's experiments, including that of the crossed prisms, but he, like Huygens, never admitted that Newton had established his position.[27] The seeds of future enmity were firmly planted.

Mathematical Correspondence

Not all Newton's correspondence – almost our sole guide to his life – was equally optical and polemical. He exchanged letters with Collins about mathematics; to him, indeed, Newton imparted in December 1672 his method of tangents (presumably with tacit permission to pass it on to others):

> This Sir [he continued] is one particular, or rather a Corollary of a Generall Method which extends itselfe without any troublesome calcu-lation, not onely to the drawing tangents to all curve lines whether Geometrick or mechanick or how ever related to streight lines or to other curve lines but also to the resolving other abstruser kinds of Problems about the crookedness, areas, lengths, centers of gravity of curves &c. Nor is it (as Huddens method de maximis et minimis & consequently Slusius his new method of Tangents as I presume) limited to aequations which are free from surd quantities. This method I have interwoven with that other of working in aequations by reducing them to infinite series.[28]

We can understand that in this single sentence Newton was giving his first hint to Collins of the *method of fluxions* (differential calculus) which he had "interwoven" with the *inverse method* (integral calculus) of which Collins already knew something from *De Analysi*. Newton then went on to make one of his rare allusions to Isaac Barrow, writing that he had casually told Barrow "that I had such a method of drawing Tangents" about the time when Barrow was printing his [geometrical] lectures "but some divertissement or other hindered me from describing it to him."

As for Sluse's "new method", it was to be communicated to Oldenburg by Slusius in January 1673, and Collins thereupon, through Oldenburg, informed Sluse of Newton's procedure. This was Newton's first appearance on the international scene of mathem-atics.[29]

Collins almost succeeded in bringing Newton into direct contact with James Gregory, the two men exchanging through Collins cordial letters about the theory and practice of reflecting telescopes; later Collins gave Gregory more news of Newton's mathematical investi-gations. Gregory may have visited Newton in Cambridge after a trip to London, in September 1673.[30] With Henry Oldenburg, Secretary of the Royal Society, Newton exchanged many letters. He was an active intermediary in the optical discussions and was also responsible for introducing Leibniz into the mathematical world of Newton and

Collins; he sent many books to Newton in Cambridge and, by means of Huygens's gift of *Horologium oscillatorium* to Newton, brought these two back into cool cordiality. Oldenburg also corresponded with Newton about such diverse topics as the cultivation of cider-apples in East Anglia and the construction of ear-trumpets. Perhaps Oldenburg's greatest achievement was to keep Newton calm when a certain Francis Hall (or Line), a Jesuit, challenged the accuracy of his experiments as late as September 1674. About a year later Newton was relaxed enough to tell Oldenburg that though it was "against the grain to put pen to paper any more" on the subject of optics, still he would send to the Royal Society "one discourse [more] by me of that subject, written when I sent my first letters to you about colours of which I then gave you notice".[31] And indeed on 21 May 1672 Newton had written to Oldenburg of his intention to attach to his reply to Hooke's 'Considerations', still delayed, a "discourse" of the "Phaenomena of Plated Bodies",

> concerning which I shall by experiments first show how according to their several thicknesses they reflect or transmit the rays indued with severall colours, and then consider the relation which these thin transparent Plates have to the parts of other naturall Bodies, in order to a fuller understanding of their colours also. And this I purpose to send because it most properly apperteines to the former discourse of light, being a declaration of the different reflectibility of the severall sorts of rays, as that was of their different refrangibility.[32]

Evidently Newton was so far encouraged by the reception of his various communciations to the Royal Society that he was willing to extend their range. And against the various criticisms, public and private, of Newton's new optics one can set the approval of the grandees of the Royal Society, and of others, such as the astronomer Richard Towneley, who thought Newton's theory of colours so admirable that a Latin account of it ought to be printed in the *Philosophical Transactions*.[33]

The Colours of Thin Plates

The new paper is usually known as the 'Discourse of Observations'. Newton gave it no title; in fact, as the Royal Society eventually received it at the end of 1675 it had become the second part of a long double document, of which the first part is headed "An Hypothesis explaining the Properties of Light, discoursed of in my several Papers."[34] The "Hypothesis" Newton had written hastily "having a little time this week to spare" in order to illustrate the 'Observations', despite an earlier resolution not to circulate any such "Hypothesis of Light & colours, fearing it might be a means to ingage me in vain disputes". In the same covering letter Newton made some friendly overtures to Robert Hooke, on the subject of refraction at an interface between glass and "vacuum", but Hooke was not to be mollified. After the reading of the "Hypothesis" at the Royal Society's meeting (16 December 1675) his only recorded comment was 'that the main of it was contained in his *Micrographia*, which Mr. Newton had only carried further in some particulars'.[35]

To begin with the older 'Discourse' of 1672, in turn based upon the essay already mentioned, "Of the coloured circles twixt two glasses", composed probably between 1668 and 1672;[36] it comprises twenty-four "observations", that is, experiments, followed by their interpretation culminating in nine propositions. It is almost the final development of Newton's study of interference colours going back to 1665, and it passed without major change into *Opticks*, Book II.[37] Whether or not this study was originally inspired by *Micrographia* Newton made no allusion to that book, taking his start from an experiment of his own, found in the notebooks, of two prisms tied together tightly so that colours were produced by the thin film of air between the glass surfaces (cf. figure 3.1). Later, he would devastatingly vindicate himself from the charge of borrowing from Hooke in his "Hypothesis".[38]

Exact measurement and closer analysis than he had given before of the 'Newton's rings' formed by films of transparent materials are the chief feature of the 'Observations'. Newton ends with dimensions in terms of millionths of an inch, derived from actual measurements in hundredths! As before, using a lens of large curvature pressed upon a flat glass surface as his experimental tool, Newton now observed that the coloured rings were seen (in alternation) by transmitted as well as reflected light. Soap-bubbles formed larger rings because the film is thinner. After setting out the experiments in detail and establishing the simple periodicity of the diameters of the rings (following from the thicknesses of the film where the rings occur, being as the series of odd numbers) Newton set out a grid, its defining points fixed by this series, relating the colours of the successive rings in their various orders to the thickness of the film.[39] By interpolating actual measurements into this grid Newton then constructed a table of the thicknesses of films of air, water and glass for seven orders of coloured rings. His example is neatly contrived to score a point: Newton chose an observation from *Micrographia* where Hooke noted

> that a faint yellow plate of Muscovy glass [mica], laid upon a blue one, constituted a very deep purple. The yellow of the first order [of rings] is a faint one, and the thickness of the plate exhibiting it, according to the table, is $5\frac{1}{4}$, to which add $9\frac{1}{2}$, the thickness exhibiting blue of the second order, and the sum will be $14\frac{3}{4}$, which most nearly approaches $14\frac{4}{5}$, the thickness exhibiting the purple of the third order.[40]

In another place Newton returned to a familiar theme, emphasizing once again the difference between Hooke and himself as natural philosophers:

> all the production and appearances of colours in the world are derived, not from any physical change caused in light by refraction or reflection, but only from the various mixtures or separations of rays, by virtue of their different refrangibility or reflexibility. And, in this respect it is, that the science of colours becomes a speculation more proper for mathematicians than naturalists.[41]

No research could more forcefully exemplify this point than that in the 'Observations', where precise measurement and calculation generate definite, verifiable predictions. (Hooke could no more emulate Newton in this than he could fly – though it is true he did also claim to know twenty ways of flying.) Modern evaluations indicate, moreover, that Newton's results – not gained without pains nor without adjustments for causes of error – were astonishingly correct. As Westfall has pointed out, Newton's triumph over the qualitative looseness of Hooke's hypotheses (and even over the acute but less pertinaceous Huygens) could not have been more complete.[42]

At length, after further detailed analysis, Newton pronounced a general conclusion, indicating both the total symmetry latent in optical phenomena and the universality of the propositions he had stated in the first letter concerning the nature of light and colours:

> It is necessary therefore that every ray have its proper and constant degree of refrangibility connate with it; . . . and what is said of their refrangibility may be understood of their reflexibility; that is, of their dispositions to be reflected, some at a greater, and others at a less thickness of thin plates or bubbles, namely, that those dispositions are also connate with the rays, and immutable, . . . By the precedent observations it appears also, that whiteness is a dissimilar mixture of all

— 133 —

colours, and that [natural] light is a mixture of rays endowed with all those colours.[43]

But this was not yet the conclusion of the 'Discourse'. Newton had still to display the third strand in his optical investigation, to define what it is in the physical structure of bodies that causes them to appear tinged with colour.

Particles and Light

In what is perhaps the most daring of his theoretical arguments in optics, Newton showed how the optical behaviour of bodies throws light upon the nature of their "least parts" or component corpuscles. His conclusions were very different from those generally accepted. Refraction and reflection were not opposed, but complementary properties in matter, for bodies that refract best also reflect best. The "least parts" must be transparent, and the spaces between them filled with air or water or perhaps some "subtiler medium" (aether). Only large spaces and corpuscles could be opaque to light rays. Rays actually striking material corpuscles must be "stifled and lost" hence light cannot be reflected from them but only from the 'empty' spaces between; and this contention is related to the fundamental point that the transparent corpuscles "according to their several sizes, must *reflect* rays of one colour, and *transmit* those of another". Finally, treating the corpuscles as fragments of thin film, Newton computed the size of surface corpuscles from the colour of the surface: "Thus if it be desired to know the diameter of a corpuscle, which being of equal density with glass, shall reflect green of the third order; the number $17\frac{1}{2}$ shows it to be about $17\frac{1}{2}/1000000$ parts of an inch."[44] Newton allows that it is difficult to guess the order to which a colour should be assigned, since they repeat in successive orders of the rings, and his reasoning about the choice is merely analogical. However, it is certain that blackness must be the colour of the smallest of all particles. (And light-particles themselves, it is obvious, must be far more minute than these.)

Newton concluded with a well-known, over-sanguine prediction: "it is not impossible, but that microscopes may at length be improved to the discovery of the corpuscles of bodies, on which their colours depend." A magnification of 500x or 600x might suffice to see the biggest corpuscles, but to see further into "the more secret and noble works of nature within those corpuscles, by reason of their transparency" Newton held to be impossible.[45]

Newton's Aetherial Hypothesis, 1675

The "Hypothesis explaining the Properties of Light, discoursed of in my several Papers" preceding these 'Observations' opens with a rare biographical allusion; Newton makes it plain that he had not long before met Robert Hooke face-to-face for the first time. This can only have happened during that visit to London (9 February to 19 March 1675) whose fruit was the patent allowing Newton to remain both a Fellow of Trinity and a layman. On 18 February he was admitted to the Fellowship of the Royal Society and presumably stayed for the meeting. Hooke recorded his attendance at Gresham College on the next Wednesday (25 February) also, and he was almost certainly there again on 11 and 18 March. For the latter day Hooke had been asked 'to have ready . . . the Apparatus necessary for the making Mr Newton's Experiments' but instead Hooke talked about the diffraction of light – perhaps there was no sun. On the former day (probably) Newton heard Hooke expatiate upon 'his thoughts . . . That light has a vibrating or tremulous motion in the Medium'.[46] So, he recorded in the "Hypothesis", he had with pleasure understood that Hooke had "accommodated his hypothesis to this my suggestion of colours, like sounds, being various, according to the various bigness of the pulses [waves]." Such an hypothesis is plausible, Newton admitted, but he himself preferred another in which the rays of light are assumed to be "small bodies emitted every way from shining substances" which, when they impinge upon any material surface, "must as necessarily excite vibrations in the aether, as stones do in water when thrown into it".[47]

This is the first open indication that Newton's theory of light involved more than the streaming of tiny particles from radiant bodies; he had, so to speak, compromised with the aetherist wave-theorists in order to account for the transmission-reflection duality and its visible consequences in 'Newton's rings'.[48] This all-inclusive hypothesis (for, as Newton himself said, it left "little room for new ones to be invented") was heuristic only, however; it was not to be taken as a definitive proposal. It postulated, in the common mechanical tradition of the age, an aether much like air but "far rarer, subtler and more strongly elastic". (Deceptive evidence of the passage of light and magnetism through an experimental 'vacuum' and the cessation of a pendulum's motion in such a vacuum almost as rapidly as in air spoke for such an aether's existence.) A puzzling draft by Newton, perhaps written in the 1670s, shows him exploring the mechanical character of air, and the relation of air to aether, and to the magnetic and electric effluvia present in both.[49] Again, as we shall see, in a well-known letter to Robert Boyle of 1679 Newton speculated very freely about the possible physical functions of such an aether. Here, in the "Hypothesis", Newton suggested that electric attraction and repulsion might be aetherial effects – his briefly stated experiments on these electric motions, difficult to repeat, greatly intrigued the Royal Society – and his flow of thought carried him into an almost poetic passage on the cosmic significance of the finest constituent "spirit" of this aether, supposed to be not pure and simple but complex:

> so may the gravitating action of the earth be caused by the continual condensation of some other such like aetherial spirit, not of the main body of phlegmatic aether [!], but of something very thinly and subtilly diffused through it, perhaps of an unctuous or gummy, tenacious or springy nature, and bearing much the same relation to aether, which the vital aereal spirit, requisite for the conservation of flame and vital motions, does to air. For, if such an aethereal spirit may be condensed in fermenting or burning bodies, or otherwise coagulated in the pores of the earth and water into some kind of humid active matter, for the continual uses of nature, adhering to the sides of those pores, after the manner that vapours condense on the sides of a vessel; the vast body of the earth, which may be every where to the very center in perpetual working, may continually condense so much of this spirit, as to cause it from above to descend with great celerity for a supply; in which descent it may bear down with it the bodies it pervades with force proportional to the superficies of all their parts it acts upon; nature making a circulation by the slow ascent of as much matter out of the bowels of the earth in an aereal form, which, for a time, constitutes the atmosphere; but being continually buoyed up by the new air; exhalations and vapours rising underneath, at length (some part of the vapours, which return in rain, excepted) vanishes again into the aethereal spaces, and there perhaps in time relents, and is attenuated into its first principle: for nature is a perpetual worker, generating fluids out of solids, and solids out of fluids, fixed things out of volatile, and volatile out of fixed, subtil out of gross and gross out of subtil; some things to ascend, and make the upper terrestrial juices, rivers and the atmosphere; and by consequence, others to descend for a requital

to the former. And, as the earth, so perhaps may the sun imbibe this spirit copiously, to conserve his shining, and keep the planets from receding further from him. And they that will, may also suppose that this spirit affords or carries with it thither the solary fewel and material principle of light; and that the vast aethereal spaces between us and the stars are for a sufficient repository for this food of the sun and planets.[50]

Like such nineteenth-century English scientists as William Thomson, Joseph Larmor and Oliver Lodge, Newton also speculated that

Perhaps the whole frame of nature may be nothing but various contextures of some certain aethereal spirits, or vapours, condensed as it were by precipitation, much after the manner, that vapours are condensed into water, or exhalations into grosser substances, though not so easily condensible; and after condensation wrought into various forms; at first by the immediate hand of the Creator; and ever since by the power of nature; which, by virtue of the command, increase and multiply, became a complete imitator of the copies set her by the protoplast. Thus perhaps may all things be originated from æther.[51]

So far as we know this is the first outpouring of cosmic speculations from Newton's pen. There is no distant antecedent to the "Hypothesis", since "De aere et aethere" can hardly be seen in that light. In these lines the exact experimenter and ingenious mathematician has been supplanted by a fanciful, organismic philosopher as daring and bizarre in his imagination as Robert Fludd or Henry More. The clockwork universe of Cartesian mechanical philosophy has made room for the animistic world of Paracelsus. It would be difficult not to see here (with Westfall) the misty influence of recent reading in chemical and Platonist authors. If we except a few strange pages about comets in Book III of the *Principia* nothing similar was released by Newton to an audience until he published the later queries in *Opticks*, from 1706 onwards. For, having demonstrated his formidable talents for wild speculation to his colleagues in the Royal Society, Newton firmly refused to permit either "Hypotheses" or 'Observations' to be printed. Accordingly, they were hidden from the world at large until 1756.[52]

It is not surprising, in the context of his age, that Newton also identified the fine "spirit" of the aether as the true 'animal spirits' by which the muscular motions of animals are controlled by the brain through the nerves. The same notion was revived in the concluding words of the second edition of the *Principia* of 1713 – a time when Newton was again interested in aetherial hypotheses.

But to return to Newton's hypothesis of light: it is that light may be "multitudes of unimaginable small and swift corpuscles of various sizes, springing from shining bodies at great distances one after another; but yet without any sensible interval of time".[53] Aether-waves as light were rejected by Newton because waves cannot travel only in straight lines or form deep shadows; moreover, he could not connect with waves the alternating reflection and transmission of light in transparent films. This second point seems strange to us, who regard 'Newton's rings' as strong evidence for the wave hypothesis. Newton saw the aether-waves, created by the impact of light upon matter, simply as modulating the reaction between matter and light. To explain the varying "bigness" of the waves, determining the reflection/transmission intervals, Newton was now compelled to introduce the idea of varying bigness in the light particles provoking the waves. The essence of this theory survived into *Opticks* save that there Newton finds the waves, not in an aether, but in the stream of particles.

In the *Principia* years later Newton was to present an account both geometrical and mechanical of reflection and refraction based on the assumption of matter's exercising an attractive force upon the particles of light. This Section XIV of Book I had been anticipated in the "Hypothesis" where, however, instead of attractive force the active part is played by an aether less dense within matter than outside it.[54] It is similar with diffraction, a newly discovered optical effect first mentioned by Newton in the "Hypothesis" (with a sharp jab at Hooke again!) where it is also explained by aether gradation. In the *Principia* attractive force explains the bending of light in diffraction.[55]

Clearly Newton wasted few of his ideas, but optics as well as mechanics shows how greatly his ideas were modified in their working-out, probably in 1679–80. Then Newton departed for ever from the Cartesian type of aetherial universe, the plenum, to enter the universe of forces where space is a vacuum. For this reason the "Hypothesis" is an invaluable document in showing us much of the physical thought of Newton before this transformation in it had occurred, giving us our first sight of a Newton who is more than a mathematician and experimenter.

Newton and Hooke – and Lucas

There seems to have been little reaction on the part of the Royal Society to this massive double communication, read aloud *in toto* by its Secretary at successive meetings. More interest was shown in Newton's primitive electrostatic experiment and in Huygens's placing a lamb's lungs in the air-pump than in Newton's new optical investigations and speculations. True, on Oldenburg's initiative Hooke again made apparatus ready, but nothing was done (2 March 1676). Oldenburg was very much Newton's friend, seeing in him an ally against Hooke, who had sworn enmity to the Secretary over the matter of fitting springs to the balances of watches. He was not slow to encourage Newton's wish to vindicate himself from the charge of borrowing Hooke's hypothesis of waves. Newton's first answer (21 December 1675) is reasoned enough: he owed less to Hooke than Hooke owed to Descartes, and Hooke had been unaware of "the two main experiments without which the manner of the production of those colours [of thin films] is not to be found out" when writing *Micrographia*. Newton was anxious to "avoyd the savour of having done any thing unjustifiable or unhansome towards Mr Hooke".[56] A second letter, three weeks later, breathes a sharper sense of wrong done but is mainly concerned with other critics than Hooke. It stimulated Hooke to make a direct conciliatory approach to Newton, blaming Oldenburg as a 'kindle-cole': 'Your Designes and myne [he wrote] I suppose aim both at the same thing which is the Discovery of truth and I suppose we can both endure to hear objections, so as they come not in a manner of open hostility, and have minds equally inclined to yield to the plainest deductions of reason from experiment.'[57]

How little men know their own natures! No more than Hooke could Newton 'endure to hear objections'. Nevertheless, he replied in fittingly self-depreciatory terms, 'you defer too much to my ability for searching into this subject. What Descartes did was a good step. You have added much several ways, & especially in taking the colours of thin plates into philosophical consideration. If I have seen further it is by standing on the sholders of Giants.'[58]

Concluding, Newton sought guidance from Hooke about an astronomical observation proposed by him, Newton having missed Hooke when he called at Gresham College (presumably in October 1675, when he was away from Cambridge). Hooke received the letter but, with astonishing insensitivity, never responded. The two men did not correspond again for nearly two years.

Much has been made of these incidents. One finds in Newton's answers to criticism throughout the main optical controversies impatience, exasperation and a conscious superiority no doubt maddening to recipients of his letters, however justified. Yet in an age when the defence of intellectual property was essential to success, and strong writing on either side of a point in dispute usual, Newton's ill manners rarely seem to exceed the norms. Newton's taunt that Hooke had failed to measure the thickness of films "& therefore seeing I was left to measure it my self I suppose he will allow me to make use of what I took the pains to find out" was not unjustified in the circumstances. Westfall's complaint that 'Newton owed his very knowledge of diffraction to Hooke's discourse' is ill-founded, since Newton already had on his shelves Fabri's book on the matter, and had read it.[59] No one would claim Newton to have been the most gracious of men; friendly commonplaces such as "Pray present my humble service to Mr Boyle when you see him" are rare enough with him.[60] The English Jesuit at Liège, Anthony Lucas, a late critic of the first optical letter, particularly aroused Newton's ire, because he seemed to accuse Newton of falsifying his experiments and demanded that Newton satisfy his own doubts. When Newton refused to correspond with him further it was because Lucas had seemed to cheat by continually shifting his ground and was discourteous: "I am called Quarreller & that for nothing els but saying — this he calls a Quarelling with the number of his Objections & every letter since has grown more sharp than the former. This it seems troubles him that he having committed an error in matter of fact cannot get off."[61] Here Newton clearly felt himself the abused party!

Such recent scholars as Richard S. Westfall speak of Newton's paranoia, his wish to humiliate antagonists, and even his 'complete loss of control that is compatible with a [nervous] breakdown'. Such judgements perhaps ignore differences in styles of rhetoric between Newton's age and our own. A tendency to classify Newton as clinically ill might perhaps be diminished by greater familiarity with other examples of coarse and 'brutal' language in seventeenth-century controversies.[62]

Another Possible Book?

However tedious controversies about optics may have become to Newton in the late 1670s, he continued to plan fresh publications – without result. Having formerly given up the extended treatment of his prismatic investigations, rejecting his optical lectures as unsuitable for this purpose, it was rational that he should not wish the 'Observations' to appear as a disconnected paper in the *Philosophical Transactions*. He told Oldenburg that he had thought of combining that piece with a new, parallel account of his prismatic work in the same format, but was puzzled to know how to connect them and also to bring in some "additions" that he proposed.[63] Perhaps as part of this exercise Newton (I assume) had printed at Cambridge proof-sheets of a new edition of his first optical letter to Oldenburg, with explanatory notes.[64] The same may be linked to a phrase in a letter to Hooke, after the Secretary's death in September 1677: "Mr Oldenburg being dead I intend God willing to take care that [his own correspondence with Lucas] be printed according to his mind, amongst some other things which are going into the Press."[65] Some time before this, Collins had heard a rumour that the engraver David Loggan, then in Cambridge preparing his *Cantabrigia Illustra* (1690) – in which the plate of Great St Mary's church is dedicated to Newton – had drawn a portrait of Newton to adorn 'a book of Light, Colours [&] Dioptricks which you [Newton] intend to publish'. The rumour may have sprung from confusion; certainly no such portrait survives.[66] Nor does any further evidence of the proposed book.

Fires, like that which recently destroyed a part of Hampton Court Palace, were all too easily caused by an overset candle. Several tales of fire are connected with Newton and his mythical lap-dog, Diamond. Long after, in 1726, Newton related to his nephew-in-law, John Conduitt, how a candle left too long unattended had burnt papers concerned with both optics and mathematics. Signs of burning are indeed to be found among Newton's papers. Humphrey Newton too heard something of such an accident. But that Newton himself should have in consequence 'run mad' and been 'not himself for a month after' will be credited only by those who see Newton as a half-demented magus. Major loss of material, in view of the rich manuscript legacy still existing and Newton's known proclivity for drafting and redrafting, is unlikely. If Newton had composed a major treatise on optics in the late 1670s, as some tales would have it, it is strange that the fire so consumed it that no trace now remains. If the puzzling proof-sheets that turned up some years ago *are* traces of such a book, it is far less likely that all else was consumed by fire, than that Newton (as on other occasions) thought better of his intention to go into print.[67]

6

Life in Cambridge, 1675–1685

With the freedom granted to all Lucasian professors to decline holy orders, Newton knew that so long as he retained that office he was assured of the tenure of his fellowship at Trinity College till death or resignation – in fact, he would hold it for another twenty-six years. He did not resign his Cambridge posts until December 1701, still not a Senior in his college though non-resident since 1696.[1] Further, to the best of our knowledge, Newton had carried out no teaching since the Revolution.

His appearances in the public life of his college and university are poorly documented. In March 1673 he joined his relative Humphrey Babington and others, Masters of Arts, in signing a protest against the heads of houses exercising their customary but doubtful powers of nominating two candidates for the post of Public Orator, for the Senate's election. The protest failed. At an unknown date Newton attempted to secure from the commissioners of taxes for Cambridge an exemption from a property tax, on the grounds that the revenue from the Lucasian professorial estate did not belong to his college. (Later, from 1688 to 1695, Newton himself was to be one of these commissioners, a mark of his eminence in the university; the vice-chancellor was one of the body *ex officio*.)[2] In 1676 Newton gave £40 – then a fairly comfortable annual salary – towards the projected new library building proposed by Barrow at Trinity, since famous as the

Wren Library closing the rear court of the college. This gift was followed by a loan of £100 made about the end of 1679; Newton also presented a number of books to the library.[3] In this year also he gave his vote for the Master of Trinity Hall, Sir Thomas Exton, to represent the university in Parliament; he voted for a new registrary and two successive librarians. The second of these, John Laughton, another Trinity man, was or became a close friend of Newton's.

That Newton was considered as a man of some weight and good sense in other colleges than his own seems to follow from a rather strange letter now in Queens' College, Cambridge, but addressed to the Master and Fellows of St Catharine's, about 1677. This contains a carefully quantitative estimate of the probable darkening effect, within Queens' College chapel, of new buildings being erected by St Catharine's on the other side of Queens' Lane, separating the two colleges. Newton reckoned that a fortieth part of the presently available light would be lost, while a darkening five times greater than this would only have caused two or three weeks' extra use of candles in the Queens' chapel in each year. At this time St Catharine's, a poor foundation, was being almost completely rebuilt; along with some heads of houses and many others Newton bought an annuity from the college for £25, which in fact was never paid him so that later he made it into a free gift.[4]

Edleston found a couple of anecdotes of experiences at Cambridge related by Newton in old age from the presidential chair of the Royal Society. Twice he recalled cutting the heart of an eel into three pieces, in the college kitchen, noting that each continued to pulsate at the same time, the pulsation being at once extinguished by a drop of vinegar though unaffected by saliva. On another occasion he told how a violent clap of thunder damaged two windows in a room adjacent to his own, and drove splinters from the floor into the ceiling. The accompanying lightning flash was exceedingly bright. Fortunately Newton was not to be, like Professor G. W. Richmann of St Petersburg, a scientific martyr to his electrical experiments.[5]

Newton's Friends

Among Newton's friends in Cambridge, nothing more is to be said of Isaac Barrow, welcomed by Newton as Master of Trinity in 1673, but not (so far as is known) regretted by him on Barrow's death (aged only 47) in 1677. His successor as Master was John North, also (it is said) a friend of Newton's. A new senior friend of his at this time or later was John Ellis, a Fellow of Gonville and Caius College, of which he was to become Master in 1703. As Vice-Chancellor pro tem he was knighted by Queen Anne together with Newton when she visited Cambridge in 1705. The Cambridge Platonist Henry More has been previously mentioned as possibly known to Newton (at least by name) since Grantham days; whatever their acquaintance before 1680, it emerges clearly into the light for a moment in a letter of More's, written in that year, describing his conversation with Newton in More's chamber. He portrays Newton as having 'a singular Genius to Mathematicks, and I take him to be a good serious man'. As to Newton personally, he wrote that 'the manner of his countenance . . . is ordinarily melancholy and thoughtfull, but then [in the course of their talk] mighty lightsome and chearfull.'[6] The conversation was all about the interpreting of biblical prophecy, especially in the *Revelations* of St John. On this they failed to agree; clearly More felt that Newton should stick to mathematics and leave exegesis to himself. Though the letter is important as showing Newton's staking a claim in this theological field as early as the late 1670s, it bears no obvious signs of close intimacy between the two men. In 1685 Newton invited More, by then a septuagenarian, to join the proposed Philosophical Club in Cambridge. More declined;[7] his death came only two years later.

Another numbered among Newton's Cambridge friends, on slight enough evidence, was the first official teacher of chemistry at Cambridge, a Veronese, John Francis Vigani. One can imagine that Newton was happy to find an expert with whom he could discuss chemical matters, and he made a careful study of Vigani's *Medulla Chymiae* (1683). He appears to have arrived in Cambridge only in 1682 and according to Humphrey Newton 'gave much delight and Pleasure of an Evening' to his master. Catherine Barton, Newton's niece and later companion, recorded that their friendship (if such we are to call it) was broken when Vigani tried to tell Newton a story about a naughty nun. But this must have been long before Catherine joined Newton's household in London.

The great experimental, philosophical and chemical influence upon Newton from his undergraduate days was of course that of Robert Boyle. They neither met nor corresponded (it seems) before Newton was admitted to the fellowship of the Royal Society on 18 February 1675. Probably their correspondence began in the following year, though the first extant letter in it is one from Newton to Boyle of 28 February 1679; and even that is unique in this period save for one from Boyle to Newton written in 1682. The rest are lost. The two men had common interests in chemistry, the nature and structure of matter, and natural religion, but their ideas were often very different. Newton – for reasons not now obvious – was sceptical of Boyle's understanding of supposedly alchemical processes and his fundamental concept of matter came to be very different from Boyle's. But he always greatly respected his older colleague.

The Letter to Boyle, 1679

Newton's well-known letter, or rather paper, of February 1679 addressed to Robert Boyle is a document wholly different in spirit and method from the mathematical principles of natural philosophy that Newton began to express with such mastery less than six years later.[8] It is for this reason that the letter is of considerable interest, providing as it does a coherent picture of the physical universe as Newton conceived it before he had become conscious of the power of mathematico–mechanical reasoning, founded upon the concept of force. It enables one to perceive how great was the transmutation in Newton's thinking that had to take place between this time and 1684, a transmutation of which we have little direct evidence. The 'Letter to Boyle' is a non-mathematical texture of hypotheses. Its opening makes it plain that Boyle and Newton had been in correspondence for some time, and freely opened their minds to each other. Newton confesses that his ideas about the foundation of physics "are so indigested that I am not well satisfied myself in them", and so would not commit them to paper if he had not bound himself by a promise to do so. For, "especially in natural Philosophy . . . there is no end of fansying." Newton's "fancies" in this letter are clearly related to those in the "Hypothesis" sent to the Royal Society in December 1675; he expected Boyle to recall that unpublished paper. As before, Newton postulates pores in all material substance, filled by an aether which stands "rarer in those pores then in free spaces, & so much the rarer as the pores are less". The density gradient of this aether causes the refraction of light, the cohesion of flat plates, and surface tension: "that a fly walks on water without wetting her feet". In particular, it accounts for the resistance offered by bodies to pressure into close contact (as with powders, for example) and their resistance to separation when such contact has been effected. It also explains some properties of the atmosphere: "for I conceive the confused mass of vapors and exhalations which we call the Atmosphere to be nothing els but the particles of all sorts of bodies of which the earth consists, separated from one another & kept at a distance by the said principle" (p. 291). It is, therefore, a repulsive principle which Newton next invokes to account for the action of solvents. If a substance is wetted by a solvent penetrating its superficial pores, the presence of the solvent (by removing the aether gradient) nullifies the cohesion of the surface particles, which drift away into the fluid. Solution also alters the colours of fluids, since this arises from the magnitude of the particles composing it, "as I think you have seen described by me more at large in another paper" (that is, the "Hypothesis"). Newton now goes deeper into the problems of solubility and insolubility: if water will not dissolve metals, this is because

> there is a certain secret principle in nature by which liquors are sociable to some things & unsociable to others. Thus water will not mix with oyle but readily with spirit of wine [alcohol] or with salts . . . But a liquor which is of itself unsociable to a body may by the mixture of a convenient mediator become sociable [with it] (p. 292).

So the addition of acid [or salts] enables water to dissolve metals. Newton does not tell Boyle what sociability is, or note his echo of Hooke, but the next paragraph makes it clear that it is identical with the no less mysterious 'affinity' of the next generation. By supposing that substances have greater or less "sociabilities" one for another, the chemist can explain how the decompositions and recompositions of substances occur (Newton puts it here much as he would later do in Query 31 of *Opticks*, only with the differential aether gradient serving instead of the differential attraction in that book.)

When violent, such chemical reactions cause ebullition, whereby the particles of bodies enter the atmosphere. Heat may have the same effect, if the particles of the substance are small enough. This gives a hint that gold, for example, could only be made volatile by breaking its normal corpuscles into fragments – a further hint to us, perhaps, accounting for Newton's chemical interest in the volatility of metals. But the small particles of water, when raised into a vapour by heat, soon condense into a liquid again as they cool, whereas "the grosser particles of exhalations raised by fermentation keep their aerial form more obstinately, because the aether within them is rarer" (p. 294). Now, Newton argues, we live in a "true permanent air" arising from the bowels of the earth; therefore these aerial particles contain little aether, and therefore also they are very dense; possibly therefore "the true permanent air may be of metallic origin" since metallic particles are densest of all. Such air Newton regards as "unactive . . . affording living things no nourishment if deprived of the more tender exhalations and spirits that float in it", confirming the hypothesis of its metallic origin.

As a final speculation, Newton adds (with unconscious irony) that a continuous variation of "subtility" in aether itself might be the cause of gravity, the finest aether being found deep in the interior of bodies, the gross sort in empty space. In a quasi-Aristotelian argument Newton suggests that a heavy body suspended in the gross aether above the Earth would endeavour to descend towards it, in order to regain the more subtle aether suitable to its own dense composition.

Other evidence, to be mentioned later, makes it clear that even at this late date Newton, for all his early computations, was still attached to the Cartesian system of celestial vortices of aether. The letter to Boyle equally proves that Newton in his "fancies" still regarded the aether as the prime subject for physical speculation, when causes are sought. In this he shared the general sentiment of his age. What is more surprising than Newton's neo-Cartesianism – more casual than the speculations that Leibniz had published some years before (no doubt unknown to Newton) but still of the same natural-philosophical genus – is his willingness to invoke a new, truly occult, principle of "sociability" that appears, on the face of it, quite outside the mechanistic tradition exemplified by Descartes, Boyle and Leibniz. What does this mean? Newton had already appealed to this new principle in his "Hypothesis".[9] As he makes plain, it is not a question of mechanical properties like size, shape and motion. It seems to be a strictly chemical notion.[10]

David Gregory

Another person, of a very different type, who was to play a major role in Newton's life till his own death in 1708, aged 49, was the Scot David Gregory. Nephew of the far greater mathematician, James Gregory, to whose chair of mathematics at Edinburgh he succeeded in 1683, Gregory secured the Savilian professorship of astronomy at Oxford through the recommendations of Newton and John Flamsteed, the Astronomer Royal of England, Newton first heard of David Gregory when sent by him in June 1684 a fifty-page mathematical pamphlet (*Exercitatio geometrica*, 'A geometrical exercise')[11] recently cobbled up by Gregory from his uncle's papers. There, too, he had found Collins's letters:

> Sir [he wrote] I perceive by severall letters from Mr Collins to my Uncle, from whose remains this [pamphlet] is for the most parte taken, that your selfe have of a long time cultivated this methode, and that the world have long expected your discoveries therein. and I hope that if yee doe me the honour to glance it over yee will see that I have according to justice acknowledged the same.[12]

Gregory begged Newton for his opinion of his 'exercitation' and for an account of Newton's 'methode of turning the root of ane equation to ane infinite series, which is infinitly troublesome and tedious to me'. 'Thus forewarned, Newton was well-prepared for Gregory's following sketch of the principles of exact algebraic integration and their exemplification in problems of the quadrature and rectification of curves and the mensuration of their solids and surfaces of revolution.' Moreover, to continue D. T. Whiteside's exposition of the situation, Newton in reading on discovered that 'Gregory devoted the remainder of his tract to elaborating (with numerous well-chosen examples) two of the three methods of reducing quantities to series . . . which Newton had set down in 'Reg. III' of his *De Analysi* fifteen years before.' Moreover – unless John Wallis (or presumably Newton himself) should come out with it first – Gregory promised to explain soon his uncle's method of extracting the roots of a literal equation. 'The challenge thrown out to Newton himself was unspoken but no less real: publish or be published.'[13]

Whether Newton replied to Gregory is not known; it may be that he chose the discourteous alternative of silence. What he did is clear from his unpublished papers: he sat down to prepare the explanation of his mathematical processes, from which – with the stimulus of a slight hint from Collins – James Gregory had (as Newton supposed) derived – "a measure of the power of his intellect" – the work now published by nephew David. But Newton did not propose to compose a new work from the whole cloth, rather to patch a commentary to a printing of the letters exchanged with Leibniz in the years 1676–7 (thereby, of course, establishing considerable priority over anything David Gregory could now produce).

> By publishing these [letters] I shall certainly oblige the reader more than if (after what Mr Gregory has done) I should write up the whole subject afresh, and especially so since in them is contained Leibniz's

extremely elegant method, far different from mine [NB!] of attaining the same series – one about which it would be dishonest to remain silent while publishing my own.[14]

If this unfinished draft, entitled *Matheosis universalis specimina* ('Specimens of a universal system of mathematics') had ever been completed and printed, would it have precipitated or stifled the quarrel between Newton and Leibniz over the invention of the calculus? After the printing of the letters, Newton proposed six chapters, of which he wrote four, by way of additional explanation. He also prepared an edited version of one of his own letters to Leibniz. Having stopped, he then began afresh with a quite new draft of the supplementary chapters under the title "The computation of series". This too was abandoned unfinished.[15]

After this wholly abortive encounter, Newton was again approached (in a very different vein) by Gregory on 2 September 1687, to be thanked by him in a memorable phrase 'for having been at pains to teach the world that which I never expected any man should have knowne . . . you justlie deserve the admiration of the best Geometers and Naturalists, in this and all succeeding ages.' Acutely, Gregory remarked on the new mathematical techniques required for a theoretical physics of the sophisticated Newtonian kind: for Newton's predecessors 'had never found out a Methode of determining geometrically what it is which represents and measures physicall qualities'.[16]

It is doubtful whether the two men met before the Revolution, or indeed before Gregory's appointment at Oxford. Since, after the Revolution, Newton spent much time in London they then had many opportunities for conversation, of which Gregory made good use. Meanwhile, I may here add that not only was Gregory an ardent and sincere admirer of *Philosophiae naturalis principia mathematica*, as we have just seen, so that one might call him the second Newtonian – Edmond Halley obviously being the first – but he rapidly recruited others, all Scots of his acquaintance: Archibald Pitcairne, a fellow-student; John Craige, perhaps a pupil (obviously acquainted with Newton before Gregory was); John Keill, another pupil; George Cheyne, and so on. It was through Gregory that it came about that so many early Newtonians were Scots: Edmond Halley and William Whiston were the great exceptions among the mathematical Newtonians.

John Craige merits a further word here since he was one of the first British mathematicians deliberately to seek out Newton at

Cambridge. In the course of a few days' visit in the summer of 1685 he was allowed by Newton not only to see early draft pages of the *Principia* but to examine (probably) the *De methodis serierum* of 1671 also. Newton did not impart to Craige the 'prime theorem' of quadrature formerly communicated to Leibniz (nor the letters of which it was a feature) but from Craige's account to him, in Edinburgh, of Newton's paper Gregory was able to establish this theorem for himself and proposed (much to Craige's embarrassment) to print it as his own discovery. In fact, it was Pitcairne who issued it in an almost invisible tract. Perhaps Newton never heard of this; at all events, the Revolution soon swept the matter from his mind. Craige was still apologizing for Gregory thirty years later.[17]

John Collins

Newton's first mathematical friend outside Cambridge, John Collins, whose correspondence with James Gregory in the early 1670s had been the *fons et origo* of this little incident, died in November 1683. He had done more than any man to encourage Newton to work and publish in mathematics, and almost everything known publicly – little enough – of his mathematical researches by men like James Gregory, John Wallis, Michael Dary and the continental mathematicians reached through Henry Oldenburg's international network had been by Collins's instrumentality. In the eight years from 1675 to Collins's death Newton wrote him only two surviving letters, both in 1676; Collins wrote eight extant letters to Newton. Others are evidently lost. The correspondence continued to be largely about mathematics. Newton asked Collins to assure Dr John Pell that no overlap existed between his own methods for solving algebraic equations and Pell's work on the solution of numerical equations, which Pell should publish without delay. (Who was talking?) Newton's second letter concerned Collin's correspondence with German mathematicians; it is terse and discouraging:

> You seem to desire that I would publish my method & I look upon your advice as an act of singular friendship, being I beleive censured by divers for my scattered letters in the *Transactions* about such things as no body els would have let come out without a substantial discours. I could wish I could retract what has been done, but by that, I have learnt what's to my convenience, which is to let what I write ly by till I am out

of the way. As for the apprehension that M. Leibnitz's method may be more general or more easy than mine, you will not find any such thing.[18]

Newton had no intention of being further involved at Collins's behest.

Unwittingly, Collins had already laid the foundations of the future dispute about the origins and first discovery of the calculus which was to bedevil the last three decades of Newton's life. Collins was not only a man who had made it his purpose to make the investigations of mathematicians in Britain and in Europe mutually known to each other, and to have them critically examined by each other: he had a strong desire to vindicate the particular excellence of the British mathematicians, hence his continued fruitless efforts to persuade Newton into print. More than once we find him asking Newton his opinion of the writings of others, not always in a field of obvious interest to Newton. In June 1674, for example, he sent to Newton a copy of *The genuine use and effects of the gunne demonstrated* (1674) by one Robert Anderson. By this time, the idea that projectiles describe parabolic trajectories, first put forward by Galileo, was well established and coming into practical use; Anderson employed it, not considering the effect of air resistance upon real projectiles.[19] Not so Newton; he replied to Collins: "Mr. Andersons book is very ingenious, & may prove as usefull if his principles be true. But I suspect one of them namely that the bullet moves in a Parabola. This would be so indeed were the horizontal celerity [speed] of the bullet uniform, but I should think its motion decays considerably in the flight." To express this, he made the arbitrary proposal that (in a horizontal projection) the horizontal intercepts of the projectile's position after successive equal times should not be in an arithmetical progression increasing – as the parabolic theory required – but rather in a geometrical progression decreasing. Ten years laters – Newton forgot nothing – he returned in a far more sophisticated way to this problem of resisted projectile motion in his very first essay on mechanics. Then shortly afterwards, drafting the present Book II of the *Principia*, he would embark upon a treatment of more general problems in fluid mechanics, adding the resistance of pendulums and ships to those of projectiles.[20]

Leibniz enters the Story

Collins was presumably a poor Latinist. As Oldenburg brought foreign mathematicians into the network of the Royal Society, Collins formed the habit of roughly drafting material for letters to them in English, leaving Oldenburg to render the same into Latin. Thus a portion of Oldenburg's foreign correspondence was really Collins's. G. W. Leibniz, for example, introduced himself to Oldenburg by letter in July 1670; he came late to mathematics, a topic first appearing in his correspondence with Oldenburg only in February 1673, when Leibniz was visiting London and had unfortunate brushes with Robert Hooke and John Pell. Oldenburg sent a copy of his letter of self-vindication vis-à-vis Pell to Newton.[21] Later in that same year his ingenuity in the treatment of series began to be evident. When correspondence was resumed after a considerable break, Leibniz had so far advanced (5 July 1674) that he had found an infinite series expressing the ratio between the radius of a circle and its circumference (π).[22] Such a series aroused great interest in England, though it was not new there, and soon the mathematical content of letters to Leibniz was being provided by Collins. Leibniz, of course, was now based in Paris where Christiaan Huygens was his mentor, but soon another mathematician arrived in London from Germany. Ehrenfried Walter von Tschirnhaus also made a great impression upon both Collins and Wallis (whom he visited at Oxford);[23] he was to be of some importance in the life of Leibniz but affected Newton little.

Intelligence of Newton's mathematical achievements was first conveyed to Leibniz by Oldenburg (perhaps in this case independently of Collins) in a rather dampening letter of 8 December 1674:

What you relate about your success in the measurement of curves is very fine, but I would like you to know that the method and procedure for measuring curves has been extended by the praiseworthy Gregory, as also by Isaac Newton, to any curve whatever, whether mechanical or geometrical, even the circle itself; in so much that if you have given the ordinate of any curve, you can by this method find the length of the curved line, the area of the figure, its centre of gravity, the solid of revolution, its surface, whether erect or inclined, and the segment of revolution of the solid of revolution, and the converse of these; and moreover, given the quadrature of any arc, to compute the logarithmic sine, tangent or secant without the natural ones being known, and conversely.[24]

Leibniz was not at all put down by this formidable catalogue and Collins evidently felt it proper to impress him with considerably greater detail in reply.[25] Further correspondence was much concerned with Gregory, but it included mention, for example, of Newton's logarithmic rulers for finding the roots of equations mechanically.[26] Things went slowly until on 2 May 1676 Leibniz, having received the cognate pair of arc-sine series from Collins–Oldenburg, which he thought 'remarkably ingenious' asked if he might be sent the demonstration of them.[27] In response, it seems, to this superficially simple request Collins began to prepare a long and confused compilation of British mathematics over the last years, more heavily weighted with Gregorian than Newtonian material, known (after Newton's example) as the 'Historiola'. This was never despatched to Leibniz in Paris (though Newton was to suppose that it had been) but Leibniz went through it during his visit to London in October 1676.

Collins, for a combination of reasons, delayed long in reporting to Newton anything of Leibniz's week or ten days' visit to London. And all Collins told him at last on 5 March 1677 was that Leibniz had 'imparted some papers, whereof I hope ere long to send you transcripts' and that the two mathematicians had talked together about topics 'taken out of two of Mr Gregory's letters'; he also relayed to Newton mathematical information from Leibniz's letter to Oldenburg, written from Amsterdam on his way to Hanover from London. Of Newton's own concern in Leibniz's visit he prudently said nothing. He thus suppressed the fact that he had allowed Leibniz to read Newton's letters to himself, *De Analysi* and the 'Historiola' (a compilation of which Newton then knew nothing). Leibniz did not, it seems, set eyes on the 1671 tract, *De methodis serierum et fluxionum*, nor on its partial predecessor, the 1666 tract on fluxions, though we may imagine that as Leibniz made his copious notes in Collins's study these even more precious documents cannot have been far away.[28] But since Leibniz during his hard day's labour evinced no interest in any passages relating to Newton's fluxional technique, perhaps if had been able to glance through these others papers he would have been equally unimpressed. This is something Newton would never have believed!

To go back to the early summer of 1676, at some time in May Oldenburg sent a letter to Newton, now sadly lost, in which among other matters to do with optics he gave Newton 'particulars' of Leibniz's diverse mathematical letter of 2 May, just mentioned. We cannot say what words Oldenburg used but they were enough to impel Newton to compose, by 13 June 1676, a long letter or rather a short mathematical tract, for Oldenburg to send to Leibniz. This Newton would later call his "First Letter" [*sc.* to Leibniz]. It opens thus (translating):

> Although Mr Leibniz's modesty (in the excerpts from his letter that you recently sent me) credits us [Britons] with much [achievement] with regard to a certain speculation about infinite series, concerning which there has already been some talk, yet I doubt not that he has himself already found out not only (as he affirms) the method of reducing quantities of any kind into series of that nature, but also various short-cuts perhaps similar to our own, if not even better.[29]

Clearly, Newton was not to be outdone by Leibniz in formal modesty.

The "First Letter", which has nothing to do with fluxions, is one of the more generous productions of Newton's life. Before hearing from Oldenburg he can have known little or nothing of Leibniz. He was under no obligation to provide more than a brief polite reply; indeed, Oldenburg had already informed Leibniz (in good faith, though incorrectly) that Newton had in the press a work on these topics. Instead, Newton wrote Leibniz ten dense pages of mathematical explanation drawn from the opening of his 1671 tract, *De methodis serierum*. These again state Newton's three basic processes for the reduction of quantities to series; in addition, for the first time he expressed the Binomial Theorem that he had discovered about ten years before. There was much else, then at the frontiers of mathematical science, some taken from Newton's letters to Collins.[30]

By mid-June Collins had completed the 'Historiola' and an 'Abridgement' of it wholly in English. Oldenburg translated the latter document into Latin, added Newton's "First Letter" at the end, and sent the lot to Leibniz in Paris on 26 July 1676. (Why his part of the process took so long is unknown; other circumstances delayed Leibniz's receiving the papers until 16 August.) Leibniz replied at some length the very next day.[31] Full of compliments – 'Your letter contains more numerous and more memorable things concerning Analysis than many thick volumes published on that subject' – and gratitude to those who had chosen 'to place so many extraordinary reflections with me', Leibniz particularly remarked that 'Newton's discoveries are worthy of his genius, which is abundantly evident from his optical experiments and reflecting telescope.'[32] But the bulk of Leibniz's letter was concerned with his own mathematical accomplishments; in one phrase Newton found the basis for a later claim that Leibniz did not at this time comprehend calculus.[33]

Newton wrote few letters in the late summer and autumn of 1676, not even (by the record) acknowledging Collins's copy of Leibniz's letter, posted to Cambridge, though he did write to Oldenburg, rather quaintly, about the encouragement of cider manufacture in East Anglia, explaining that the famous English cider-apple, the Red-streak, did not do well there.[34] Between 10 September (when he read Leibniz's reply to the "First Letter") and 24 October (when the "Second Letter" was despatched to Oldenburg) Newton was wholly occupied in composing this even longer mathematical tract than the "First".

It was a substantial, highly compressed essay on the construction and use of infinite series, whose autobiographical portions are accurate enough, as the private manuscript record confirms; otherwise it is a paper 'hard to master in its mathematical subtleties' deriving, and in part copied, from *De methodis serierum et fluxionum* (1671). Newton prepared future trouble for himself by expressing Problems 1 and 2 of that tract in two so-called 'anagrams' (lists of letter-frequencies in the sentences); could Leibniz have read them they would (as we now know) have taught him nothing new. But Newton's claim would have been clearly asserted. 'In all other respects the tone of Newton's letter is one of friendly helpfulness, even in his criticism of certain 'oversights' in Leibniz' previous letter: indeed he may have intended to give Leibniz a full account of his fluxional calculus at a subsequent date.'[35]

This summary of the position by the master of Newtonian mathematics is unexceptionable; certainly in 1676 Newton felt no distrust of Leibniz and to devote six weeks to the enlightenment of an unknown foreign colleague is no mean act of generosity. Leibniz for his part, acknowledging 'Newton's truly splendid letter', added 'I shall read it once and again with the care and attention that it in truth both merits and requires.'[36] This letter of Leibniz's was not written until June 1677: a copy of it by Collins did not reach Newton until 30 August. Oldenburg's private measures to create an international postage service to Germany had almost failed. Within a few days of Newton's receiving Leibniz's reply, Oldenburg himself was dead, and Newton lost any opportunity to continue the correspondence, had he so wished. We shall return to it in a later chapter.

Collins vainly urged Newton to print these latest formulations of his now ten-year-old mathematical inventions.[37] We have seen already that Newton's own short-lived thoughts of using the letters in print in 1684 (to forestall David Gregory) came to nothing. Unknown to Newton, Collins had despatched copies of both letters to his friend

John Wallis at Oxford, whose identical advice to Newton was equally to fall upon deaf ears. Wallis did not easily submit to the defeat of his patriotic aspirations, however. After Collins's death, while preparing his *Treatise of Algebra both Historical and Practical* (1685) he put into it English versions of passages from both the "First" and the "Second Letters", noting that they were 'full of very ingenious discoveries and well deserving to be made more publick'.[38] In this rather lame fashion, in another man's book, Newton's mathematical genius was first displayed in print. Another decade passed; then Wallis, well advanced in years, began to issue all his mathematical works in stately Latin folios. Once more he thought of Newton's letters. At last, in (probably) July 1695, Newton acquiesced, himself correcting Wallis's copies, and giving him the nowadays well authenticated account of the origin of his methods:

> As to the time of my finding the method of converging series, the exactest account I can give of it is this, That in the year 1664 between Michaelmas & Christmass I borrowed & read your works & found the intercalation of your series that winter. For in the notes I then took out of your *Arithmetica Infinitorum* I find this intercalation set out for squaring the circle & the method of reducing quant[it]ies into converging series by division & extraction of roots & thereby of squaring all curves. And then (that is in the beginning of the year 1666) I retired from the University into Lincolnshire to avoyd the plague.[39]

Yet four more years were to elapse before the letters at last saw full light, in their original language, in the third volume of Wallis's *Opera mathematica*.

Newton and the Royal Society

It would be unfair to blame the decease of his correspondence with Leibniz wholly upon Newton, who was not perhaps filled with unmitigated sorrow because the Royal Society's channels of communication with foreigners now silted up, Robert Hooke, the Society's new Secretary, being happy to have it so. Leibniz, too, remained silent for years after his letters of June 1677 had vanished into a void, though he knew of Oldenburg's death. This event indeed marks a period in Newton's relationship with the Royal Society. Apart from the publication of the *Principia* – and here the essence was a personal relationship between Newton and Halley – he had little contact with that body from 1679 until his own election as President in 1703. Before 1677, few Fellows other than Oldenburg and Collins corresponded with Newton; in the next ten years he wrote far less to their successors, Hooke, Flamsteed and Halley. Other than his friend Francis Aston, Edmond Halley and again Flamsteed, few Fellows or none bothered to visit him in Cambridge, apart from those normally resident there. Oldenburg had brought Newton into the Society, cajoled him into remaining a Fellow, soothed his irritability, and kept him in the swim of scientific affairs. Of seventy-three surviving letters written by Newton between January 1672 and Oldenburg's death about five and a half years later, fifty-one were addressed to him. In contrast, from the next five and a half years of Newton's life only twenty-four letters *in toto* are now extant. Newton might complain that philosophy was a litigious lady, regret Mr Oldenburg's "ways" beguiling him into correspondence, and long for the peace denied him by publication and polemic, but it is obvious that many such cries of woe were the products of momentary annoyance. Had Newton's letters to Oldenburg not been written (and published in the *Philosophical Transactions* by him) both contemporaries and posterity would have been robbed of much knowledge of Newton's investigations and ideas about the natural world. Newton himself would have lost much by the absence of the Secretary's sympathy and stimulation. In no period of his creative scientific life was Newton less isolated than in those years between 1672 and 1677, and again from 1684 to 1687, when he was reassured by a co-operative friend.

Robert Hooke

Yet strangely it was Oldenburg's successor, the ambitious Robert Hooke, who gave Newton the crucial stimulus that transformed his life and reputation. Hooke, the Royal Society's Curator of Experiments and Gresham Professor, was an investigator of great manual skill, practical inventiveness in the mechanical sciences and fertile physical intuition. He had constructed the first laboratory air-pump ever made, produced the finest book of microscopical observations of the time, and freely laid before his colleagues in the Royal Society his neo-Cartesian interpretations of a great variety of phenomena: heat and light, rock strata and fossils, animal physiology, the movement of bodies on Earth and in the heavens. Wider-ranging than Newton in experimental science, he could not – as he admitted – approach

Newton in mathematics. Here he was hardly more than a competent practitioner of elementary geometry. An excessive confidence in his own optical theories had brought Hooke into collision with Newton in 1672; Hooke's dismissive reaction then, and again in 1675 when Newton's long papers of 'Hypotheses' and 'Observations' were read to the Royal Society, was never forgotten nor entirely forgiven by Newton. Officially, however, all was smooth when Newton at the end of 1677 approached Hooke, now Secretary, about the printing of the letters exchanged (through Oldenburg) with Anthony Lucas, the last of Newton's adversaries in optics; for (he wrote) "Mr Oldenburg being dead I intend God willing to take care that they be printed according to his mind amongst some other things which are going into the Press."[40] He also thanked Hooke for repeating before the Society Newton's basic experiment on the formation of the prismatic spectrum, confirming the account of which Lucas had been sceptical, and paid Hooke a warm compliment. In return, Hooke politely assured Newton of the full success of the experiment but regretted his inability to help Newton further.[41]

The making of this experiment, on 27 April 1676 in fact, had been an important event. At last here was independent verification that Newton's fundamental claim about the colour and shape of the solar spectrum, so often challenged by the Royal Society's foreign correspondents during the last four years, was correct. It is probable that Henry Oldenburg had pressed for this to be done, so that he could assure such sceptics that Newton's account had been formally confirmed *coram publico* by the Society whose motto was *Nullius in Verba*. Three weeks before a committee (whose members were Sir Jonas Moore, Dr Croone, Dr Abraham Hill, Dr Nehemiah Grew and Mr Robert Hooke) had been nominated by the Society to take charge of the performance of Newton's experiment. It is worth quoting Oldenburg's minute for the decisive day: 'The experiment of Mr. NEWTON, which had been contested by Mr. Linus and his fellows at Liege, was tried before the Society, according to Mr. NEWTON's directions, and succeeded as he all along had asserted it would do.'[42]

Newton himself, of course, was not present; he was almost certainly in Lincolnshire rather than Cambridge. It is likely, indeed, that Newton did not visit the Royal Society again until several years after his admission in 1675.

Death of Newton's Mother, 1679

His absences from Cambridge in these years were few and brief. In 1676 he missed four days only (27 May to 1 June); in 1677, however, he left the college on four occasions, once for a month's absence. On 26 March in this year he left for London, where he remained till about mid-April; a letter records his failing to meet John North – the new Master of Trinity in succession to Barrow – at his lodgings in Chancery Lane. They were to have discussed his brother Francis North's new book on music, about which Newton wrote instead.[43] In 1678 he spent most of May at Woolsthorpe, whence he wrote to Hooke requesting him to deliver two letters in London.[44] In 1679 occurs a gap of some nine months in Newton's correspondence, and in this period he was away from Cambridge during most of the time from 15 May to 27 November. The reason is obvious: his mother was dying and, after she had been buried at Woolsthorpe on 4 June, he remained to clear up her affairs and secure possession of the estate to himself. Writing to Hooke on 28 November 1679 Newton explained that he had temporarily deserted philosophy: "For I have been this last half year in Lincolnshire cumbred with concerns among my relations till yesterday when I returned hither [Trinity]; so that I have had no time to enterteine Philosophical meditations or so much as to study or mind anything els but Countrey affairs." After this very acceptable excuse Newton unnecessarily launched into one of those mildly paranoiac tirades to which he had formerly treated Henry Oldenburg:

> And before that, I had for some years past been endeavouring to bend myself from Philosophy to other studies in so much that I have long grutched the time spent in that study unless it be perhaps at idle hours sometimes for a diversion; which makes me almost wholy unac-quainted with what Philosophers at London or abroad have of late been imployed about . . . And having thus shook hands with Philosophy, & being also at present taken of with other business, I hope it will not be interpreted out of any unkindness to you or the R. Society that I am backward in engaging my self in these matters, though formerly I must acknowledge I was moved by other reasons, to decline as much as Mr Oldenburg's importunity & ways to engage me in disputes would permit, all correspondence with him about them.[45]

The allusion to Oldenburg's extension of the right to reply was unkind. Newton himself had initiated fresh correspondence with the Royal Society about optical matters after 1672 and he had more than once manifested not only readiness to respond to objections against his optical papers, but satisfaction at having opportunities to do so. We may guess that Newton arrived back in Cambridge annoyed by his long detention in the country, and perhaps was provoked by Hooke's manner. Later he told Halley that this letter to Hooke (to which I shall have occasion to return) "refused his correspondence" and that Newton "expected to heare no further from him, could scarce perswade my self to answer his second letter, did not answer his third . . ." Moreover, Newton had been (it seems) engrossed in chemical experimentation in the period before his mother's death.[46]

The evidence bearing on Newton's feeling for his mother, and hers for him, is minimal and though we may plausibly infer that Newton now assumed the somewhat doubtful rights of the lordship of the manor of Woolsthorpe with satisfaction, we do not know what his emotions at her death may have been. Ignorance of the actual situation is of course no impediment to psychological fiction masquerading as reasoned interpretation. To Frank E. Manuel 'Newton's mother is the central figure in his life', his Oedipal relationship with her accounting for all the complexities and contradictions in his character. It is true that according to one anecdote Newton, who had as a child purposed burning their house about his mother's and stepfather's ears, attended his mother in her fatal fever (perhaps typhus or typhoid?)

> with a true filial piety, sate up whole nights with her, gave her all her Physick himself, dressed all her blisters with his own hands, and made use of that manual dexterity for which he was so remarkable to lessen the pain which always attends the dressing, the torturing remedy usually applied in that distemper with as much readiness as he ever had employed it in the most delightfull experiments.[47]

As a matter of record, Newton could not have so tended his mother for more than a week; but one suspects embroidery. One may ask too whether the anecdote, if true, justifies Manuel in asserting that this 'fixation upon his mother may have crippled Newton sexually'. Though Freud's reification of Greek mythology followed the pattern of Newton's own exercises in ancient history, we may now leave Clio to repose unquestioned upon the psychiatrist's couch.

Exchanges with Hooke, 1678–80

After this long digression, it is time to return to Newton's correspondence with Robert Hooke and Anthony Lucas in 1677 and 1678. It seems that the fire already mentioned, perhaps occurring in January 1678, destroyed the copies of Lucas's letters in Newton's possession. He then sought to obtain fresh copies from the originals in the Royal Society's keeping, for printing in his proposed publication. In the event, Lucas himself sent two copies directly to Newton, and one by Hooke.[48] We need concern ourselves no further with Newton's final letters to Lucas about the disputed optical experiments, since Newton before long refused to receive anything further from the man; but we must note that Newton and Hooke exchanged further cordial letters about other matters in May and June 1678. Particularly, Newton enquired in Lincolnshire on Hooke's behalf about the longest piece of level ground to be found outside the Fens, which was said to be on Lincoln heath.[49]

Thereafter silence, perhaps because Hooke abandoned his plan. It was he who broke the silence, but about a year and a half later, again writing in official style, and rather absurdly opening his letter with the words: 'Finding by our Registers that you were pleased to correspond with Mr Oldenburg . . .' as though every reader of the *Transactions* in Europe did not know this!

> I hope therefore [Hooke continued] that you will please to continue your former favours to the Society by communicating what shall occur to you that is philosophicall, and in returne I shall be sure to acquaint you with that we shall Receive considerable from other parts or find out new here. And you may be assured that whatever shall be soe communicated shall be noe otherwise farther imparted or disposed of then you yourself shall prescribe.

Both these promises were, at least as to the letter, soon broken by Hooke, who went on to blame 'some' for misrepresenting himself to Newton. Further in the same vein he wrote:

> For my own part I shall take it as a great favour if you shall please to communicate by Letter your objections against any hypothesis or opinion of mine, And particularly if you will let me know your thoughts of that of compounding the celestiall motions of the planetts by a direct motion by the tangent & an attractive motion towards the centrall body, Or what objections have against my hypothesis of the lawes or causes of Springinesse.

The remainder of his letter was a summary of philosophical news, including Flamsteed's 'confirmation' of the parallax 'of the orb of the earth'.[50]

Newton answered Hooke with the letter already discussed, intended to put him off; however, "in complement, to sweeten my Answer" as he later recalled for Halley, Newton added "the experiment of Projectiles (rather shortly hinted then carefully described").[51] This was putting it too lightly: "the fansy of my own about discovering the earth's diurnal motion" by a physical experiment (surely Newton was aware of Hooke's *Attempt to prove the motion of the Earth* by astronomical observation), illustrated by a sketch, is quite precisely expressed. Newton had reasoned (correctly) that, assuming the diurnal rotation and negligible air-resistance, a heavy body placed as far as possible above the Earth's surface would have had a circumferential velocity eastwards minimally greater than if it stood upon the ground below. Accordingly, when released to fall from the height, the body "will not descend in the perpendicular line AC, but outrunning the parts of the earth [beneath] will shoot forward to the east side of the perpendicular describing in it's fall a spiral line ADEC."[52] An experiment to demonstrate such a deviation might be easily made from the top of a tall building or within a deep well, the point perpendicularly below the suspended heavy body being first established before its release by methods that Newton detailed with some ingenuity. If by trial the falling body showed an eastwards deviation, the Earth's rotation would be proved. Newton ended his letter with a reiteration of his "affection for Philosophy being worn out, so that I am almost as little concerned about it as one tradesman uses to be about another man's trade or a countryman about learning" (what strange comparisons), explaining again "that I have of late & still do decline Philosophicall commerce but only out of applying my self to other things". After which, quite unconsciously, he belied himself by adding in a postscript that "Mr Cock [an instrument-maker] has cast two pieces of Metal for me in order to a further attempt about the reflecting Tube [telescope] which I was the last year inclined to by the instigation of some of our Fellows."[53]

The scientific significance of this curious letter – which also reveals so much of Newton's complex nature – is in its diagram and the relevant phrase "describing in it's fall a spiral line ADEC", for Newton showed this spiral curling in less than a complete revolution into the centre of the Earth, where the projectile evidently came to rest. Now this was indeed careless, especially as he had said nothing about the resistance (or none) experienced by this remarkable object

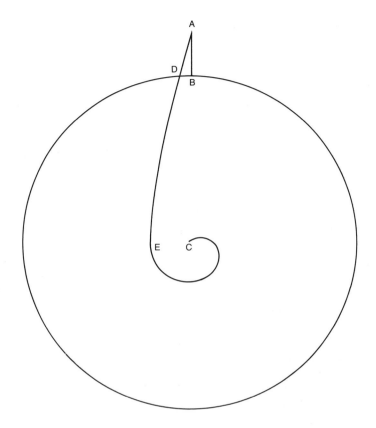

FIGURE 6.1   The spiral line of descent to the centre, after the sketch in
Newton's letter to Hooke, 28 November 1679

in falling through the air and the Earth itself; it is as though Newton
were illustrating the ancient Aristotelian notion that (could it do so) a
weight would descend to the Earth's centre with increasing speed,
and there instantly stop. Hooke was not slow to note this carelessly
imagined trajectory and at once scored a point by writing to stress
Newton's error, of which he also advised the Royal Society. In so
doing he set Newton's foot on the road to the *Principia*, a road to be
traced in a later chapter. True, Newton himself would at first hardly
be aware of the massive significance of what he was about to
accomplish in these days of late December 1679 and early January
1680, for in his current mood of "begrutching" time spent on
philosophy he put his papers away and paused on this new, fruitful

track. It required Halley to bring from him the admission that he had thrust aside the barrier seeming to block the way to the reconciliation of mechanics and astronomy.

Other Scientific Correspondence

The mention just now of a letter of Newton's about music has indicated that Newton's correspondence in the pre-*Principia* epoch was not wholly about mathematics and optics. Though in that letter to John North Newton depreciated his own skill in musical theory, he sensibly insisted that sound could not travel through a vacuum and that notes of the same frequency, travelling different distances, do not necessarily strike the ear in phase.[54] A stranger kind of physics was envisaged by a Dr Joshua Maddock, who imparted to Newton a mathematical theory of anti-light, rays of darkness. Newton told him that his theory was correct, on the assumptions stated; what he had worked out was new and would advance ordinary optics, if the existence of rays with such properties should ever be demonstrated experimentally.[55] But far more important was the emergence of Newton as a mathematical astronomer. We know from his reading that he studied the subject, though he never attempted to acquire competence as a precise observer. Positional astronomy was a subject in which Newton perforce relied upon specialists, above all (as we shall see) John Flamsteed. Most of Newton's own observing was of comets, but still relying upon others equipped with the proper instruments for precise positions.

The first letter linking Newton with mathematical astronomy had a Lincolnshire origin. Arthur Storer, son of a Woolsthorpe farm tenant, a boy thrashed by Newton, was the brother of the girl who (according to Stukeley) once touched Newton's heart. Storer was about to emigrate to Maryland when in August 1678 he sent to Newton by his own uncle Humphrey Babington (whom we have supposed to be Newton's early patron) tables of the motion of the North Star in altitude and azimuth.[56] These were soon followed by like tables for the Sun. Seemingly Newton, acknowledging the former table on 11 September 1678, found some fault in Storer's calculation. Later Storer sent Newton from Maryland details of the motions of the comets seen in December 1680 and August 1682; the latter was to be Halley's comet. No reply from Newton is recorded.[57]

Cometary astronomy was also the topic of a major series of letters exchanged between Newton and Flamsteed, passed through the hands of James Crompton, a Fellow of Jesus College contemporary with Newton.[58] A conspicuous morning comet had been seen in November 1680: 'I concluded,' wrote Flamsteed percipiently, 'that having past the Sun it would appear after his seting in December,' as indeed it did and stayed visible in the evening into January 1681. As Flamsteed explained to Halley – for the two astronomers were still on cordial terms at this time – he imagined that the Sun attracts to itself all the celestial bodies 'that come within our Vortex, more or less according to the different substance of their bodyes and nearenesse or remotenesse from him', this attraction bending the formerly straight path of the comet into a curve. The next element in Flamsteed's speculation is incoherent: approaching the Sun the comet turns towards it an opposite magnetic pole, and while also continually revolving with the vortex, is now repelled from the Sun. Hence in his view the comet's path was bent around *before* reaching the Sun and its (ultimately straight) line of departure from the vortex (figure 6.2(a)). He guessed that a comet might be 'some planet belonging to a vortex now ruined: For Worlds may die as well as men.' The tail might be the product of its humid atmosphere, much like smoke from a chimney on a moving ship.[59]

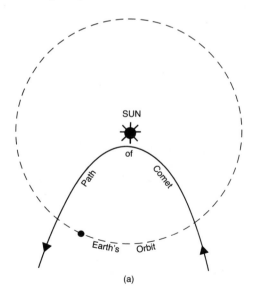

FIGURE 6.2   (a) Flamsteed's cometary hypothesis, after Newton's sketch;

John Flamsteed

Though we do not know precisely how Newton became apprised of this theory of Flamsteed's, his letter of 28 February 1681 obviously responds to it, incidentally alluding to the motion of projectiles debated with Hooke a year before. By this time the astronomer had been established at Greenwich Observatory for some six years. Only four years junior to Newton he had slowly made a name for himself. Oldenburg, Collins and other Fellows of the Royal Society began to encourage him from 1669. He had been – and remained – a critic of Newton's theory of light and colours. Travelling home to Derby from London by Cambridge in 1670, to enrol at Jesus College, he took the opportunity to meet both Barrow and Newton. Four years later, spending the period 29 May to 13 July in Cambridge (during which he took the MA degree which would in due course allow him to proceed to holy orders) he attended one of Newton's 'Lectures on Algebra' and retained a paper of notes he was then given. Why, when Flamsteed's name was again brought to Newton's notice in 1681 he did not write directly to him is not clear.[60] Newton could not understand how a comet, which upon Flamsteed's hypothesis must at some point be moving upon a line aimed directly at the Sun, could

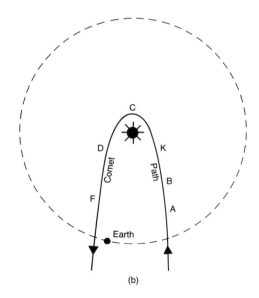

(b)

(b) Newton's revision of it. Note that in both hypotheses the comet's path near the Earth's orbit is almost rectilinear

avoid destruction by falling into it. "The only way to releive this difficulty [he wrote] is to suppose the Comet to have gone not between the Sun & Earth but to have fetched a compass about the [Sun] as in this figure" (figure 6.2(b)). It seems to me a fair guess that Newton at this point was devoting to Flamsteed's cometary theory – itself a good step in the right direction since he rightly identified successive morning and evening comets as the same body following a curved path rather than two (as Newton and all astronomers had supposed) – the mechanical way of reasoning into which he had been jolted by Hooke. His revision of Flamsteed's hypothesis, putting the Sun at the 'focus' of an elongated semi-orbit (see figure 6.2(b)) formulates the basic idea of his mature theory of comets. Moreover, Newton went on:

> though I can easily allow an attractive power in the Sun whereby the Planets are kept in their courses about him from going away in tangent lines, yet I am the lesse inclined to beleive this attraction to be of a magnetic nature because the Sun is a vehemently hot body & magnetick bodies when made red lose their vertue.

He emphasized in addition that if a celestial magnetic attraction were polarized, as Flamsteed supposed like Kepler before him, it would not serve their turn, for if the comet were attracted by the Sun, it could not be again repelled from it. Moreover, Newton doubted Flamsteed's identification of the seeming pair of comets of 1680/1 as a single one, but we need not dwell on Newton's strongly mustered objections, in part invoking the appearance of the tail, about which Newton had strictly cross-examined a "schollar of our College" who claimed to have seen the comet in November, near the star Spica Virginis.[61] Evidently the comet aroused considerable interest in Cambridge: Humphrey Babington had seen it with the tail extended over King's College chapel between 5 and 6 a.m. (Newton gives a little sketch of this) and he himself had studied it several times with a seven-foot telescope, using a pair of compasses to ascertain its distance from neighbouring stars.

Noting in passing the wealth of detail about the positions and probable orbit of the comet to be found in the discussion between Newton and Flamsteed, we turn to a fascinating draft of (perhaps) April 1681, in which Newton strengthened his arguments that the Sun, being white hot all through, could not possess (ferrous) magnetism and that the directive force of a magnet is stronger than its attractive force. Nor could he agree with Flamsteed's primitive notions about angular momentum, which were contrary to "the laws of motion". For once, Newton reverted to boyhood's amusements:

> Upon a spell [a stick or bar of wood] or bridge such as schoolboys play with, lay a large ball one hemisphere of which is white the other black. Either hemisphere lying upwards, strike the end of the bridge to make the ball rise & if the ball receive not any circulating motion from the stroke you will see that hemisphere which is laid upwards continue upwards as well falling as rising. If I did not know the event of the experiment by the reason of it, yet could I guess at it by what I have observed of a handball tossed up.

The unconscious echo of Galileo's attitude to thought experiments is also amusing. Equally, Newton found Flamsteed's idea of force to be at fault; should the comet in its course reach a place where "the magnetick repuls continually [urges] the Comet to go from the Sun, [it] would make [the Comet] go away faster & faster continually". After applying further mechanical reasoning against Flamsteed's hypothesis, Newton summed up his position by writing:

> But all these difficulties may be avoyded by supposing the comet to be directed by the Sun's magnetism as well as attracted, & consequently to have been attracted all the time of its motion, as well in its recess from the Sun as in its' access towards him, & thereby to have been as much retarded in its recess as accelerated in his access. & by this continuall attraction to have been made to fetch a compass about the sun in the line ABKDF, the *vis centrifuga* at C overpow'ring the attraction and forcing the Comet there notwithstanding the attraction, to begin to recede from the Sun.[62]

This seems to have been the end of the matter, though in only a year or two more Newton would be making friendly demands upon Flamsteed for precise astronomical data about planets and satellites. The letter that the astronomer certainly received from Newton in April 1681 makes it clear that though Newton was ready to consider hypothetically the mechanics of the orbit of the comet(s) of 1680–1 supposed to be one and the same, he considered them in reality to be two, and saw no strong objection to the Keplerian hypothesis that comets move in straight lines.[63]

Newton on Celestial Motion

Thus Newton, while analysing clearly the components of a novel treatment of planetary motion, chose to endorse none of them as physically plausible. Equally, he made no comment upon Flamsteed's accepting the Cartesian celestial vortex as valid – probably at this stage he did so himself. Hence Newton, with many of the elements of a potential revolution in his hands, preferred to remain a conservative. He was by now (as we shall see later) aware of Hooke's hypothesis of "an attractive power in the Sun, whereby the Planets are kept in their courses about him from going away in tangent lines" – Newton's words are a close recollection of Hooke's formulation – but in 1681 solar attraction did not commend itself to Newton as a likely cosmic hypothesis, and when in 1684 he re-examined mathematically the idea of a central attractive force, he did not approach the problem in Hooke's way. Why he rejected the Kepler–Flamsteed concept of solar attractive force is quite clear: it was because that force was magnetic and polarization was one of its characteristics. Possibly Newton thought Hooke's force to be of the same kind. When Newton considered the possibility of solar force of a different kind, a non-polarized, gravitational force acting upon the planets and drawing them towards the Sun and treated this as a serious physical hypothesis, he saw that his former objections to the Kepler–Flamsteed hypothesis vanished at once, and along with them went the need for a celestial vortex to swirl the planets and comets around their orbits.

As we shall see, Newton had in 1680 solved the mathematical problem of determining the orbit traced by a body moving relative to an inverse-square law centre of attraction. But we have no reason to suppose that at this time he imagined his solution to have such astronomical validity as to be applicable to the celestial motions. His letters to Flamsteed prove that he did not, at least so far as comets were concerned. I do not believe that in those discussions with Flamsteed, just summarized, Newton deliberately concealed the possibility of viewing the Sun as the centre of a non-polarized force of attraction, gravity in short; and the consequence following from this that bodies approaching the vicinity of the Sun must, being attracted towards it, necessarily describe a path – a conic – including the Sun within itself. The latter assertion would have provided the best possible answer to Flamsteed's hypothesis placing the Sun externally to the orbit, had Newton been aware of it. Moreover, I cannot believe that a mathematician acquainted with celestial mechanics would have failed to grasp the identity of the comet(s) of 1680–1.

Newton's mathematical papers confirm his interest in comets. At some point he attempted to define a rectilinear path for the comet of 1680–1. He observed the great comet ("Halley's") of 1682 and at another time collected data about a number of comets. In the same manuscript he at last took over the revolutionary position that he had formerly rejected, that is, he accepted the comet's path as closed or open with the Sun within, possessing a gravitational power far greater than that of any other body in the solar system. These manuscripts are undated, but by the autumn of 1684 Newton was certainly convinced that planets and comets are governed by the same laws of motion. Not until 1685 did he admit to Flamsteed that the same comet had appeared twice, in late 1680 and early 1681.[64]

Christ's Hospital

Other letters from these years illustrate Newton's increasing involvement in business of various kinds. Less than a year after the correspondence between Newton and Flamsteed about comets they encountered each other again over the appointment of a teacher of mathematics to Christ's Hospital. Originally a London foundation of King Edward VI, new arrangements made in 1673–4 with government money provided for the support of forty King's Scholars who were to be trained in mathematics and navigation to the advantage of the Royal Navy or the merchant marine. This plan to improve navigation at sea was heartily promoted by such men as Christopher Wren and Samuel Pepys (Secretary of the Admiralty), both future Presidents of the Royal Society. For the post of mathematical master to the King's Scholars Newton recommended (upon consultation by his friend John Ellis in 1682) a younger Fellow of Trinity, Edward Paget. He wrote to Flamsteed and Collins to enlist their support for Paget and sent a testimonial to the governors of the Hospital. Despite Pepys's fear of Paget's lack of experience in navigation the academic mathematicians prevailed: Paget was appointed. After some years he became drunken and idle, but one of his last useful acts was to draw up a revised curriculum for the boys, endorsed by Newton, Wallis and David Gregory, and later adopted. After Paget's resignation in 1694 Newton was again consulted on the choice of a successor and his nominee (the unrelated mathematician Samuel Newton) was approved; in subsequent years Newton sometimes examined the King's Scholars.[65]

Thomas Burnet

A glimpse of Newton's biblical studies is furnished by a letter to Thomas Burnet, a Fellow of Christ's College, in January 1681. An older contemporary of Newton, Burnet was about to publish *Telluris theoria sacra* ('The sacred theory of the Earth'), a book in which he attempted to interpret Moses's narrative of the creation in a rational and philosophical manner in order to account for the present state of the globe and its future destruction. Evidently towards the end of 1680 he asked Newton to review his text before it went to press. Newton did so, but his first letters to Burnet are lost, save for a quoted fragment. A letter of defence by Burnet exists, as well as the single surviving (but undated) letter by Newton in reply.[66] It is mainly of interest as an introduction to Newton's style in the exegesis of ancient documents. Of course he found no strangeness in Burnet's assumption that Moses is our sole authority on cosmogony, nor in his belief that the creation was accomplished in six days, though these Mosaic days he interpreted (in good company) allegorically: "If you would have a year for each days work you may by supposing day & night was made by the annual motion of the earth only, & that the earth had no diurnal motion till towards the end of the six days." But, "I must profess I know no sufficient naturall cause of the earth['s] diurnal motion." In general, Newton argued that Moses's account, terse, intelligible to the vulgar yet complete as to essentials, could not be bettered; Moses wrote neither as a poet nor as a philosopher but as a plain man speaking to the multitude. To explicate Moses, however, Newton resorted to scientific analogies or to homely ones such as "if beer be poured into [milk] & the mixture be let stand till it be dry, the surface of the curdled substance will appear as rugged and mountanous as the Earth in any place." Particular interest attaches to Newton's answer to Burnet's question: is the Earth a perfect sphere or axially elongated as the French suppose? Newton favoured the former alternative: "And my chief reason for that opinion is the analogy of the Planetts. They all appear round so far as we can discover by Telescopes, & I take the earth to be like the rest. If it's diurnal motion would make it oval that of Jupiter would much more make Jupiter oval." As indeed it does; Newton was inevitably ignorant of observations of the oblate shape of Jupiter already imparted in August 1673 by G. D. Cassini to Flamsteed (but by him denied). Again, within a few years Newton would reverse his opinion, with a full mechanical justification.

William Briggs

Burnet, a man much attacked in years to come and active in his own vindication, is well known to history and was obviously on friendly terms with Newton, but we know nothing of their relationship.[67] Another man whom Newton addressed as his "honoured friend", and with whom he proposed to discuss the theory of vision at their next meeting, is similarly shadowy. This was a physician, William Briggs, a contemporary of Newton and Fellow of Corpus Christi College, but practising in London. In 1676 Briggs had published *Ophthalmographia* (second edition, 1685), while in March 1682 a paper of his on 'A new theory of vision' was read at the Royal Society. He sent a copy of this to Newton, who wrote back on 20 June 1682.[68] He approved Briggs's ideas about the pairing of the fibres in the optic nerve and the perception of (almost) the same scene by each eye, but found other minor points to criticize here and in a subsequent letter of 12 September, hastily written because of "Sturbridge Fair friends". Vision had been of great interest to Newton since he first opened his notebooks, as it had been to Kepler and Descartes, and so it is not surprising to find him remarking upon the eye of the fish and the chameleon. He urged Briggs to print his *New theory* in Latin, which was done in 1685; to this version was prefixed a commendatory letter by Newton praising Briggs's skill in optics and anatomy; "Your skill and dexterity in the dissection [of the eye]," he recalled, "formerly afforded me no little enjoyment."[69]

Universal Arithmetick

Newton was now a figure of substance in the world of Cambridge and London, a man whose judgement scholars in other fields sought and respected. To none did he complain that they made him a slave to philosophy. Meanwhile regular teaching continued, with a few seeming gaps in the record. Whether he at any time lectured upon mathematical geography, as might be suggested by his edition of Varenius (p. 110) and his gesture of support for the English geographer John Adams, is speculation.[70] What is certain is that he claimed to have given ninety-seven lectures upon algebra between October 1673 and October 1683, depositing the fair copy about 1685 in rough accord with the statutes of his chair. In the later 'authorized' printed book of the lectures in English (1720) they make 237 small but closely printed pages, that is about two and a half pages per lecture, which does not seem heavy work.[71] But the relationship of Newton's performances in the Schools to the revised manuscript deposited in the University Library is far from straightforward (as is invariably the case with Newton). Flamsteed, the only recorded auditor of just one of these lectures, confirms that it was given on material rehearsed by Newton in the deposited text, but not on the same date. Moreover, Newton assigned lectures to the Michaelmas term of 1679 when he was almost totally absent from Cambridge after his mother's death. D. T. Whiteside argues from the handwriting that the text was written continuously and rapidly, presumably after the nominal closing date. The preparatory materials for the deposited text have for the most part vanished.[72]

Nor can anything positive be put forward to account for Newton's choice of topic. Perhaps, justifiably, he wished to adopt a more basic approach than before – if he did so, he was far from realizing his object. He could have followed a more obvious and practical line from trigonometry to its applications in cartography, astronomy and navigation. It may be that his work on Kinckhuysen, of which he made use in the lectures, led him into the same branch of mathematics. His success as an expositor to beginners must be regarded as mixed. Newton himself began (in 1684?) a complete re-casting of the deposited text, and when (a generation later) he at last issued his own printed revision of it in 1722 he undertook a thorough overhaul. But even so, Newton could not eradicate the 'larger faults of poorly conceived construction and muddled notions of what a novice would find both comprehensible and useful'.[73] For all that, under the printed title of *Universal Arithmetick*, the doubly revised 'Lectures on Algebra' were many times reprinted in the eighteenth century (both in Latin and in English) and found their way into gentlemen's libraries.[74] Leibniz, unhesitatingly divining their author beneath the cloak of anonymity, gave them a long review in the *Acta Eruditorum* of Leipzig in 1708. Written thirty years before, he noted, and now deservingly printed by William Whiston, he assured the reader that 'you will find in this little book certain particularities that you will seek in vain in great tomes on analysis.' His close associate, Johann Bernoulli, despite some adverse remarks paid Newton the compliment in 1728 of basing his own course on the elements of algebra upon Newton's text.[75] Perhaps partly in consequence of Newton's recent death, in Britain too the book began about this time to arouse greater interest than when it was first issued in 1707.

Universal Arithmetick, like so many of Newton's writings, remained for ever incomplete. The final page of the deposited text promises a new chapter following 'De Reductione fractionum et Radicalium ad series convergentes' ('On the reduction of fractions and roots to converging series') but nothing follows. Innovations in the technique of analysis were noted by Leibniz in his review and he also remarked on Newton's preference (shared by himself) for classical geometric methods. Newton's 'guiding doctrine that algebra is "universal arithmetick"' – that is, an arithmetic that operates generally without specifying particular numbers – 'embroiders a theme stated briefly in an opening phrase of his 1671 treatise on infinite series and fluxions' but to the stratospheric levels of that treatise the lectures of course did not rise. Even so they were far too difficult for beginners in mathematics like Thomas Horne, Fellow of King's College and perhaps the only individual known to have signed a letter to Newton 'your humble and thankfull pupill'. Possibly Horne attended the lectures.[76]

William Whiston

The William Whiston mentioned by Leibniz who, much against Newton's wishes, published (as an anonymous work) *Arithmetica universalis, sive tractatus de resolutione et compositione arithmetica* ('Universal Arithmetic, or a treatise on arithmetical resolution and composition') in May 1707, was, like his friend the physiologist Stephen Hales, a member of the first generation of Cambridge students to emulate Newton's method and principles. He went up to Cambridge in 1686, claimed to have attended one or two incomprehensible lectures by Newton on his *Principia*, and was elected a Fellow of Clare Hall in 1691. After taking orders he left Cambridge for a while, returning in 1700 when chosen by Newton to be his deputy as Lucasian Professor. About a year later, upon Newton's resignation and commendation, Whiston succeeded him.[77] Aberrant theology was to be his downfall. While Newton and their common friend Dr Samuel Clarke kept private their doubts about Trinitarianism, the Creed and the Thirty-nine Articles, Whiston sought publicly to amend the errors of the Anglican faith; for this he was summoned before the heads of houses in the university and dismissed from his post in 1710. Well before this, however, Whiston (who had read and presumably copied the 'Lectures on Algebra' in the University Library) had in 1706 arranged for their printing and publication in

London. Their appearance was to be for a time delayed by Newton, who in the end reluctantly acquiesced; he did not set his own improved version before the public until fifteen years later, in 1722.

Other Mathematical Work

At the start of this chapter, when describing Newton's relations with younger mathematicians, I by no means rehearsed all the occasions on which he enlightened others by correspondence (like Horne), sometimes through Collins until his death in 1683. And besides this, amounting in Gregory's case (1684) to a fair body of work, not to say the 'Lectures', Newton in the decade 1674–84, comprised within the fourth volume of his *Mathematical Papers*, tackled many varied areas of mathematics with all his invariable skill and power of innovation. Some of this material was reworking of previous accomplishments, other parts new. Some notion of the range of Newton's interests may be given merely by repeating the chief headings assigned by D. T. Whiteside:

> I. Researches in Algebra, Number Theory and Trigonometry
> (*c*.1675–84)   200 pages
> II. Researches in Pure and Analytical Geometry
> (*c*.1678–80)   175 pages
> III. The Geometry of curved Lines   101 pages

And to these should be added the material, already discussed, concerning David Gregory.[78]

A few points of biographical interest may be drawn from this mass of technical mathematics. We obtain another rare glimpse of Newton as a teacher, his pupil being Henry Wharton of Gonville and Caius College (admitted in 1680), who probably joined a 'select Company in [Newton's] own private Chamber' to receive instruction in mathematics. His tutor was that John Ellis, a friend of Newton's, already mentioned in connection with Christ's Hospital. Wharton in 1683 received a copy of a summary of trigonometry written by Newton but never printed.[79] Another abortive labour in the same branch of mathematics at about this time was Newton's completion and preparation for publication of a *Trigonometry* by one St John Hare, who cannot be positively connected with Newton or even identified. Accordingly, one cannot explain Newton's undertaking this thankless – and, once more, abortive – task. His own comments, however, do contain original ideas about the treatment of the trigonometrical functions.[80]

A different kind of mathematical interest is revealed in Newton's late study of Greek mathematicians – not indeed in their own language but in recent translations and restorations of lost works that became available in the 1670s. D. T. Whiteside's opinion is that Newton had by no means, before this time, acquired a profound understanding of the advanced mathematical problems tackled by the Greeks. Among the results of his late studies of Pappus, Apollonius and Euclid's 'Porisms' were examinations of the 'solid locus' problems of the Ancients, and a short paper demonstrating errors in Descartes's geometry. Newton was now convinced that Descartes had wrongly depreciated the mathematicians of Antiquity; henceforward he began increasingly and deliberately to obscure his own indebtedness to Descartes in both mathematics and physics.[81]

Yet another undated study seemingly of this period is a short draft essay on the general cubic curve, the ancestor of Newton's printed treatise on the classification of the cubic curves.[82] But more significant, perhaps, was a long paper on *Geometria curvilinea* ('The geometry of curved lines'), another intended exposition of the fluxional calculus. This is connected with the anti-Cartesianism just mentioned. Newton never took it beyond the first Book and so it was to be another forgotten fragment, having only faint echoes in *De quadratura curvarum* ('On the quadrature of curves'), written in 1693, published along with *Opticks* in 1704. Part of it provided the base for the 'fluxions scholium' inserted – needlessly, one might think – into Book II of the *Principia*. This relation provides a *terminus ante qua non* for this undated fragment. Newton's purpose in writing it is clearly stated in the opening paragraphs. Cartesian analysis is clumsy, "intolerably roundabout" and sometimes unphilosophical:

Observing therefore that numerous kinds of problem which are usually resolved by [an algebraic] analysis may (at least for the most part) be more easily resolved by synthesis, I have written the following treatise on the topic. At the same time, since Euclid's elements are scarcely adequate for a work dealing as this does, with curves, I have been forced to frame others. He has expounded the foundations of the geometry of straight lines. Those who have taken the measure of curvilinear figures have usually viewed them as made up of parts infinitely small and numerous. For my part I shall consider them as generated by growing, arguing that they are greater, equal or less according as they grow more swiftly, equally swiftly or more slowly from their beginning. And this swiftness of growth I shall call the fluxion of a quantity. So when a line is described by the movement of a point, the speed of the point – that is, the swiftness of the line's generation – will be its fluxion.[83]

This concept Newton regards as the most natural and simple that can be. The elements of its geometry were to be presented in the first Book of this work, problems on the derivation of fluxions in the second Book, while the third would deal with the inverse problem of the derivation of quantities from their fluxions (that is, integration). The nature of curves in general was to be the subject of the fourth Book, together with the construction of problems by the intersections of curves. What a masterpiece of a new geometry might have been written! But Newton went so far as to outline eleven problems in Book II, then laid down his pen.

As his introduction hinted, the method is Euclidean: Definitions, Axioms and Postulates precede the series of thirty Propositions in Book I. Here Newton (since his quantities are denoted by lines, not algebraic symbols) employed the abbreviation "fl.", that is, "fl.AB" is the fluxion of the line AB. And here also for the first time he set down the 'classical' definition of a fluxion (to be printed in the *Principia* scholium): "Fluxions of quantities are in the first ratio of their nascent parts or, what is exactly the same, in the last ratio of those parts as they vanish by defluxion [that is, flow to zero magnitude]." This was as far as he could go in distinguishing the fluxion from the seventeenth-century concept of the indivisible or infinitesimal.

Just as the *Geometria curvilinea* materializes from nowhere, so too its fading into inanition is without explanation. According to D. T. Whiteside's suggestion,[84] the last, tailing-off pages show signs of the working of Newton's highly irritable temperament – the essay as it developed contained less of elegance and perfection than was tolerable to him and the idea of working through sets of problems no doubt bored him. To guess what other concerns in biblical study or chemistry may have seemed more real and enticing is idle. We can only say that the record reveals Newton's abandonment of mathematical research for a few years from the late 1670s until his interest – or self-interest – was awakened again by David Gregory in 1684.

7

The Chemical Philosopher,
1669–1695

In his translation of Hermann Boerhaave's *Elements of Chemistry* (1741) Peter Shaw wrote: 'It is by means of chemistry, that *Sir Isaac Newton* has made a great part of his surprizing discoveries in natural philosophy,' an opinion which becomes more understandable, if no less extravagant, when related to Shaw's wide claim that 'chemistry, in its extent, is scarce less than the whole of natural philosophy.'[1] Such a high view of chemistry would not have been expressed a century earlier, indeed the rise of chemistry as a department of natural philsophy had taken place in Newton's lifetime. In this rise the writings of Robert Boyle had played a principal part, exercising (as we have seen) considerable influence over Newton himself.[2] Many considerations lead me to believe that Newton's chemical atomism was Boyle's corpuscular chemistry revised, made more precise and rendered more complete, but also more deeply speculative. However, as Shaw correctly states, the Queries in Newton's *Opticks* had done much to enhance chemistry's reputation as a branch of theoretical science, the science of matter.

When Newton's chemical interests first took shape, *Opticks* and its Queries were still half a century in the future. It would be rash indeed to extrapolate the sophisticated atomist theory of chemical reaction found in them back into Newton's initial experiments of the late 1660s. No positive statements can be made about their date, nor about what book or which individual may have inspired Newton to attempt this kind of investigation. Dr B. J. T. Dobbs dates to 'About the year 1667 or 1668' a dictionary of chemical terminology compiled by Newton, and therefore implying a fairly early stage of his interest at that time, but elsewhere she contemplates the possibility that Newton might have begun alchemical experimenting as an undergraduate or a schoolboy![3] Like many other writers on Newton, she solves problems about the origins of Newton's various interests by making Newton's older 'friends' responsible: 'it is suggested here that it was through Barrow and More and their friends, as his relationships with them developed in the 1660's and 1670's, that Newton was put in touch with the larger scene of English alchemy.'[4] Since, in fact, no document throws light upon Newton's relations (if any) with Barrow and More during the first decade mentioned, except with regard to Barrow and the mathematical sciences at its close, by which time (as we know) Newton's interest in chemistry was already established, Dr Dobbs's suggestion is one that can be neither confirmed nor denied. True, we have more evidence on the mutual relationships of Barrow, More and Newton during the 1670s, though it is very meagre; none of it relates to chemistry or alchemy. However, while admitting that Barrow's activity as 'an active experimenter during the 1660's has not been recorded', Dr Dobbs infers from Barrow's supposed friendship with his Trinity colleague John Nidd (see p. 28) and his mention in 1654 of those in Cambridge who 'fear not to espouse the cause of the noble Goldmaking Stone with lavish faith, whether it be fable or history' that Barrow himself was practically engaged upon this same search.[5] Further, she admits that, supposing Barrow to have realized 'that the young Newton's interests turned in the direction of natural philosophy as well as mathematics . . . It does not seem at all unlikely that Newton had been encouraged by Barrow to use whatever equipment the [Nidd] group had in the 1660s' . . . The evidence is hardly definitive.' It would be plainer to say that no evidence exists to support any of this 'reconstruction'. Dare one reiterate that nothing is known as a fact about Barrow's awareness of Newton's existence before the summer of 1669? And that the story of the 'Nidd laboratory' is wholly fictitious? One may accept Dr Dobbs's speculation that the period of Barrow's greatest intellectual influence upon Newton was before 1669 – though his greatest influence upon Newton's career was to be in that year and later – but why go on to affirm dogmatically that 'sometime between 1664 and 1669 Barrow had given Newton's mind a firm set towards experimentation, induction, and mathematizing in "philosophizing"?'[6] Such an affirmation belongs not to history but to a romance of discipleship.

The same author detects in Barrow a 'sympathy for the Neoplatonic philosophical position as opposed to strict Cartesian mechanism' which may have stimulated Newton towards his 'reading Neoplatonic alchemy before the decade of the 1660's was out'.[7] However this may be – and I do not mean to follow these cobwebby contentions further – Henry More, the Cambridge Platonist, is universally recognized as exerting a far greater influence than Barrow upon these aspects of Newton's thought. More loved mystical philosophies, and he was certainly a friend of a Cambridge alchemist, Ezekiel Foxcroft (a Fellow of King's College) and also of an even more mysterious physician and philosopher, Francis Mercury van Helmont. More made these two men acquainted at his own table in Christ's College. But van Helmont was in no ordinary sense an alchemist, while More was certainly inclined to consider alchemical pretensions as fables rather than history. More judged chemists – not excepting Robert Boyle – to be pitiful philosophers and there is no word of evidence to suggest that he ever undertook chemical experimentation of any sort, or read any authors in the tradition of operative alchemy. It is therefore difficult to see how, even allowing that (as I have previously supposed) he may have encountered Newton at Grantham, or at Cambridge during the 1660s, he could have impelled Newton along a path which he himself regarded as unpromising.[8]

I have argued elsewhere that because of the paucity of evidence relating to Newton's life in the 1660s, from which (in the present context) we learn only that Newton had read More's *Immortality of the Soul* (1659) and was convinced by More's advocacy of atomism, we should be sceptical of imaginative attempts to make Henry More a close friend of the young Newton and a major formative influence upon his natural philosophy.[9] On the one point they are known to have discussed, years later, their ideas (by More's testimony) were far apart; on all points of scholarship their mental tempers were very unlike. We know that Newton rejected the fundamental feature of More's idea of the universe, the Spirit of Nature. That, in common with many contemporary scholars, they both held to the notion of *prisca sapientia* (the first ancient sages did not know more than we do but derived correct principles to which the moderns have returned) proves nothing about More's relationship to Newton.

Newton's Chemical Furnaces

For relief, let us return to the certainty of Newton's purchase, already mentioned, of chemicals and his building of a furnace, either in the summer of 1668 or a little later. With these events – which I take to mark Newton's first venture into practical chemistry – may be associated his early 'chemical dictionary', from which Dr Dobbs quotes the following passage, here modernized:

> *Furnace.* As (1) the Wind-Furnace (for calcination, fusion, cementation etc.) which blows itself by attracting [!] the air through a narrow passage [flue]; (2) the Distilling-Furnace by naked fire, for things that require a strong fire for distillation. And it differs not much from the Wind-Furnace, only the glass[-vessel] rests on a cross-bar of iron under which is a [stoke-]hole to put in the fire, which in the Wind-Furnace is put in at the top. (3) The Reverbatory-Furnace where the flame only circulating under an arched roof acts upon the body. (4) The Sand-Furnace where the vessel is set in sand or sifted ashes heated by a fire made underneath. (5) The *Balneum* or *Bain-marie* where the body is set to distill or digest in hot water. (6) The *Balneum Roris* or *Vaporosum* [vapour-bath] in which the glass hangs in the steam of boiling water. Instead of this may be used the heat of horse-dung (called *venter equinus* [horse-belly]); that is [alternatively], brewer's grains, wheat-bran, saw-dust, chopped hay or straw, a little moistened, close pressed [together] and covered. Or it [the material] may in an egg-shell be set under a hen. (7) Athanor, Piger Henricus [Lazy Henry] or Furnus Acediae for long digestions [the vessel] being set in sand heated with a turret full of charcoal which is contrived to burn only at the [bottom], the upper coals continually sinking down for a supply. Or the sand may be heated [by an oil-]Lamp. And is [then] called a Lamp-Furnace. These are made of fire-stones, or bricks.[10]

No word of the philosopher's stone here! Much of the practical information could have been derived from technological treatises of the sixteenth century, Biringuccio's *Pirotechnia* (1540) or Agricola's *De re metallica* (1556); such furnaces are illustrated in these books. Water-baths, steam-baths and perhaps sand-baths were easily found in the kitchen. That Newton was already interested in the old-established literature of mining and metallurgy appears from his bizarre (and derivative) letter to Francis Aston (18 May 1669) where the young traveller is adjured to

> observe the products of nature in severall places especially in mines with the circumstances of mining & of extracting metalls or mineralls

out of their oare and refining them and if you meet with any transmutations out of one species into another (as out of Iron into Copper), out of any metall into quicksilver, out of one salt into another or into an insipid body &c) those above all others will bee worth your noting being the most luciferous & many times lucriferous experiments too in Philosophy.

The next item is "The prizes of diet & other things"! The "transmutations" mentioned are not exactly those of the alchemist, for Newton shortly after asks Aston to find out "Whither at Schemnitium [Schemnitz = Selmeczbanya] in Hungary (where there are Mines of Gold, copper, Iron, vitrioll, Antimony, &c) they change Iron into Copper by dissolving it in a Vitriolate water which they find in cavitys of rock in the mines."[11] (The "transmutation" is real, a surface deposition of copper.)

In the same manuscript dictionary Newton does indeed list such alchemical terms of art as "magistery", "elexir" and "projection" for definition, leaving some for ever blank. But the great preponderance of the definitions are (in Dr Dobbs's words) 'so precise and operational that one could draw from this single manuscript a fairly full description of the state of straightforward chemical knowledge and practice as it stood in general in 1667[?], excepting medical preparations.'[12]

And so it is with a certain proportion of Newton's manuscript material to be considered in this chapter, and the tiny fraction that was released into print in his lifetime (chiefly, the essay *De natura acidorum*, 'On the nature of acids'). Like the chemistry of the Queries in *Opticks* the treatment is clear, intelligible and matter-of-fact. Apart from the technical terms of art occurring in the manuscripts (not in the printed material), these examples of Newton's writing upon chemistry contain not a vestige of alchemy. We have to remember that though such a name for a copper compopund as "volatile Venus" may seem redolent of alchemy, it is simply a name, no more mysterious intrinsically than 'butter of antimony' or 'sugar of Saturn [lead]'; such names constituted the only chemical language available to Newton. With the great mass of Newton's alchemical transcripts the case is far otherwise.

Newton's Chemical Records

Having set the scene, and leaving aside the question of the origin of Newton's interest in (al)chemical investigation as likely to be for ever insoluble, we may next examine those manuscripts which uniquely throw light upon Newton's experiments with his chemicals and furnaces. These are Cambridge University Library MS Add. 3973 and 3975. The former is a bundle of loose sheets, the latter a bound notebook. The sheets seem to precede the notebook, in which descriptions of the same experiments are usually fuller. The two manuscripts essentially cover the same work on the same dates. The earliest date marked in them is 10 December 1678, the latest February 1696, when Newton was on the eve of leaving Cambridge for ever. Clearly, records for the decade or so before the existing notes are lost, perhaps purposefully destroyed. Newton's friends knew of his chemical researches some time before these manuscripts begin. Collins wrote to James Gregory (29 June 1676): 'Mr Newton intends not to publish anything, as he [had previously] affirmed to me, but gives in his [optical] Lectures yearly to the [University] public Library, and prosecutes his Chimicall studies and Experiments.'

Four months later, Collins explained to Gregory that he had not bothered Newton with mathematical letters for nearly a year, Newton 'being intent upon Chimicall Studies and practises, and both he and Dr Barrow &c beginning to think mathematicall Speculations to grow at least nice [pedantic] & dry, if not somewhat barren.'[13] As we have seen, Newton had at this point at least six years of chemical experimentation behind him, perhaps more.

For some years (1680, 1686–90, 1694) in the period covered by MSS 3973, 3975 no chemical work is recorded, but the gaps may be only apparent as many experiments are reported without dates. Some periods of particularly intense activity appear: December 1678 to January 1679; May to June 1681; May to August 1682; April and May 1686, the last perhaps a relief from the intense preparation of the *Principia*, then approaching completion. Of course, one cannot reliably analyse these notes as though they constituted a precise laboratory diary, though the manuscripts do have something of this character.[14]

As with other notebooks containing Newton's personal researches, MS. 3975 starts with reading notes compiled under such headings as "Of fire, flame, the heate & ebullition of the hearte & Divers mixed liquors, & Respiration" or "Of Salts, & Sulphureous bodyes, & Mercury & Metalls". Some notes are in English, others in Latin without any apparent significance in the distinction. Much material is drawn from Robert Boyle's *New Experiments Physico-Mechanical touching the Spring and Weight of the Air* and especially the *Origine of Forms and Qualities* (1666).[15] With a title by George Starkey, *Pyrotechny asserted* (1658), however, one enters the different world of Newton's alchemical reading – material well-known to Boyle also, of course. As in other notebooks, reading notes continue without a break into reports of Newton's own experiments. Here is one such, lightly modernized:

> Munday June 26 [1682] Regulus of copper 8 [parts], serpens non destillata [perhaps mercury sublimate, not distilled] 1 [part], destillata 1 [part]. Of this without being melted 15 parts, sal ammoniac 20 parts, there remained in the bottom [after heating] 3 parts. The sublimate was white and with water gave a very white precipitate not readily dissolvable in Aqua Fortis, fusible in a great heat like antimony and something more volatile. Out of the water nothing more was precipitated by salt of tartar so that all that sublimed besides sal ammoniac was precipitated before by water alone. The white sublimate 18 parts, salt of iron 9, left in the bottom 7 parts, so that it carries up by 2/7, and perhaps the salt was not thoroughly dried before.[16]

This was the first of seven similar experiments made on the same day. All are factual and quantitative. Most used salts of metals, some used ordinary metals (iron, copper, lead, bismuth), others again "ores" of metals, perhaps as more naturally vigorous than the smelted substance; all involved antimony in the "regulus". Newton and contemporary chemists supposed that when antimony (or rather, its ore, stibnite) was melted with another metal, the often crystalline "regulus" of metal found at the bottom of the cold crucible contained the second metal. In fact it did not – regulus of iron, regulus of copper and so on were all antimony metal. Newton in each case examined the regulus carefully; it was just metallic antimony but he thought there was much more to it than this. Crystalline features of the metal as it cooled in the crucible struck him as having deep meaning. At one point he melted various proportions of "regulus of iron" with copper, examining the structure of the solidified mixture: nine and a quarter parts of the former to four parts of the latter "gave a substance with a pit hemisphericall and wrought like a net with hollow work as twere cut in;" again, eight and a half parts to four gave "no pit but a net worke forme spread all over the top, yet more [deeply] impressed in the middle;" two parts of regulus to one of copper "gave a net work but not so notable as the former, and so did Regulus iron 5 [parts], copper 2". From this Newton judged that the higher proportions of regulus to copper were preferable, presumably for obtaining striking crystalline appearances.[17]

The Star of Antimony

The crystalline 'star' of the regulus of antimony – of the 'pure' metal, that is – had long been esteemed by (al)chemists, as it was by Newton.[18] He would have read 'Basil Valentine's' warning that many have 'spared neither labour nor expense to bring about its preparation. But very few have succeeded in realizing their wishes. Some have thought that this Star is the true substance of the Philosopher's Stone. But this is a mistaken notion, and those who entertain it stray far afield from the straight and royal road.'[19] He discovered that a complicated regulus made with stibnite, iron and ores of iron, copper, tin lead and bismuth, purified by saltpetre "had a glorious Star" and that it was "wrought with network". Regulus is also the name of a celestial star (α Leonis) in the constellation called Leo.

Alchemical writings are replete with high-sounding, unexplained allegorical names: the Caduceus of Mercury, Diana's Doves, the Dragon, the Green Lion, often also names of astronomical significance, whose meaning the tyro had to learn from his master or discover through his own labours, as Newton tried to do. So Newton writes:

> Dissolve the volatile Green Lion in the central salt of copper and the Green Lion is distilled from this spirit. The Blood of the Green Lion, Venus, the Babylonian Dragon interposing his poison in all things [but] overcome by the beating [of the wings] of Diana's Doves, the Bond [Fetter] of Mercury.
>
> Neptune with his trident leads philosophers into the academic garden. Therefore Neptune is a mineral, watery solvent and the trident is a water ferment like the Caduceus of Mercury, with which Mercury is fermented, namely, two dry Doves with dry ferrous copper.
>
> Certainly the Caduceus of Mercury is a double vitriol [double sulphate] fermenting white natural stibnite. For these metallic principles are not readily melted, and so each clings more strongly to itself (as appears from the regulus of iron and the net) than it does to Mercury (as appears from the fermentation of the regulus with mercury.[20]

I do not pretend to understand these Latin sentences in the experimental notes here rudely done into English. But it is evident that Newton is trying to decode the symbolism of his alchemical sources. As we shall see more fully later, he believed that fable, myth and prophecy enciphered the wisdom of the Ancients, their knowledge of Nature and of human history. *Serpens*, the serpent, another name of a constellation, is one of these esoteric terms. In a passage of general elucidation of alchemical language quoted by Dr Dobbs Newton wrote:

> The Dragon kild by Cadmus is the subject of our work, & his teeth are the matter purified.
> Democritus (a Grecian Adeptist) said there were certain birds (volatile substances) from whose blood mixt together a certain kind of Serpent ([symbol for mercury]) was generated which being eaten (by digestion) would make a man understand the voice of birds (the nature of volatiles how they may be fixed)
> St John the Apostle & Homer were Adeptists.
> Sacra Bacchi (vel Dionysiaca) [the rites of Bacchus (or Dionysus)] instituted by Orpheus were of a Chymicall meaning.[21]

As before, Newton decodes for his own benefit – seemingly in a rather arbitrary way – the esoteric language of alchemy, and gives to the *prisca sapientia* of Homer and St John a chemical significance. In this passage the serpent is taken to be mercury. Elsewhere, as in the extract quoted previously, the serpent is an agent for the production of the regulus, as Newton also explains in yet another quotation that I take from Dr Dobbs: "Still more often this Regulus or star may be led through the fire with stony serpents (that is, with mercury sublimate, for stony serpents, that is, vitriol & Saltpetre are present in the sublimate) so that at length it is completely consumed and joins itself with the serpent."[22] This sentence is in fact taken from the *Triumphal Chariot of Antimony* of 'Basil Valentine', but the explanatory words in parenthesis are Newton's own. They point to vitriol (copper sulphate) and saltpetre – both crystalline – as the 'stony serpents' employed with sublimate of mercury. Possibly, therefore, serpents were a group of substances performing the same function, rather than a particular substance. Who can tell whether Newton's decoding is 'correct' or not?

Practical Notes

By no means all the experiments noted in the chemical manuscripts relate to alchemical dreams. There is for example this note on "Venetian sublimate":

> Venetian sublimate is made of mercury 2 parts, refined silver 2 parts, vitriol calcined to red 1 part and salt decrepitated 1 part. The Hollanders

sophisticate it with arsenic. The sophisticated is in long splinters and turns black with oyle of tartar dropt on it. But the true turns yellow and is in little grains like hempseed.

A similar account appears in the chemical dictionary. In another place Newton describes the preparation of ether, with no obvious context. A little more doubtful (though perfectly plain and rational in itself) is this quantitative account of the preparation of *mercurius dulcis* (mercurous chloride) from the sublimate of mercury (mercuric chloride):

> Mercury sublimate 4, mercury 4, sublimed together into mercury dulcis and a little mercury adhered to the top of the glass. This mercury dulcis put in with its weight of fresh mercury would imbibe none of it but left the mercury running. So that mercury sublimate 4 will imbibe but mercury 3 or $3\frac{1}{2}$. Note that mercury dulcis is much less volatile than mercury sublimate.

Equally without conceptual context or explanation of their significance are paragraphs like the following:

> If sal ammoniac be dissolved in aqua fortis to make aqua regia, and the menstruum distilled, the aqua fortis in a gentle heat comes over first and leaves the sal ammoniac behind, the same in weight and vertue as before: so that the sal ammoniac is not altered or destroyed by the Aqua Fortis until the menstruum be imployed in dissolving gold or some other body.[23]

There is no obvious explanation either of Newton's concern to discover that alloy of common metals which should have the lowest melting-point – unless this was part of another search for a 'running mercury'. To attain this end, he experimented with proportions of lead, tin and "tinglass" (bismuth): "Lead two parts, tin 3 parts, tinglass 4 parts melted together make a very fusible metal which in summer will melt in the Sun."[24] (Newton's metal is today defined as a composition of lead 5, tin 3, bismuth 8, melting-point 94.5° F.) Newton also employed both "tinglass" and "spelter" (zinc) in the making of a regulus.

What Metals are made of

It is possible to form any notion of the objects of Newton's procedures, pursued with so much attention to quantitative proportions, after a manner unheard of in the previous tradition of alchemy? Or rather traditions, for there were several different schools, more or less mystical, and by no means united in determining the true path to success in the Great Work. We may begin by recalling that the (al)chemists had developed a theory of metals rejecting the ancient Aristotelian scheme of four universal elements of matter. They invoked two principles. All metals, sufficiently heated, become liquid: one, mercury, is always fluid at room temperatures and special alloys (such as those Newton investigated) liquefy at temperatures less than that of boiling water. The chemists believed therefore that metals contain a principle of liquidity which they called *mercury* after the fluid metal. But metals are also hard and crystalline, and in some cases a sample will fracture rather than bend: they were therefore said to have a principle of firmness or fixity, called *sulphur*, supposedly earthy but also 'fat' and unctuous, hence also the principle of flammability. To these two principles some added *salt* to make a trinity.[25] In theory, therefore, by changing in some way the particular form of one or both of these principles in one metal, it should be possible to transmute that metal into another. (Sometimes this seemed to happen naturally, as in the central European streams Newton had heard of.) The processes for bringing about such transmutations of metals were variously understood by different (al)chemists, their operations and hopes being further confused by such extraneous issues as the possibility of hitting upon some noble medicine. Because the twin illusory objectives of making gold and abolishing death were always closely allied, a good step towards the former was judged to make progress towards the latter also. In practice, many old alchemical hands seem to have boiled and sublimated, fused and distilled, any combination of materials that came to hand, organic and inorganic, fair or foul, without any conceptual scheme but perhaps (as Newton did) making what they wished of allegory and myth.

One identifiable line of approach was to try to isolate the pure 'mercury' of metals (never found in Nature) so that it could be recombined with 'sulphur' to make a new metal. One method was to reduce the mineral ore or some compound of a metal so as to obtain the metal itself, which might be identified with 'mercury'. If antimony ore were added the result might be a regulus of antimony metal. Alternatively, a metallic compound might be sublimed (volatilized), that is, particles driven off by heat would condense into a solid powder in a cool part of the apparatus. Sublimate of mercury (mercuric chloride), volatilizes readily and condenses in this way, hence its name. If a metal were heated with this substance the chloride of this metal might be formed, and be sublimed, while the mercury in the sublimate would be released and reappear as liquid metal. Alchemists frequently believed that they had obtained no ordinary quicksilver but a 'philosophic mercury' newly extracted by the process from the metal that had been heated with the sublimate of mercury.[26]

Much later – but we cannot tell when he gained this new depth of understanding – Newton wrote of the business of metallic chlorides in the Queries in *Opticks*:

> And is it not also from a mutual Attraction . . . that when Mercury sublimate is sublimed from Antimony, or from Regulus of Antimony, the Spirit of Salt lets go the Mercury, and unites with the antimonial metal which attracts it more strongly, and stays with it till the Heat be great enough to make them both ascend together, and then carries up the Metal with it in the form of a very fusible salt, called Butter of Antimony, though the Spirit of Salt alone be almost as volatile as Water, and the Antimony alone as fix'd as Lead?[27]

If we substitute for "Spirit of Salt" (HCl) – whose composition was unknown to Newton – its component, chlorine, we have an essentially modern view of a replacement reaction $HgCl + Sb \rightarrow Hg + SbCl$. Newton's new prosaic theory of differential attractions (later called affinities) between the component corpuscles of compound substances had no more need to invoke 'philosophic mercury' than to disguise volatile substances as birds.

One of Newton's experiments of the former type was directly modelled, Dr Dobbs points out, on a process of Boyle's described in the second edition of *Certain Physiological Essays* (1669):[28]

> In Aqua fortis [nitric acid] 2oz. dissolve mercury 1 oz. or as much as it will dissolve. Then put an ounce of Lead laminated or filed into it by degrees and the lead will be corroded dissolving by degrees into mercury and besides there will fall downe a white precipitate like a limus being the mercury praecipitated by the sulphur of lead. Out of an ounce of lead may bee got 1/3 oz. of mercury. If the remaining liquor bee evaporated there remains a reddish matter tasting keen like sublimate. The same liquor will extract the mercury of tin.

(The "limus" was in fact the oxide of tin, white lead, formerly used in paint.) We know that any metallic mercury produced by this process comes from the nitrate produced in the first step; Newton hoped that it was 'philosophic mercury' out of the lead. This Boyle had denied, yet even he was inclined 'to look upon it [in 1669], as somewhat differing from common Mercury, and fitter than it for certain Chymical uses'.[29] Newton did correctly appreciate that some mercury had gone into the "limus". After making the same experiment with tin, he repeated it again with copper, whereupon the blue tint of the solution told him that it held the copper. This made him less confident of the source of the fluid mercury: "I know not whither that mercury came out of the liquor or of [the] copper, for the liquor dissolves copper." These experiments too were to figure in Query 31 of *Opticks*, now mustered to make the point that "the acid Particles of *Aqua fortis* are attracted more strongly . . . by Iron, Copper, Tin and Lead, than by Mercury."[30] The theory of differential attractions had made it clear to Newton that the mercury did come from the nitrate solution.

In Dr Dobbs's opinion, Newton hereafter exploited only the approach by volatizing the chlorides of metals. One more example from the many experiments recorded in the notebooks will suffice here:

> Sublimate of Venus (a chloride of copper?) made with sublimate of antimony, dissolved and philtred to separate the antimony and dried and mixed either with iron filings or with spar would not rise in a second sublimation but stayed behind with the iron or spar and made the spar of a keen tast. The design was to separate the sal ammoniac from the salt of copper but the sal ammoniac did not fasten on the spar nor much on the iron, but rose alone without the copper. And if Spar and sal ammoniac were taken alone, the sal ammoniac rose from the spar without being destroyed by it.[31]

As in this passage, Newton frequently used his tongue as an analytical tool. The experiment is unusual in that its rationale is to some extent recorded.

Newton's Objectives

If I understand Dr Dobbs correctly, she takes Newton's alchemical researches to have been directed to two principal issues, allied but distinct: (1) interpretation of the fabulous writings of the alchemists by rendering them into plain language; such interpretation might be by the normal scholarly means – comparison of different authors' texts, reference to classical mythology – or by making chemical experiments; (2) preparation of a true 'philosophic mercury' and (to a lesser extent) elucidation of such alchemical mysteries as Jupiter's eagle and Saturn's daughters (presumably derivatives of tin and lead respectively) not to say the 'fat and scaly silvery flickering little fishes' which the philsopher is to catch in his net by experimental means and following the methods decoded from the corpus of texts. I shall turn in a moment to the literary aspects of Newton's chemical studies, but first a word more may be said about his particular interest in mercury and antimony.

Robert Boyle, though supposing in 1669 (as we have seen) that the 'running mercury' produced in chemical reactions was not identical with the quicksilver obtained from mines, had within ten years become quite disillusioned on this point and convinced that such mercury had been present among the reagents in the first place. Obviously Newton too had come to this view by 1706 and probably long before; indeed, such an alchemist as George Starkey in *Ripley reviv'd* (1677–8) had adopted it.[32] However, the following note of Newton's may date from the 1680s: "this [mercury] drawn out of bodies hath as many cold superfluities as common [mercury] hath, and also a special form & qualities of the metals from which it was extracted, which makes it more remote from the philosophical [mercury] then the common [mercury] is." Here, *pace* Boyle and Starkey, Newton still supposes the mercury from metallic compounds to be different from common mercury and so (because of the tainting of the metals) he now proposed to use common, mineral mercury as a source of 'philosophick' mercury.[33]

He returned later to volatilization. Meanwhile, he did an enormous amount of work on the regulus of antimony, only lightly outlined above. Dr Dobbs's supposition is that in this area Newton was much influenced by the writings of the alchemist Michael Sendivogius, which he transcribed (he also owned several versions of one book by Sendivogius – see Harrison, *Library*, nos 445, 1192, 1485). Sendivogius allegorically described a 'Steel' or magnet which Newton decoded as being antimony metal. Perhaps then he attempted experimentally to draw the 'philosophic mercury' out of other metals by the magnetic pull of the star of the regulus. He was aiming to decompose common metals into their principles, mercury and sulphur, by 'using antimony as a matrix which was rather different from the Sendivogian natural philosophical one but much like the parallel alchemical one which Sendivogius offered to his readers'.[34] The antimony should presumably have become imbued with the 'mercury' of the gold, copper, lead, etc. which could then by some other process be transferred to another metal, to give it the properties of the first.

It would be strange if there were not interconnections between the body of Newton's alchemical manuscripts, even though these are largely transcriptions, and his experiments as we find them in the notebooks. The following passage, though clearly a programme for an experiment rather than an experiment, reads very like those he actually performed:

> To understand the following Treatise [by Sendivogius] know the Author intends you should melt by spoonfulls the pouders of [stibnite, tartar and nitre] in an hot crucible. Then shake it [so] that the [Regulus] may fall to the bottom: Which freed from its dross shines like [tin] after the expulsion of [lead]. Then melt it four times with half so much [iron], still seprating it from the gross dross. Then you will have the [star regulus] of [iron] which melt with so much [copper] till both are catched in a fine net, & you have the Philosoph[ic] flying [gold] and [star] & [star of copper].[35]

Here is another such programme:

> note that for Jupiter's eagle, lead must perhaps be otherwise prepared then for copper, viz by adding to lead in his first or second working a little of the two mineras of tin [Jupiter], or rather a little of the eagles salt, or minera, that all may putrefy together. Thus you shall have copper the daughter of Saturn [lead] and Jupiter's eagle.[36]

To this Newton adds a long experimental note of "Quaeres".

Newton's Chemical Library

I shall now turn to Newton's literary labours in the (al)chemical field. So far as we know, he began his book-collecting in 1669 with the "*Theatrum Chemicum*". This book could have been either of two, both recorded as present in Newton's later library: (1) *Theatrum Chemicum Britannicum*, edited by Elias Ashmole (1652; Harrison no. 93) or (2) *Theatrum Chemicum*, edited by L. Zetzner (Strasbourg 1659–61; Harrison no. 1608). The latter work, six quarto volumes much studied by Newton, is the more likely.[37] Both are collections of alchemical tracts, Ashmole's all by British authors, the Zetzner collection continental and in Latin. Altogether Newton possessed by Harrison's count some 138 books on alchemy and 31 on chemistry, making combined some

10 per cent of his library. For comparison, the number of his mathematical and scientific books *in toto* was 369, and that of his books on theology 447.[38] Therefore if we reckon mathematics with science, theology and (al)chemy as his three principal interests, the last was the least well represented in his library. On the other hand, we might guess these ratios to be distorted because Newton would have been unlikely to receive many presentation copies of alchemical books, and their total number published was probably fewer.

Newton possessed many books now reckoned by scholars as pioneer works of modern scientific chemistry (besides his collection of twenty-four Boyle titles) such as Jean Beguin's *Tyrocinium chemicum* (Amsterdam, 1669 – a late reprint), Nicolas Lemery's *Course of Chemistry* (London, 1698, a late translation) and Andreas Libavius's *Alchymia* (Frankfurt, 1606 – not a work of alchemy as that term is now used).[39] Some of these books were published only after Newton had left Cambridge and presumably ceased his experimental investigations. Moreover, his library included some books on industrial metal-working, notably Georg Agricola, *De re metallica* (Basel, 1621, a reprint showing signs of use) and two other books by the same author, and Alvaro Alonso Barba's esteemed *Art of Metals* (in two English versions, 1670 and 1674), a book (originally published in Spanish) as significant for New World metallurgy as Agricola was for that of the Old World.

De re metallica is notable too because it is one of the few non-mathematical books directly linked with Newton in a secondary account of him. In his recollections of life with Newton in Trinity College during the 1680s Humphrey Newton recollected that 'he would sometimes, though very seldom, look into an old mouldy book which lay in his elaboratory, I think it was titled *Agricola de Metallis*.' The form suggests that Humphrey may have read only the title on the spine. He remembered nothing of Newton's alchemical books, but he had more memory of his great namesake as a chemical philosopher experimenting than as a mathematician or a physicist, relating how active Newton was

> especially at Spring and Fall of the Leaf, at which times he us'd to imploy about six weeks in his Elaboratory, the Fire scarcely going out either night or day, he sitting up one Night as I did another, till he had finished his Chymical Experiments, in the Performance of which he was the most accurate, strict, exact: What his aim might be I was not able to penetrate into, but his Pains, his Diligence, at those sett Times made me think at something beyond the Reach of humane Art and Industry.[40]

Humphrey, no doubt taking a commonplace view, nevertheless opined that 'The transmuting of metals [was] his chief design, for which purpose antimony was a great ingredient.'[41]

Though Humphrey fair-copied the *Principia* he barely mentioned doing so in his reminiscences! This unheeding assistant, though recalling the 'Chymical Materials' (that is, apparatus) with which Newton surrounded himself, 'which was very little made use of, the Crucibles excepted, in which he fused his metals', tells us nothing of Newton's busy pen transcribing alchemical writings. The resulting mass of manuscripts, purchased by Lord Keynes in 1936 and by him presented to King's College, Cambridge, must next be considered.[42]

Newton's Alchemical Transcripts and other Papers

Keynes eventually acquired nearly half the 120 lots of alchemical papers sold by Sotheby's in 1936; some still remain unlocated, presumably in private hands, but a great many have passed into academic libraries, especially the Jewish National and University Library, Jerusalem.[43] A large part of the mass consists of copies taken from well-known alchemical writers: William Bloomfield, 'Hermes Trismegistus' (by attribution), Raymund Lull, Michael Maier, Thomas Norton, Sir George Ripley, Michael Sendivogius, William Yarworth (an alchemist certainly in touch with Newton after his removal to London) and many more. Newton had nothing of Paracelsus nor of either van Helmont. Most, but not all, of this transcribed corpus is in Newton's hand; some papers bear notes or comments by him, as we have seen. He compiled several long lists of chemical authors and no doubt tried to see all their writings. He made two alchemical dictionaries of terms and processes. He collected many recipes and accounts of the preparation of metals. Besides long transcripts, he made series of extracts and notes. There are translations, both made by Newton himself and by others. One transcript not written by Newton bears at the end considerable additions in Newton's writing, introduced as follows: "Here follow several notes and different readings collected out of M. S. communicated to Mr F. by W. S. in 1670, & by Mr F. to me 1675."[44] M. S. might well be Michael Sendivogius; W. S. is not known to me; Mr F. might be Ezekiel Foxcroft of King's, already mentioned. Newton later (after 1690) referred to "Mr F." as the translator of an allegorical book published in 1690, *The Hermetick Wedding*; Foxcroft's name is given as that of the translator on the title-page.[45] It is therefore certain that Newton received alchemical material from others, who might (to hypothesize) have included Robert Boyle. Whether these others constitute an 'alchemical circle' of which Newton was a member may be doubted and there is no evidence to determine its membership (if it existed). We have seen that John Collins had some knowledge of Newton's chemical interests by 1675; that Boyle knew of them is obvious; so did Newton's young friend Nicolas Fatio de Duillier (from 1689 onwards). A curious relic survives to show that the secret, if such it was, did not lie in their hands only: "On Munday March 2d or Tuesday March 3 1695/6, A Londoner acquainted with Mr. Boyle and Dr. Dickinson making me a visit, affirmed that [in] the work of Jodochus a Rhe [an alchemist] with [vitriol] twas not necessary that the [vitriol] should be purified but the oyle or spirit might be taken as sold in shops."[46] At the date of this note Boyle had been dead for several years. Edmund Dickinson (1624–1707) was an Oxford MD who became physician to two kings; he is said to have entertained Charles II with experiments in a chemical laboratory just beneath the royal bedchamber. He was certainly an alchemist.

To be known as an 'alchemist' in the late seventeenth century was less bizarre than it would be in our own time in which, however, astrology seems to flourish. One can hardly insist too often that the appellation was of broad significance: if a 'chymist' was not commercially engaged in the metal trades, nor in preparing medicines, he was likely to be an alchemist. There were few 'chemical philosophers' besides J. B. van Helmont, long dead, Robert Boyle and Isaac Newton, and none who was not fascinated by the nature of metals. At that time there was no reason of theory why the metals should be judged elementary (in our sense of the word), that is, judged to be pure and simple substances that could not be divided into component parts, as most minerals can. Boyle certainly never viewed the metals in this way. Nor perhaps was there a compelling *a priori* reason for regarding as absurd the many plausible stories in print of the transmutation of base metals into gold by the 'powder of projection'. The constant scarcity and high price of the precious metals through the ages was perhaps the best argument against their truth. Collins and his friends do not seem to have regarded Newton's chemical interests as in any way extraordinary.

What is an Alchemist?

An alchemist may be a person who:

(1) practices chemistry (in the modern sense), chemistry and alchemy being etymologically the same word. Or
(2) seeks for the philosopher's stone, an extra-natural substance transmuting metals and immortalizing humans. Or
(3) studies the writings of chemists and alchemists (sense 2) with the object of understanding their meaning and repeating their processes. He may not necessarily be certain that all or any of these writings are truthful and valid.

It is my impression that within the learned tradition of alchemy the task of decoding older writings and recomposing their meaning in fresh allegorical texts was at least as important as experimental work with furnaces and metals. The former was the normal business of scholars. It is obvious that Newton pursued this scholarly investigation with tremendous vigour, just as (by similar methods) he pursued parallel investigations in ancient history and mythology, and in the foundations of Christianity. Only in relation to the alchemical tradition could Newton test his developing interpretation by the exact experiments Humphrey Newton observed.

In the other fields of scholarship he investigated, as we shall see later, Newton prepared systematic treatises, some of which were published before his death, some posthumously, others not at all until recent times.[47] In relation to alchemy he composed no such treatise so far as we know. (There is a not unlikely tradition that at some point in his mature life Newton destroyed many papers by fire; an alchemical treatise might have been among them.) Sherwood Taylor interpreted as Newton's original composition a manuscript which is rather a compilation or patchwork of quoted passages compiled by him.[48] More recently Dr Dobbs proposed Keynes MS 18 (King's College) as a product of Newton's own thought, opining that 'the weight of evidence' falls 'heavily on the side of its being Newton's'. Richard S. Westfall has taken the same view.[49] Despite their plausible arguments this manuscript, the *Clavis* or 'Key', has proved to be a Latin version of a text written in English by 'Eirenaeus Philalethes' (almost certainly George Starkey, an early Harvard undergraduate then living in London) and sent by him to Robert Boyle. It was written in late April 1651, at a time when Boyle and

Starkey were jointly investigating the alchemical literature, much as Newton was to do twenty years later.[50] Whatever the significance of the parallels between this document and other alchemical texts in Newton's handwriting, it is certain that Newton must be the derivative, not the originating author. To say that Newton never wrote an alchemical treatise is by no means to overlook the many passages of alchemical commentary and explanation that he wrote, sometimes in high-flown allegorical language, sometimes at some length. But we do not have to read these as conveying the notion that Newton believed the material he was examining to be true.

I am far from sure that Newton ever claimed to have made any solid progress towards the Great Work along the path so obscurely delineated by the alchemists, or affirmed that their goals were attainable. For we can now be confident that in the apparently decisive passage of Keynes MS 18 beginning 'I know whereof I write, for I have in the fire manifold glasses with gold and this [philosophick] mercury . . .' not Newton but Starkey was speaking.[51] (Humphrey commented upon the rarity of Newton's use of glass vessels in his experiments.) In 1676 Newton wrote a letter to Henry Oldenburg indicating a degree of scepticism about the whole business. Robert Boyle had printed in Oldenburg's *Philosophical Transactions* a relation of how a certain 'mercury' that he had prepared (by a process not disclosed) when mixed with gold made a hot amalgam. Newton did not find "any great excellence in such a [mercury] either for medical or Chymical Operations", proposing a hypothetical mechanical reason for the heating of the amalgam. For all that, he begged Boyle not to reveal the secret of this 'mercury', because of possible "immense dammage to the world if there should be any verity in the Hermetick writers" arising therefrom, "there being other things beside the transmutation of metalls (if those great pretenders bragg not) which none but they understand". These sentences suggest no great confidence in the reality of the cosmic mystery Humphrey imagined Newton to be seeking; they allow only the possibility of it.[52]

Boyle, Newton and Locke

Greater experience does not seem to have increased Newton's credence in the "great pretenders" he studied so assiduously. After Boyle's death (30 December 1691) John Locke – yet another searcher into metallic mysteries – was appointed with Edmund Dickinson and Daniel Cox to inspect his papers, presumably with a view to selecting some for publication.[53] Copies of two processes that Newton wished to see were sent him by Locke with an offer of further copies. (The single part-process surviving describes a purification of mercury by repeated mixing with sulphur and soap.) In reply, on 2 August 1692, Newton wrote Locke a strange letter, returning to the dubious heating of gold with a 'mercury', just described, and emphasizing his reluctance to have anything to do with Boyle's processes. It is clear from this letter and other evidence that Newton and Boyle had been in regular correspondence since 1676 (little of this survives), but Newton revealed to Locke that Boyle, the more experienced invest-igator, had not played quite fair with him in concealing steps in his processes. Nevertheless, Newton was sure that because Boyle's processes could not succeed in multiplying gold, Locke would waste his time in trying to recover them. He ended his letter with the following powerfully negative judgement:

> In diswading you from too hasty a trial of this R [recipe] I have forborn to say any thing against multiplication in general because you seem perswaded of it: tho there is one argument against it which I could never find any answer to, & which, if you will let me have your opinion about it, I will send you in my next.[54]

Unfortunately, there is no next extant.

To me, this a strong indication that Newton had never felt firm confidence in the alchemists' reveries of turning all things to gold, though he may well have hoped (rather than believed) that they knew some chemical processes of importance. Alternatively, one would have to suppose that Newton had only recently become sceptical, or that he was lying to Locke.

However that may be, no one supposes that Newton wished to enrich himself with fairy gold. As Richard S. Westfall writes: 'Noth-ing whatever in the vast corpus of Newton's alchemical manu-scripts even hints that gold making, in the vulgar sense of that phrase, ever dominated Newton's concern . . . Truth and Truth alone held that power over him.'[55] That, in my view, makes Newton somewhat less than an alchemist 'in every sense of the word', as certain scholars have termed him.

Perhaps to write, as my wife and I did thirty years ago, that 'Newton was not in any admissible sense of the word an alchemist' was going too far.[56] To call Newton a magician because he applied quantitative chemical experimentation to the study of alchemical writings is going too far in the opposite direction. At the most, Newton was an alchemist in the third sense defined above. If Newton's researches into the literature of alchemy – where probably no one has gone further – demonstrate again his belief in the tradition of esoteric learning born in Antiquity, disguised in allegory and symbolism, they also illustrate his concomitant belief that the genuine knowledge carried within that tradition under disguise is rational, factual and (so far as Nature is concerned) experimental. There was nothing magical in alchemy as Newton conceived it, no method of acting upon material substance by supernatural powers. We also wrote, thirty years ago, that Newton sought *facts* in his experimental study of (al)chemy:

> the correlation between his own experiments and alchemical literature was significant only in so far as it indicated to Newton the direction in which he should pursue the experiments which yielded him facts. One might say that alchemical writings were for Newton only a means to an end – the end being the elucidation of chemical entity and structure.[57]

In other words, knowledge, not gain. This still seems to me a correct judgement.

The Chemical Hierarchy of Matter

The concept of structure may well be important here. In recent years Dr Karin Figala has developed in detail the relationship between Newton's study of alchemical processes directed towards the supposed transmutation of metals and his hierarchical, atomic concept of matter. (Alchemy was of course notably hierarchical in its concepts: some substances were taken to be 'higher' or 'purer' than others; Dr Figala also points to the hierarchical series of metals among the alchemists – lead, tin, copper, iron – repeated inversely in the replacement series of Query 31.)[58] In her analysis, Newton is found to have established as it were a table of nobility for substances; at the top is gold, the most noble and most dense of substances, water (as usual) being taken as the standard of comparison. The less dense a substance is, obviously, the less is its content of solid matter in proportion to the empty space within its volume: 'one can even draw up a kind of pedigree of chemicals which stretches from water *via* the acids, salts and different metals to the noblest, nearly perfected, gold.'[59] In the later editions of *Opticks* Newton explained in some detail how such a hierarchical system of matter might work, so that the proportion of truly hard substance in a given volume of 'solid' matter might be very small.[60] Dr Figala's account of how this system related to older traditions of thought is considered in Appendix A.

It is obvious that thirty years of study and the writing of one million words must have left their mark on Newton's mind. Precisely what kind of mark these alchemical investigations made is in debate between historians. Some attribute a profound effect upon Newton's late philosophy of Nature to this influence, others are sceptical. At any rate, we know that when Newton left Cambridge for London in 1696 he took with him not only a 'parcel of Mathematical instruments' but also a 'parcel of Chymical glasses' for both were mentioned in the inventory of his estate. Meanwhile, we turn now to Newton's life in Cambridge during these middle years, when chemical investigation was to him so important an occupation.[61]

8

The Mathematical Principles of Natural Philosophy, 1679–1687

Hooke and Newton, 1679–80

Recently D. T. Whiteside has published an authoritative article on the evolution of the *Principia* from 1664 to 1686, following hard upon a large volume of facsimiles of the relevant documents.[1] My account here, therefore, may be confined to bare essentials, especially as so little is known of Newton's personal life in the period of most intense effort devoted to his great work, that is, from the summer of 1684 to the late spring of 1686, when his surviving correspondence with Edmond Halley opens. Of the dozen extant letters from this period of some twenty-six months, no fewer than nine concern Flamsteed's putative provision of astronomical data for use in the *Principia*.

We must go back to chapter 6, where we left Robert Hooke enticing Newton (against his desire) into a philosophical correspondence. In a rejoinder of 9 December Hooke returned to Newton's carelessly considered spiral of descent, ending at the centre of the globe, as sketched in his letter to Hooke of 28 November 1679 (see figure 6.1).[2]

Regretting Newton's renunciation of philosophy as 'a little Unkind' and disclosing that he had read Newton's reply to his first approach at a meeting of the Royal Society, Hooke agreed, as had other Fellows then present, that a weight would fall to the east of its point of departure, not the west.[3] Not realizing that Newton had supposed the observer of his spiralling trajectory to turn with the Earth, Hooke utterly rejected it: 'my theory of circular motion,' he wrote, 'makes me suppose it would be very differing and nothing at all akin to a spirall but rather a kind [of] Elleptueid . . . I conceive the line in which this body would move would resemble an Elleipse.' Hooke's figure makes his meaning more explicit (see figure 8.1(a)). Supposing the body to fall in a gap between two hemispheres starting from A (apogee) without experiencing any resistance, it would reach no nearer to the centre than G (perigee); but if the movement of the weight were resisted (Hooke told Newton) it would indeed spiral inwards to the centre. If the descent began away from the equatorial plane, the 'elleipse' thus described would lie in a southeasterly plane 'but more to the south then the east' (again putting Newton right).

In this, given his as yet unstated assumptions, Hooke was roughly correct, supposing now the observer to be at rest away from the Earth. Hooke did not explain how he came to his 'orbit' beyond saying that it and 'many other considerations . . . are consonant to my Theory of Circular motions compounded by a direct motion and

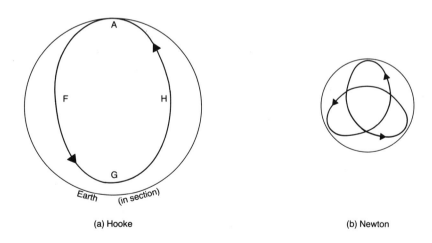

(a) Hooke

(b) Newton

Figure 8.1   (a) Hooke: the 'elliptical' path of a falling body; (b) Newton's cusped path an attractive one to a Center.' It is natural to imagine that Hooke, having qualitatively (and to a minor extent, experimentally) explored his notion of this combination of undefined central force and tangential motion, guessed that the freely falling body under the improbable circumstances of this exercise would fall like a planet, that is, move in a roughly elliptical path. It is certain that he could not have computed such a trajectory, nor was he ever able to do so.

Once more Newton had to make a partial concession to Hooke, on the basis now not of analogical reasoning but of a mathematical analysis of what Newton took to be the original problem. Hooke's statements were correct, he conceded, because if the gravity of the falling body were "supposed uniform it will not descend in a spiral to the very center but circulate with an alternate ascent & descent made by it's *vis centrifuga* & gravity alternately overbalancing one another." That such a continuous, awkwardly curved orbit would loop round neatly into an "Ellipsoeid" Newton denied: the perigee would not lie on the diameter AD but to one side, and the form of the curve would be complex, as in Newton's figure (see figure 8.1(b)). Hooke's letter, Newton declared, had put him "upon considering thus far the species of this curve [and] I might add something about its description by points *quam proxime*". Perhaps, as Whiteside suggests, Newton realized that no exact, finite solution is attainable to the problem when gravity is "the same at all distances from the center".[4] At any rate, Hooke this time was happy to agree but equally ready to change to the ground he had chosen.[5] He had from the first (he wrote) proceeded upon no such supposition, but rather 'my supposition is that the Attraction always is in a duplicate proportion to the distance from the Center Reciprocall, and Consequently that the Velocity will be in a subduplicate proportion to the Attraction and Consequently as Kepler Supposes Reciprocall to the Distance.'[6] On this quasi-astronomical supposition, 'Which I conceive doth very Intelligibly and truly make out all the Appearances of the Heavens', Hooke reasoned that the diameter through the origin would be the fixed apse-line of the trajectory. And, he writes, it is because of the analogy between this physically fanciful trajectory and the celestial motions that he has pursued the matter with Newton, for he had never really believed that the inverse-square law would apply to a body falling (impossibly) towards the centre of the Earth, since the attraction to the centre would diminish as it fell. But if the problem with which he began, 'that of compounding the celestiall motions of the planetts of a direct motion by the tangent & an attraction towards the centrall body' could be solved, the solution will be of great Concerne to Mankind, because the Invention of the Longitude by the Heavens is a necessary consequence of it: for the composition of two such motions I conceive will make out that of the moon . . . This Curve truly Calculated will shew the error of those many lame shifts made use of by astronomers to approach the true motions of the planets with their tables.

Because Newton was later to wipe out from his published writings any trace of his indebtedness to Hooke, and was in private to ridicule Hooke's pretensions to priority in the concepts of celestial mechanics founded upon attractive force, it is but just to record that no historian now doubts that the first page of Hooke's crucial letter of 6 January 1680 fairly summarized thoughts developed over some fifteen years.[7] When Newton wrote disparagingly to Halley "Mr Hook . . . will prove the last of us three that knew it", meaning that Newton himself and Christopher Wren had preceded Hooke in discovering the inverse-square law of gravitational force, he may have been right; for we need not suppose that Hooke had long kept this discovery to himself when he wrote to Newton about it.[8] Further, Newton was on the strongest ground (as Hooke himself allowed) in asserting that Hooke had made no progress at all from these firm but over-general principles to the mathematics of orbital motion. This fact indeed Newton was to make a further cause of complaint against Hooke (in writing later to Halley):

> he [Hooke] has done nothing & yet written in such a way as if he knew & had sufficiently hinted all but what remained to be determined by the drudgery of calculations & observations, excusing himself from that labour by reason of his other business: whereas he should rather have excused himself by reason of his inability. For tis plain by his words he knew not how to go about it. Now is not this very fine? Mathematicians that find out, settle & do all the business must content themselves with being nothing but dry calculators & drudges & another that does nothing but pretend & grasp at all things must carry away all the invention as well of those that were to follow him as of those that went before.[9]

No doubt with some such resentful feeling as this in his mind, Newton left Hooke's letter of 6 January 1680 unanswered. Nor was he softened by a further plea from Hooke on the 17th, that he solve this curve. Hooke would not perceive that his letters had not, after all, been written in vain until Halley brought Book I of the *Principia* to the Royal Society in May 1686. To him, after Hooke's protestations, Newton told his version of the exchanges just considered:

The summe of what passed between Mr Hooke & me (to the best of my remembrance) was this. He soliciting me for some philsophicall communications or other I sent him this notion. That a falling body ought by reason of the earth's diurnall motion to advance eastward . . . and in the scheme [diagram] wherein I explained this I carelessly described the Descent of the falling body in a spirall to the centre of the earth: which is true in a resisting medium such as our air is. Mr Hooke replyd it would not descend to the center, but at a certaine limit returne upwards againe.

Newton paid Hooke the compliment of supposing that his assurance was based upon a mathematical demonstration, and therefore

> I then took the simplest case for computation, which was that of Gravity uniform in a medium not Resisting . . . and in this case I granted him what he contended for, and stated the Limit as nearly as I could. he replyed that gravity was not uniform, but increased in descent to the center in A Reciprocall Duplicate proportion of the distance from it. and thus the Limit would be otherwise than as I had stated it, namely at the end of every intire Revolution, and added that according to this Duplicate proportion, the motions of the planets might be explained, and their orbs defined. This is the summe of what I remember.[10]

And quite fair to Hooke, one might think. Some six weeks later, again writing to Halley and with this whole bad business still pressing upon his overburdened mind, Newton thought fit to recall one further vital point:

> This is true, that his [Hooke's] Letters occasioned my finding the method of determining Figures, which when I had tried in the Ellipsis, I threw the calculation by being upon other studies & so it rested for about 5 yeares till upon your request I sought for that paper, & not finding it did it again & reduced it into the Propositions shewed you by Mr Paget: but for the duplicate proportion I can affirm that I gathered it from Keplers Theorem about 20 years ago.[11]

This single sentence is the unique evidence from Newton's pen that in January 1680 he proved the elliptical orbit to be consistent with an inverse-square central force. By "the method of determining Figures" we may, I believe, understand Problem I of *De motu* (1684) and what follows (in three Theorems and as many Problems) down to the application of the method to the ellipse.[12] The sheets upon which this work was recorded, no doubt less elaborately than in the *De motu* text itself (which was written for others' eyes), were already lost by the summer of 1684.

Why Newton did not make immediate capital of his success instead of carelessly thrusting it aside has already been hinted in what I wrote of Flamsteed and the comet of 1680–1 (page 168). A year after brutally closing down his correspondence with Hooke – and so denying him the key to the longitude, as the latter supposed – Newton was still not ready to take a simple mechanical view of planetary motion, based on the idea of attractive force in the central body; that is to say, he still intuitively rejected the concept that Hooke later told the world Newton had stolen from him. So long as Newton believed (as of course his great contemporaries Huygens and Leibniz believed to the end of their days) that matter could only be put in motion by the impact upon it of other matter, the common seventeenth-century principle that is expressed in all Newton's theoretical papers before 1684, he could neither abolish the planetary vortices nor embrace a pragmatic notion of attractive and repulsive forces divorced from aetherial mechanisms. His rejection from 1684 onwards of the common principle of mechanism in philosophy was the great intellectual adventure of Newton's life, the grand step that rendered Newtonian celestial mechanics more than a synthesis of the diverse achievements of Galileo, Kepler and Descartes. How bold this adventure was may be reckoned from the fierceness with which Leibniz criticized Newton for embarking upon it.

Halley and Newton, August 1684

We know neither when nor why Newton threw away vortices and impacts in favour of a pragmatic, mathematical mechanics of forces, first sketched in *De motu*.[13] We can only record that he embarked upon the exploration of this astonishing new world as a result of a visit paid to him in Cambridge in August 1684 by Edmond Halley, now a considerable scientific figure in London. Why this astronomer took the Cambridge Professor of Mathematics to be his man of the moment is not clear either; by implication, the two men were acquaintances, not yet close friends. No doubt Newton appreciated Halley's abilities in mathematics as he may also have admired his enterprise in astronomy. It is certain that Halley came to put a precise question to Newton because neither Wren, Hooke nor himself could connect the elliptic planetary orbit with an inverse-square law of centripetal force, or gravitation. Halley's narrative continues:

The August following [of 1684] . . . I did myself the honour to visit you [Newton], I then learnt the good news that you had brought this demonstration to perfection, and you were pleased, to promise me a copy thereof, which the November following I received with a great deal of satisfaction from Mr. Paget; and thereupon took another Journey down to Cambridge, on purpose to conferr with you about it, since which time it has been enterd upon the Register books of the [Royal] Society.[14]

Robert Hooke has left no recorded reaction to the arrival of *De motu* at the Royal Society, in which he was now less active than formerly, nor did he ever produce a demonstration correlating the inverse-square law of force with the planetary ellipse.[15]

But from Cambridge, despite Newton's confidence, Halley had returned empty-handed. According to the story told by the mathematician Abraham de Moivre, after being assured by Newton that the inverse-square law would produce an ellipse, 'Dr Halley asked him for his calculation . . . Sir Isaac looked among his papers but could not find it [that is, the sheets written in January 1680 and thrown aside], but he promised him to renew it, & then to send it to him.'[16] This proved less straightforward than Newton had expected, as many have found in like case:

> Sir Isaac in order to make good his promise fell to work again, but he could not come to that conclusion which he thought he had before examined with care. However he attempted a new way which although longer than the first, brought him again to his former conclusion, then he examined carefully what might be the reason why the calculation he had undertaken [just] before did not prove right . . . That being perceived, he made both his calculations agree together.[17]

From this, Newton composed altogether *three* demonstrations of the elliptical orbit from mechanical principles: the first in January 1680, the second (longer) in August/September 1684, and a third immediately afterwards by emendation of the mistake which had at first given him a check after Halley's departure. We have every reason to believe from Newton's language that the first and third were the same, as de Moivre affirmed, since Newton was reconstructing his lost paper for Halley when he made an error in sketching his geometrical figure. And this method of reasoning we may suppose to be the same as that which passed into *De motu* and later (in more rigorous form) into the *Principia*. But what of the second, longer way to the same conclusions, of which de Moivre wrote? It is the plausible suggestion of D.

T. Whiteside that this too survives, in the form of an English paper, 'On motion in ellipses', afterwards given by Newton to John Locke (March 1690) as a simpler proof of the *Principia*'s primary result.[18]

What Newton's occupation was during the autumn of 1684 – a period from which no correspondence survives – is evident enough. He was recovering, then several times over rewriting, always becoming more precise but also more detailed, the short tract which was first called *De motu corporum in gyrum* ("On the motion of bodies in an orbit"), then *De motu sphaericorum corporum in fluidis* ("On the motion of spherical bodies in fluids") and finally and simply *De motu corporum*. The tract received by Halley in November, in fulfilment of Newton's promise, was in a state of rapid and expansive evolution.[19] By the winter or early spring of 1685 (it seems, from Whiteside's reconstruction) Newton was well forward with a book-length manuscript, of which forty-three propositions survive. Already in the last days of 1684 Newton had begun to request from Flamsteed the astronomical data that he would need in order to prove that his theorems of mechanics held in the observable heavens.[20]

The Royal Society encourages Newton

In all this Newton received encouragement from London. The detail of what happened in the winter of 1684–5 is lost: we do not know the date of Halley's second visit to Newton, nor when he read the later *De motu corporum*, on which his comments survive. We do not know what Newton arranged with his friend Paget, nor what may have passed in lost letters between Newton and Aston, Flamsteed, Halley, Paget and perhaps others. The record shows that not until after Halley's *second* visit to Cambridge did he, on 10 December 1684, give to the Royal Society (Samuel Pepys, Esquire, President, in the chair):

> an account, that he had lately seen Mr. NEWTON at Cambridge, who had shewed him a curious treatise, *De Motu*; which, upon Mr. HALLEY's desire, was, he said, promised to be sent to the Society to be entered upon their register. Mr. HALLEY was desired to put Mr. NEWTON in mind of his promise for the securing his invention to himself till such time as he could be at leisure to publish it.[21]

Now we have learned from Halley's letter of 29 June 1686 that he had already received the tract *De motu* (brought to London by Paget) in November; was he therefore disingenuous in disguising from his colleagues that a copy of the tract was safe in his hands and could at once be laid before the Society? I do not think so. What Halley meant was that during his second visit to him, Newton had shown Halley the extensive beginnings of a large 'treatise' (*not* a tract or a paper), *De motu [corporum]*, which Newton meant soon to complete, and Halley then to register in London, as the full account of Newton's 'invention'. However, as the 'treatise' continued to expand in typically Newtonian fashion, and nothing of it arrived from Cambridge, Newton's friends in the Society decided – after a month or more – to register as an interim measure the tract *De motu*, already in London, to protect Newton's priority. *De moto corporum* would arrive in London, after many more developments and under its more famous title of *Philosophiae naturalis principia mathematica* only after the lapse of a further eighteen months.

The Short Tract on Motion, 1684

What was this tract, carefully recorded in the Society's archives, yet little read?[22] Its English version occupies about twenty-two pages in *Mathematical Papers*, volume VI, including the diagrams. It begins with three definitions and four hypotheses; the second of the latter is very roughly equivalent to the first of Newton's Laws of Motion, the fourth is a version of Galileo's rule for accelerated motion ($s\alpha t^2$). There follow two geometrical Lemmas: one of these, unknown to Newton, was stated long ago by Apollonius. The text opens with motion in a non-resisting medium. Newton proves, just as in Proposition I of the *Principia* and looking back to Hooke's old formulation, that if a body moving uniformly in a straight line is suddenly attracted to some centre – by any rule of force whatever – the curved path it then begins to trace will describe equal areas in equal times by radius-vectors drawn from the body to the centre of force. This has been justly termed the general mechanical form of Kepler's Second Law, and is the key to much that follows. No less essential is the next relation to be derived, one longer familiar to Huygens and then to Newton, that (in circles) central forces are proportional to the squares of the peripheral velocities divided by the radii. Newton notes of the fifth corollary to this relation (that is, $r^3\alpha T^2$, to us Kepler's Third Law) that it holds true among the heavenly bodies.[23]

Note that Newton from the first introduces infinitesimal arguments into these geometrical demonstrations. The general form of Kepler's

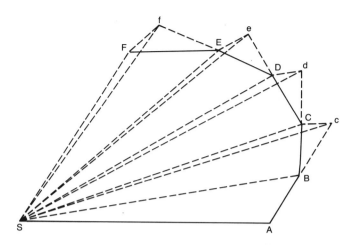

FIGURE 8.2   The proof that began it all – the generalized Second Law of
Kepler; the areas ASB, BAC , CAD &c. are proved to be equal under any
law of force to the centre S

Second Law could only be proved true of a smooth curve, rather than
a succession of straight segments, if the segments may be supposed
to be infinitely small so as to coincide with little arcs of the curve. The
v^2/r theorem similarly requires a tangent to become infinitely small.
By a similar infinitesimal argument, involving the coincidence of two
points, Newton next defines a solid which can be assigned in any
curve as the measure of the centripetal force directed to any nominated centre within the curve, such that a body will describe that
curve as the measure of the centripetal force directed to any nominated centre within the curve, such that a body will describe that
centripetal force is directed towards the centre of the ellipse, it must
vary directly as the distance for a body drawn by it to describe the
ellipse. If the centripetal force is directed towards a focus, the force
must vary inversely as the square of the distance for the same result.
"Therefore [Newton went on quietly in a Scholium] the major planets
revolve in ellipses having a focus in the centre of the Sun; and the
radius-vectors to the Sun describe areas proportional to the times,
exactly as Kepler supposed."[24]
    The remainder of this section of *De motu* – several pages – is
concerned with applying these fundamental geometro-mechanical
principles and methods to establishing the parameters of the orbits of
the planets, starting from corrected observations of positions and

times, much as Hooke had correctly but too vaguely hoped. Lastly in this section of the tract, Newton shows how similar processes might be applied to comets, in order to define their periods of orbiting the Sun and their orbital parameters, by the comparison of which astronomers might learn "whether the same comet returns with some frequency to us".[25]

The final section of *De motu* may well have been an afterthought. Newton turns to the class of motions that are resisted by the air, terrestrial motions. Projectiles are eminently in this class. He solves the problem of defining the movement of a body travelling in a straight line through a medium whose resistance to it varies directly as the speed of the body. The next case is that of a heavy body's ascent and descent [under normal gravity] in such a medium. Thirdly, after the example of Galileo, Newton combines these two in order to obtain the trajectory of a projectile shot off at any angle to the horizontal through this medium. Hints are given about how the resistance to a given ball might be established by experiment and extended by proportionality from this case to others. Surely Newton must have recalled Anderson and James Gregory and his own jejune thoughts on these problems – not that in *De motu* Newton's treatment of resisted projectile motion is other than grossly simple. But only two men in Europe were capable of rivalling or perhaps even comprehending these relatively straightforward arguments, which again employ infinitesimals: they were Huygens and Leibniz.[26]

The Fable of Fluxions

These fairly elementary conversions of geometrical elements such as lines and areas into infinitesimal quantities converging towards finite ratios as they vanish in the limit is characteristic of a method of reasoning employed extensively by Newton in the *Principia*. It has been commonly but erroneously supposed that he had used algebraic analysis and his fluxional calculus to reach conclusions that he afterwards demonstrated in the book by geometry. Towards the end of his life Newton himself encouraged this fable.

> By the help of the *New Analysis* Mr. Newton found out most of the Propositions in his *Principia Philosophiae*: but because the Ancients for making things certain admitted nothing into Geometry before it was demonstrated synthetically, he demonstrated the Propositions synthetically, that the Systems of the Heavens might be founded upon good

Geometry. And this makes it now difficult for unskilful men to see the Analysis by which those Propositions were found out.

He even projected this false notion back to the beginning: "By the inverse Method of fluxions I found in the year 1677 [*recte*, 1680] the demonstration of Kepler's Astronomical Proposition viz. that the Planets move in Ellipses, which is the eleventh Proposition of the first book of the Principles."[27] Such statements were elicited from Newton by the exigencies of his dispute with Leibniz over the invention of the calculus. They are counter to Newton's own preference for, and supreme efficiency in, geometrical analysis – he who called algebra the analysis of bunglers – including the use of geometrical equivalents of first and second differentials; they are counter to all documentary evidence, for with only one proposition in the *Principia* (Book II, Proposition 35, determining the optimum shape for bodies moving through a resisting medium) can any fluxional analysis be connected. In this case he stated his conclusions in the book without proof. Otherwise we have no reason to doubt that Newton drafted the antecedents to the *Principia* and that text just as we can trace them through the manuscripts. 'The conclusion seems inescapable. The published state of the *Principia* – one in which the geometrical limit-increment of a variable line-segment plays a fundamental role – is *exactly* that in which it was written.'[28]

Newton was not extending the methods of the classical geometers – largely known to him only through recent writers – nor was he employing an algebraic calculus of fluxions; the tool he was developing from the autumn of 1684 onwards and brought to fruition in the final text of the *Principia* was an idiosyncratic geometry in which infinitesimal increments of lines and areas perform the functions of first and second order differentials, a geometry intimately integrated with his dynamical principles. Newton employed a 'calculus' indeed, but in geometrical form, not the algorithm he had perfected in 1671. And of this too his method of series was a part, though again not expressed in algebraic terms. The singularity and power of Newton's methods in the handling of dynamical problems created difficulties in the understanding of his work both for contemporaries, whose skills went little beyond elementary geometry, and for their successors who look to find algebraic forms.

Newton's Search for Data

It might seem that in this decisive winter of 1684–5 Newton would have been solely concerned with the theoretical section of his future *magnum opus,* as the manuscripts printed by Whiteside and discussed by Cohen seem to indicate.[29] But Newton's correspondence indicates otherwise. While the mathematical propositions were piling up at a great speed – for Newton's genius was in full flow – he told his friend Aston (now a Secretary of the Royal Society) that "the examining of severall thinges has taken a greater part of my time then I expected, and a great deale of it to no purpose."[30] His problem is evident from his letters to Flamsteed; he lacked data to show that the orbits of the satellites of the two outer planets were elliptical and satisfied Kepler's Laws. "I have not at all minded astronomy of some years till on this occasion which makes me more to seek. I cannot meet with H[u]ygens' book of Saturn. Mercator & another or two which I have consulted leave me as wise as I was."[31] And in the end, in the first edition, Newton had to content himself with treating only Jupiter's satellites. To obtain really accurate information, especially with regard to the perplexing motion of the Moon, was always to remain one of Newton's chief desiderata; his search for precise facts from astronomers, from experiments directed by himself, and from the reports of travellers would continue to the last years of his life.

Another difficulty was the theory of comets. I have already mentioned Newton's construction for a straight-line portion of a comet's path traced from observations – appropriate enough when the comet is fairly remote from the Sun – and at first he seems to have thought that it would not be more difficult to define the highly inflected perihelial part of a cometary orbit than to construct the orbit of a planet, the Sun always being at the focus. Experience proved otherwise. As late as 20 June 1686 – weeks after the first Book of the *Principia* had been sent to the Royal Society – Newton wrote to Halley that its Book III still "wants the Theory of Comets . . . In Autumn last [1685] I spent two months in calculations to no purpose for want of a good method, which made me afterwards return to the first Book & enlarge it with divers Propositions some relating to Comets others to other things found out last Winter."[32] This indicates that Book I was still being remodelled during the winter of 1685–6 (if not later) and that the theory of comets as applied to observed positions was not yet finally worked out for Book III. There were experiments on the velocity of sound carried out in Book II, pendulum experiments to confirm the theory of resistance, and much else that cannot be exactly dated.

The Principia begun

What we can affirm with some confidence is that about the turn of the years 1684–5 Newton abandoned the tract *De motu* family of drafts and started upon a new group of longer texts, ending in the first-edition manuscript of the *Principia*. Some thirty years later Newton recorded (in drafts) that "The Book of the Principles was writ in about 17 or 18 months, whereof about two months were taken up with journeys, & the MS was sent to the RS in Spring 1686."[33] In a variant he wrote "beginning in the end of December 1684". Parts of two of these *De moto corporum* drafts were later deposited by Newton in the University Library as his Lucasian lectures for the years 1684 to 1687. No one now credits "Octob. 1684" as the date when Newton began to set the *Principia* before bemused undergraduates, and the whole treatment of these near-final drafts as a course of lectures has been revealed as a sham by modern analysis. We do, however, have Whiston's confirmation that Newton 'read' from the *Principia* in the Schools, perhaps in 1686.

Neither of these two drafts extends beyond Book I of the *Principia* as ultimately printed, and only scattered fragments of preliminaries to the present Book II remain.[34] In his initial plan, Newton proposed to deal with "The Motions of Bodies" in a single Book, to be followed by another dealing with "The System of the World". At an unknown date, before the summer of 1685, he decided to make a Book II "On the Motions of [Resisted] Bodies" out of some propositions on fluid mechanics formerly placed at the end of Book I, to which he added new material, and so "The System of the World" now became Book III. After Book I had been received with rapture by the Royal Society (Samuel Pepys as President affixed his imprimatur to it on 5 July 1686) Newton wrote to Halley (to quote his letter of 20 June yet again) that "the second [Book] was finished last summer [1685] being short & only wants transcribing & drawing the cuts fairly. Some new Propositions I have since thought on which I can as well let alone." Evidently these propositions were retained, and Newton also changed his mind about removing the whole "System of the World" (Book III) because of Hooke's making a quarrel: "The third [Book] I now designe to suppress. Philosophy is such an impertinently litigious Lady that a man had as good be engaged in Law suits as have to do with her. I found it so formerly & now I no sooner come near her again but she gives me warning." He also "upon second thoughts" proposed now to retain the *Principia* title for the work, "Twill help the sale of the book which I ought not to diminish now tis yours."[35]

In reply Halley begged Newton

> not to let your resentments run so high, as to deprive us of your third book, wherein the application of your Mathematicall doctrine to the Theory of Comets, and severall curious Experiments, which, as I guess by what you write, ought to compose it, will undoubtedly render it acceptable to those that will call themselves Philosophers without Mathematicks, which are by much the greater number.[36]

Newton continued in further letters to Halley to explain by references to his own past thoughts the unreasonableness of Hooke's claim to have taught him anything at all, but was much mollified by "the great kindness of the Gentlemen of your [!] Society to me, far beyond what I could ever expect or deserve", and some months later he despatched Book III to Halley. But it was now quite recast in a less popular, more mathematical mould in order to put off the incompetent and disputatious reader.[37] The original draft of Book III, after contributing much to its successor, remained among Newton's papers to be posthumously printed (1728) under the duplicate title *The System of the World*; another part-copy was deposited by Newton as the Lucasian lectures for 1687. As for Book II, it had come to London about a month before Book III;[38] it contained the "fluxions Scholium" and some of the most difficult mathematical argument in the whole volume (though again no fluxions). Indeed, Newton made more than one mistake in treating in this pioneer fashion the flow of fluids and the motion of projectiles in a resisting atmosphere, mistakes which cost him annoyance some thirty years later.

It had been Newton's wish not to publish his book before the Hilary (summer) term of 1687; the late despatch to London of Books II and III (for reasons not now clear) forced Halley to make haste in order to have the book ready before the Royal Society and all 'the Town' deserted the capital. He engaged a second printer and brought all to a conclusion by 5 July, when he wrote to inform Newton of the fact. The author received twenty copies for himself – one went to Christiaan Huygens – and by waggon forty copies which in their unbound state Halley asked Newton to dispose of to the Cambridge booksellers for five shillings each.[39] No letter from Newton to Halley survives dealing with these matters and Halley's enormous contribution to the *Principia*; Halley's labour went far beyond organization and proof-reading, though little trace of it now remains. Newton did, however, pay him a magnificent tribute in the preface to the book:

> In the publication of these things the most acute Mr. Edmond Halley, a man learned in every branch of literature, not only assisted me in correction of the printer's errors and in taking care of the engraving of the figures, but was responsible for my embarking upon these pages. For when he had sought from me my demonstration of the shape of the heavenly orbits, he persisted in a request that I communicate it to the Royal Society, which body by its encouragement and good offices then caused me to begin thinking of giving them to the public.

If not exactly historical, *e ben trovato*; none of Newton's later editors received such warm thanks.[40]

Newton himself confessed that *The Mathematical Principles of Natural Philosophy* was not a work to be read straight through from the first line to the last. It was a book to be dipped into, tasted and assimilated slowly, perhaps moving from the opening pages to the "System of the World", and when this was mastered going back to the precise justifications earlier in the book. Nineteenth-century students of mathematics in Cambridge were not expected to master more than the first few sections. In 1687 the *Principia* could find very few competent readers in the whole of Europe: Halley of course, David Gregory and John Craige, Huygens and Leibniz, Guido Grandi and Fatio de Duillier. It is not therefore a book to which easy justice can be done in a few non-technical paragraphs. It was also, in 1687, for all its grandeur, a very imperfect book. As Newton recognized, it was written far too quickly and it was hardly set before him in print before he began an extensive programme of revision. Its gravest omission had to be supplied: the dynamical theory of the Moon had defeated Newton at this stage. There were vulnerable statements that he wished to amend, demonstrations to be improved, factual evidence to be made more cogent. Yet Newton was still, in the 1690s, unconscious of some errors that he had made. The flaws in the magnificent edifice make it the more difficult to survey, without uncritical eulogy.[41]

The Structure of the Principia

The book, even in its first edition, was massive: 510 dense Latin pages (and a leaf of errata). Twenty-five preliminary pages are given to definitions and laws with their numerous corollaries and scholia, then Book I (at 210 pages much the longest) opens with eleven geometrical Lemmas. Others are introduced into the text *passim* later. The general theory of unresisted motions under the action of forces, and especially of orbital motions, is set out in ninety-eight Propositions. Despite Newton's possibly ambiguous remark, Book II on resisted motions and the mechanics of fluids in fifty-three Propositions, ending with the celebrated proof that the planets cannot be carried round by material vortices, is not the shortest in the volume (at 165 pages). Book III with no more than 110 pages is much the least in scale as well as the easiest to approach. Here Newton initially links the abstract mathematical treatments of Books I and II (embraced under the title "De Motu Corporum") with the real universe by means of a series of nine "Hypotheses", of which three are logical axioms, one a metaphysical principle ("The centre of the universe is at rest") and the remainder particular applications of Kepler's Laws. The volume concludes with the theory of comets. Though of course Newton recognized even at this stage that a comet might pursue an elliptical orbit (of great elongation) and so be seen more than once at perihelion, he preferred to treat all comets as having non-returning, parabolic orbits, for no instance of a periodic comet was yet known to him. Throughout, Book III is closely indebted to the synonymous "System of the World" (as yet in manuscript) whose arrangement and language it follows in many respects.

There are five distinct major achievements in the *Principia*:

(1) Starting from abstract conceptions of motion and force, but concentrating mainly upon a type of attractive force that varies directly as the masses of bodies and inversely as the squares of the distances between them, Newton established the fundamental importance of the conic as the path described by one body moving with respect to a fixed centre. He realized all too clearly how immensely complex the problem of path description becomes when two or more mobile bodies are involved in a system, so that (as he remarked) in a multiple family of bodies such as that of Sun, planets, satellites and comets, no planet ever traces exactly the same orbit twice.[42]

(2) In what are now probably the least read portions of Book I, Newton provided the necessary geometrical underpinning or detailed properties of conics necessary for establishing the orbits of both planets

and comets from the observations of astronomers. For a tight, rather than a loose and in-principle fit between astronomy and mechanics to be created, it was necessary to be able calculate parameters and orbital data in fine detail. If Alexis–Claude Clairaut was able (1758–9) to make a more accurate calculation of the parameters of Halley's Comet (and hence a more precise date for its return) than Halley himself had been able to do, it was thanks to improvements in these specialized mathematical techniques: there was no fresh observational evidence to go on before the comet's return. In this respect Newton laid the foundations of specialized methods that after detecting Neptune and Pluto have gone to the direction of interplanetary probes.

(3) Although a good deal of work had been done on hydraulics by the Italians and the French, of which Newton probably was largely ignorant, Book II of the *Principia* gave a new physical, and hence mathematical, basis to the investigation both of dense fluids (liquids) and tenuous ones (gases). Clifford Truesdell has remarked sternly that 'Almost all of the results are original, and but few correct.'[43] However, Newton's identification of the approximate ballistic curve – a distorted hyperbola – and his calculation from theoretical principles of the speed of sound in air were colossal achievements, victories unimaginable to the mathematicians of half a century before.[44]

(4) Newton greatly extended the mathematical 'Earth sciences' by his demonstration of the dynamic shortening of the Earth's polar diameter as compared with the equatorial diameter, and by his theory of gravitational causation of tidal movements. He understood also how the atmosphere, held to the Earth by its gravitation, becomes thinner with height, computing the gradual decline in density. The confirmation of his precise evaluations, involving great labour in collecting and correlating data, raised standards of verification to a new height of accuracy.

(5) Similarly with the heavens: Newton in 1687 was unable to bring the Moon and the satellites of Saturn within the scope of celestial mechanics, only partly because of his wanting appropriate information, but he was able to demonstrate to a high degree of accuracy how the known parameters of the orbits of the remaining planets and satellites could be derived from mechanical principles and shown to conform to those derived by astronomers from observation.

On one point Newton could not satisfy himself, nor did he ever hope to do so. While vigorously defending the uniqueness and absolute nature of time and space, Newton regarded these attributes of the universe as consequences of the divine creation; the immutability of time, space and the universe reflected the immutability of the divine nature and purpose. (But whereas time and space were changeless because God's existence creates time and space, it would not similarly be inconceivable that God's volition could destroy and remake the universe within time and space.) Therefore Newton proceeded always on the premise that the universe men investigate is an unchanging divine artefact, because it was perfect as God created it; and it remains perfect because God would not allow his creation to diminish. Since Newton understood the obvious lesson that the action of force and mechanical laws might probably tend to diminish the perfection of the universe (for example, by the planets' tending to fall inwards upon the Sun) he imagined that more than mechanical interaction, or in other words divine intervention from time to time, was necessary to maintain the stability of the creation.

God and the Universe

Newton departed in two ways from the self-perpetuating universe of the Cartesian tradition that lasted well into the eighteenth century. Firstly, he denied that motion could be conserved through all impacts: if the bodies colliding are less than perfectly elastic, part of their motion before impact fails to reappear afterwards; as we should way, it is converted into heat.[45] Accordingly, no theory of the conservation of motion in the universe could be based on a physics of collisions; for since

> the variety of Motion which we find in the World is always decreasing, there is a necessity of conserving and recruiting it by active Principles, such as are the cause of Gravity, by which Planets and Comets keep their Motions in their Orbs, and Bodies acquire great Motion in falling; and the cause of Fermentation, by which the Heart and Blood of Animals are kept in perpetual Motion and Heat; the inward Parts of the Earth are constantly warm'd, and in some places grow very hot . . . For we meet with very little Motion in the World, besides what is owing to these active Principles.

But for the existence of such principles, stars, suns and all heavenly bodies "would grow cold and freeze, and become inactive Masses; and all Putrefaction, General, Vegetation and Life would cease, and the Planets and Comets would not remain in their Orbs."[46] Secondly, there must be recruitment of the active principles themselves and since these are the ultimate causal agents within Nature their replenishment can come only from outside Nature, that is, from God. Newton specifically points to "Irregularities . . . which may have [a]risen from the mutual Actions of Comets and Planets upon one another, and which will be apt to increase, till this [celestial] System wants a Reformation" but in principle the same must be true of all motions everywhere.

That God is the author and sustainer of all things Newton decisively maintained in the final pages of the later *Opticks*.[47] The hard and solid atoms composing the material universe were

> variously associated in the first Creation by the Counsel of an intelligent Agent. For it became him who created them to set them in order. And if he did so, it's unphilosophical to seek for any other Origin of the World, or to pretend that it might arise out of a Chaos by the mere Laws of Nature.

In the General Scholium concluding the second edition of the *Principia* the same thought is repeated: "This most elegant fabric of the Sun, planets and comets could not arise save by the wisdom and power of an intelligent and powerful Being. And if the fixed stars should be the centres of similar systems all these, created by the same wisdom, fall under the dominion of the One."[48]

The logically necessary link between the divine creative and sustaining powers was very plainly argued by Samuel Clarke (1717) against the perfectionism or 'pre-established harmony' of Leibniz: '[God] not only composes or puts Things together, but is himself the author and continual Preserver of their *Original Forces* or *moving Powers*: and consequently tis not a *diminution* but the true *Glory* of his Workmanship, that nothing is done without his *continued Government* and *Inspection*.' If, Clarke continued, we follow Leibniz's idea of the universe as a great, eternal machine that has no need of attention from its creator, we embrace 'the Notion of *Materialism* and *Fate*, [which] tends, (under pretence of making God a *Supra-Mundane Intelligence*) to exclude *Providence* and *God's Government* in reality out of the World;' an idea which, taken to its conclusion, means that the universe created itself.[49]

Since human intelligence can no more comprehend how God acts upon matter to sustain its properties, and the universe as a whole, than comprehend how he created material existence from nothingness in the first place, Newton's line of thought ends in an insoluble problem. Active principles such as gravity and fermentation cannot in the last resort be 'explained' by reduction under some law of conservation or other physical principle. Throughout his working life from 1687 onwards, however, Newton tended to emphasize and re-emphasize assurances that he did not mean to exclude the possibility of the future discovery of further links in a chain of reduction before the ultimate mystery is reached. In other words, there might be other mechanical processes to be found out in the action of gravity and other principles, wholly unforeseen at present. So in *Opticks* he affirmed, speaking of attractive forces in general:

> it's well known, that Bodies act one upon another by the Attractions of Gravity, Magnetism, and Electricity; and these Instances shew the Tenor and Course of Nature, and make it not improbable but that there may be more attractive Powers than these. For Nature is very consonant and conformable to herself. How these Attractions may be perform'd, I do not here consider. What I call Attraction may be perform'd by impulse, or by some other means unknown to me. I use that Word here to signify only in general any Force by which Bodies tend towards one another, whatsoever be the Cause. For we must learn from the Phaenomena of Nature what bodies attract one another, and what are the Laws and Properties of the Attraction, before we enquire the Cause by which the Attraction is perform'd.[50]

There are several analogous disclaimers in the *Principia*: attraction (mathematical) might in Nature be impulse (physical). They are, of course, consistent with Newton's methodological principle, also asserted in *Opticks*, "the main business of natural Philosophy is to argue from Phaenomena without feigning Hypotheses, and to deduce Causes from Effects. till we come to the very first Cause, which certainly is not mechanical." But let us note that here again, after opening this door to reductionism, Newton rapidly closed it again by raising the issue of the divine design, so powerful a theme in English philosophy from Henry More and John Ray to the Darwinian debate.[51]

For myself, while recognizing that it was tactically necessary and methodologically correct for Newton to leave open this door by which further extensions of mechanistic reduction might pass into philosophy, it is difficult to believe that he viewed the likelihood of such extensions with optimism, especially during the period 1687–1710.[52] David Gregory recorded in 1705: 'The plain truth is, that he [Newton] believes God to be omnipresent in the literal sense . . . he supposes that as God is present in space where there is no body, he is present in space where body is also present . . . He believes that they [the

Ancients] reckoned God the cause of [gravity], nothing els, that is no body being the cause, since every body is heavy.'[53] Newton transferred his own predilections to Antiquity. A physical mechanism for gravity could not be excluded from the conceptual scheme of possibilities, but none seemed plausible to Newton because of the requirement that the celestial spaces should, in effect, be wholly void of matter.[54]

Forces in Nature

Although we have to turn to the extended Queries in the later editions of *Opticks* for a full view of Newton's mature thoughts about forces and atoms in their relation to the divine process of creation, Newton's commitment to both (as the active and passive bases of the physical universe) is clear in the first *Principia* (1687). What may easily be read between the lines of the text is made totally plain in the preface to the book. In a well-known passage that is itself a précis of a much longer essay on the link between the microscopic and the macroscopic in Nature,[55] he indicated that just as the gravitational force conspicuous between massive bodies arises from the gravity of the atoms composing them, so other qualitatively different forces also acting at the atomic level are responsible for qualitatively different phenomena affecting some gross bodies (as with magnetism) or all of them (as with cohesion).

Book III of the *Principia*, Newton wrote in the printed preface, is an exemplification of the general theory of motion developed in Books I and II, in an explanation of the "System of the World". He continued:

> Would that, by the same kind of argument, the rest of the phenomena of Nature might be derived from mechanical principles. For many considerations induce me to suppose that all of them depend upon certain forces by which the particles of bodies are either impelled towards one another and cohere in regular figures, or are repelled from each other and fly apart, by causes which are not yet understood. These forces being unknown, philosophers have hitherto attempted the search of Nature in vain. I hope that the principles here proposed may throw some light upon the method of philosophizing here proposed, or upon some truer one.

In a parallel passage (perhaps, as we shall see, not remote in time from this preface) Newton wrote in a draft conclusion to *Opticks*, likewise suppressed:

As all the great motions in the world depend upon a certain kind of force (which in this earth we call gravity) whereby great bodies attract one another at great distances: so all the little motions in the world depend upon certain kinds of forces whereby minute bodies attract or dispell one another at little distances . . . The truth of this Hypothesis I assert not because I cannot prove it, but I think it very probable because a great part of the phaenomena of nature do easily flow from it, which seem otherways inexplicable: such as are chymical solutions, praecipitations, philtrations, detonizations, volatizations, fixations, rarefactions, condensations, unions, separations, fermentations, . . . the reflexion and refraction of light, . . .[56]

At this stage Newton clearly believed that the mathematizable concept of atoms mutually exerting a variety of forces (each with its own characteristics) upon other atoms was adequate to account for all the appearances of Nature. That this was a belief founded upon his original mathematical success of January 1680, and strengthening rapidly with the even greater success obtained by the developing methodology of the *Principia* from 1684 onwards, seems obvious enough. A marvellous vision of how celestial mechanics and the mechanics of fluids might be patterns for, and subsequently elements in, some great structure of mathematical physics had evolved in Newton's mind. And mathematical physics would be the key to a total understanding of the fabric of Nature.

How in later life Newton came to express a sense that something more was required will be dealt with in its proper place.

9

Private and Public Life,
1685–1696

Humphrey Newton's Recollections

In the last year of the reign of King Charles II, according to his own recollection, Humphrey Newton came to Trinity College, Cambridge, from Grantham School to be sizar to his great namesake, Isaac. Neither Newton admitted or claimed relationship with the other. Humphrey remained with Isaac for five years before returning to Grantham as a country physician, during which period he lived in some intimacy with the Lucasian Professor, chiefly serving as his copyist. John Conduitt, after Isaac's death, sought in 1728 Humphrey's recollections of him, first published by Brewster in 1855. With all its obvious imperfections and silliness, Humphrey's is the only personal record of Newton's daily life from someone who had every opportunity for close and long observation.

Humphrey was much more struck by Newton's manners than by his mind, which was beyond his appreciation. His only memory of the *Principia*, after having copied it, was that some of the scholars in Cambridge to whom Newton bade him take presentation copies, declared 'that they might study seven years before they understood any thing of it'. (They might have spent seven years in worse ways.) He remembered that Newton kept a five-foot-long [refracting] tele- scope at the head of the stairs leading from his rooms to the garden below, but all he could say of Newton's study of astronomy was that 'several of his observations about comets and the planets may be found scattered here and there in a book entitled *The Elements of Astronomy* by Dr. David Gregory.' His only attempt at a personal portrait of Newton was: 'When he was about 30 years of age his grey hairs was very comely, and his smiling countenance made him so much the more graceful.' Moreover, 'His carriage then [but not later?] was very meek, sedate and humble, never seemingly angry, of profound thought, his countenance mild, pleasant and comely.' Doubtless Newton, like Henry More, found Newton's facial expres- sion somewhat melancholy, if comely (as from Kneller's 1689 portrait it certainly was) for he adds 'I cannot say I ever saw him laugh but once', at some foolish man's inquiry about the usefulness of Euclid's geometry.[1]

As a student, Newton may well have received and largely followed the same advice that was given by a later college tutor, David Waterland, in 1706:

> Never go to any Tavern or Alehouse unless sent for by some Country Friend; and then stay not long there nor drink more than is convenient. Covet not a large and general Acquaintance but be content with a very few Visitants, and let those be good . . .
> Come in always before the Gates are shut, Winter and Summer; and before Nine of the Clock constantly when your Tutor expects you at Lectures in his Chamber.[2]

At any rate, Humphrey found his master living by these same blameless rules: 'He always kept close to his studies, very rarely went a visiting, and had as few visitors, excepting two or three persons, Mr Ellis [later Master of Caius College], Mr Laughton of Trinity [Librar- ian], and Mr Vigani, a chemist, in whose company he took much delight and pleasure at an evening when he came to wait upon him.' Newton's taking the air in his garden for distraction was 'but at some seldome time' and the hall dinner attracted him only 'on some public days'; equally, the college chapel saw him rarely 'that being the time he chiefly took his repose'. But Humphrey would not have us imagine that Newton was mean and misanthropic: 'Foreigners he received with a great deal of freedom, candour and respect. When invited to a treat, which was very seldom, he used to return it very handsomely, and with much satisfaction to himself.' It is interesting that Humphrey was aware of foreign visitors to Cambridge making a point of calling upon the Professor of Mathematics. Humphrey believed that Newton was open with his purse to deserving causes and knew that he was generous to his many poor relations.

Whatever his public state – and we have seen that Newton spent money on his furniture and furnishing his table – in private his eating habits were careless. 'So intent, so serious upon his studies [was he] that he ate very sparingly, nay, ofttimes he has forgot to eat at all, so that, going into his chamber, I have found his mess untouched, of which, when I have reminded him, he would reply, "Have I?" and then making to the table, would eat a bit or two standing, for I cannot say I ever saw him sit at table by himself.' His dress was slovenly; he commonly took short sleeps or naps in his clothes, and went forth from his rooms (if not reminded) 'very carelessly, with shoes down at heel, stockings untied, surplice on, and his head scarcely combed'. We have no evidence that Newton ever washed or bathed. One could almost imagine that Humphrey was following some stereotype of the absent-minded professor: 'At some seldom times when he designed to dine in the Hall, [he] would turn to the left hand and go out into the street, when making a stop when he found his mistake, would hastily turn back, and then sometimes instead of going into the Hall, [he] would return to his chamber again.' Little food and little sleep sufficed him, but in winter he was a lover of apples. Despite the story that a pet named Diamond overset the candle that caused a fire among Newton's manuscripts, Humphrey is firm that Newton 'kept neither dog nor cat in his chamber, which made well for the old woman his bedmaker' (named Deborah) who was happy to bear away the dishes from the college kitchens that Newton left untouched.

It is certain that, unlike pious Fellows, Newton never read prayers and homilies to students in his chamber. Besides sparing the chapel, Humphrey noted that 'As for his private prayers I can say nothing of them; I am apt to believe his intense studies deprived him of the better part' – perhaps an unusual admission for those days. But he repeated that Newton was 'mild and meek' in behaviour, contrary to the assertions of some historians, a man 'without anger, peevishness, or passion, so free from that, that you might take him for a Stoick . . . taking patience to be the best law, and a good conscience the best divinity.'[3] In support of these opinions of Humphrey

Newton are memories of him in late life from various individuals remarking upon his affability, hospitality and good nature. But clearly all men did not find him so.

Humphrey wrote more about Newton's chemical experiments, in which he assisted Isaac by sitting up on alternate nights to tend the fires, than about any other theme; his recollections have been quoted in chapter 7, and need not be extended here.

After the Principia

One might imagine that the publication of the *Principia* would soon have augmented Newton's correspondence, but that was not the case: rather his exchanges with Flamsteed and Halley ceased, or at any rate are lost. David Gregory wrote a flattering letter to Newton after reading the book, assuring Newton that he deserved 'the admiration of the best Geometers and Naturalists, in this and all succeeding ages' for his success in determining 'geometrically what it is which represents and measures physicall qualities'. Gregory was still in Edinburgh and had not as yet met Newton.[4] Another reader was Gilbert Clerke, a former Cambridge man some twenty years senior, who had published mathematical books. Newton wrote cordially in reply to Clerke's first letter ("If there be any thing els material for me to know . . . pray do me the favour of another letter, or two") but answers to Clerke's later letters are lost.[5] It seems that, as Newton remarked, his difficulties were largely verbal. A third early purchaser (and careful reader) of the *Principia* was the young Scots mathematician, John Craige, who lent his copy to another Scot, Colin Campbell.[6]

Little, indeed, is known of Newton's life in the short interval between the publication of the *Principia* and the political crisis of 1688 which swept Newton into university and national affairs. It is certain that he began almost at once to rewrite his optical papers in the form that would be published as *Opticks*. His first attempt was begun in Latin under the title *Fundamentum opticae* ('The Foundation of Optics') – later rendered into English in the first part of Book I without much change.[7] In his "Advertisement" to *Opticks* Newton wrote of composing its first Book about 1687 and later in the text of making reflecting telescopes (1671) about sixteen years ago.[8] It almost seems as though Newton had suddenly decided to make up for lost time.

We have no evidence of his paying attention to pure mathematics in these years. His mathematical labours seem to have been wholly given to the improvement of the *Principia*, and may be found detailed in the latter part of volume VI of *Mathematical Papers*.

Newton defends his University

Much of the reason for this (relative) lack of learned activity obviously lies in the fact that Newton was already entering upon his most important role in the history of Cambridge University, as a particip- ant in the English Revolution. The active, Catholicizing policy adopted by James II immediately after his accession induced him to place Roman Catholics in positions of trust and power wherever possible: as army officers, judges, officers of state. Every vacancy in Oxford and Cambridge was occasion for a royal mandate nominating a Catholic to fill it. In February 1687, just as Newton was finishing Book III of the *Principia* (which book, six months later, was to be somewhat fulsomely presented to the monarch by Edmond Halley) the Vice-Chancellor of Cambridge received a royal mandate for the admission of Alban Francis, a Benedictine monk, to the degree of MA without any examination or oaths. It was evident that this monk meant to be active as an MA in the university, and that (if admitted) he would not long be unique. On 19 February Newton advised the Vice-Chancellor to disobey this mandate: "Be courragious therefore & steady to the Laws & you cannot faile."[9] Newton argued that the king's personal orders cannot override statute law, and that no man can be punished for disobedience to them if they enjoin disregard of statute; he cited the similar refusal of Thomas Burnet at the Charter- house, which had gone unchallenged.[10] Newton was possibly also the author of a printed pamphlet, *An Account of the Cambridge Case*, of which a variant text in Humphrey Newton's hand has been found among Newton's papers.[11]

Whether or not Newton's advocacy of resistance to the king was widely known and had any influence, John Peachell, the Vice- Chancellor thus caught in a dilemma, did refuse the mandate, and Newton (once more with Humphrey Babington) was linked with his defiance. With the Public Orator, Newton was deputed to inform the Vice-Chancellor of the Senate's negative view of a second royal mandate reiterating the first, and was again deputed (with seven other members of the Senate) to support Peachell when he was summoned before the Ecclesiastical Commission presided over by the dreadful Lord Chancellor, George Jeffreys. The Cambridge case was heard at the end of April 1687, as Halley was hastening the printers to complete the *Principia*. Peachell was deprived of the mastership of Magdalene College and the vice-chancellorship (though restored to both before the year's end) and the whole Cambridge party was roundly denounced by Jeffreys. While all this was happening Newton was absent from Cambridge from 25 March, presumably at first in Lincolnshire (since he told Halley on 5 April to write to him there), probably on estate business, returning to Trinity in time for the legal business in London.

At this stage, if we may believe John Conduitt's story from Newton's own recollections, he played a crucial role in stiffening the university's resistance to the king. The Vice-Chancellor's delegation was almost inclined to compromise before meeting Jeffreys's wrath in London; it was Newton, who has left many drafts concerning the defence of the university's position in his papers, who gave them courage to stand their ground.[12] After their first appearance before the Lord Chancellor, at which Peachell was ineffective, it was again possibly Newton who prepared a document legalistically and rationally putting his case; but this the Commission would not look at.[13]

Newton thus won great fame in the university as a Protestant stalwart at the very moment when, among the learned, his fame as an incomparable mathematician began to rise. He was one of three candidates for the election to the Convention Parliament and one of the two chosen by the university (15 January 1689). This Parliament, having established the joint monarchy of William and Mary and re-established the Anglican Church, was dissolved after one year and a month. Newton was not again returned to Parliament by the university until the end of 1701. He had then been for a year the Master of the Mint.

The Portrait by Kneller, 1689

Perhaps it was to celebrate his new dignity that Newton had his portrait painted by Godfrey Kneller; this is the earliest authentic portrait, since the one by Loggan of which Collins had heard in 1677 never materialized. Kneller was to paint three more portraits of Newton as the years passed, in 1702, 1720, and 1722. The first is undoubtedly the finest that exists (see appendix B), plainly showing a youngish, vigorous, intellectually alert man with all the candour of high purpose and profound originality.

Kneller had plenty of time for his work, since Newton spent fifty-five weeks of the years 1689–90 in London, taking lodgings near the Abbey and Palace of Westminster. We can only guess that Newton, as a conscientious and determined representative, would have attended dutifully at the votes that settled Church and State in England, marching consistently into the lobbies in support of a firm Protestant stand. He presumably applauded the virtual deposition of James II, the offering of the Crown jointly to William and Mary and the legislation against popery. That this legislation also condemned the terrible heresy of Arius no doubt concerned him wryly, as we shall see. No participation of Newton in the debates is known, if we except the negative point that he was not listed among those who opposed the deposition of James on 5 February 1689. An anecdote records Newton's only utterance in the House of Commons: feeling a draught, he requested that a window be closed. However, anecdote may be erroneous. Since Newton asked leave to bring in a Bill to confirm the charters and privileges of the University of Cambridge, he must have made some remarks, however terse.[14] At this time Newton was in frequent correspondence with John Covel, Master of Christ's College and Vice-Chancellor, to ensure that the university took as much advantage as possible of the fluidity of the political situation. Cambridge was slower to act than Oxford was; Newton's own proposals for new university privileges included "inhibiting of [Royal] Mandates, regulating Visitations, entituling Professors to livings annexed to their Professorships, granting one book of every printed copy to the public library of either University for ever & restoring the right of University Preachers."[15] Together with his parliamentary colleague, Sir Robert Sawyer, Newton also tried to explain to Covel the new legislation concerning oaths of allegiance. But Newton's rational arguments did not prevent some Fellows of colleges from becoming non-jurors.

Newton's sojourn in London was by no means devoid of social activity, whose traces make apparent his rise as a man of distinction. In 1679 Charles Montagu, grandson of the parliamentarian Earl of Manchester, had been admitted (at the age of eighteen) as a Fellow-commoner to Trinity College; after a time he was elected a Fellow. Under William and Mary he was to become a major statesman. For the present, the point is that he became a friend of Newton's, despite the twenty years' difference in their ages. Early in 1685 – when well into Book I of the *Principia* (as it became) – Newton wrote to his friend Francis Aston, a Secretary of the Royal Society: "The designe of a Philosophick Meeting here Mr Paget when last with us pusht forward, and I concurred with him, and engaged Dr [Henry] More to be of it, and others were spoke to partly by me, partly by Mr Charles Montague."[16] This project came to nothing, because no one was willing to be a Hooke and provide experiments for the entertainment of the rest. Three years after this Montagu resigned his fellowship to plunge into public life; he was one of the signatories to the invitation to the Prince of Orange to pursue his own and his wife's claims to the throne of Great Britain. Thereafter he was the patron, Newton the client. Probably he was instrumental in persuading William III of Newton's merits, but his great act of patronage (after Montagu had become Chancellor of the Exchequer) was still some years in the future, for Newton's first attempt to set himself up in a more exalted position than that of university professor, soon after the Revolution, proved a failure. He advanced himself as a candidate for the provost-ship of King's College, Cambridge, arguing at length that the college's statutory requirements for its provost-elect to be a Fellow of the college in priest's orders could be set aside. Whether Montagu had encouraged this false hope is not known.

But it is clear that at one stage Newton enlisted as an ally the King's fellow-countryman, Christiaan Huygens, then visiting London for the third time. There his elder brother Constantijn was William's secretary. Newton and Huygens first met at a session of the Royal Society on 12 June 1689. On that occasion Huygens, anticipating a forthcoming publication, spoke of the cause of gravity and of the double refraction in Icelandic spar. Newton too took up this latter topic. Within a few days they entered on a friendly exchange of views about the theory of resisted motion developed in the *Principia*, of which Huygens had received a copy from Newton shortly before he left Holland.[17] On 9 July Newton went to visit Huygens at Hampton Court (where he was with his brother at court); next day Huygens and Newton, together with a powerful Whig MP, John Hampden (grandson of the opponent of ship money) had an audience of the King to press Newton's case.[18] William was as willing to appoint Newton provost as anyone, indeed eager to assert his royal powers by doing so, but the protests of the college proved too strong. In August the King in Council decided to drop the matter.[19]

Fatio de Duillier

Also present at Hampton Court, it seems, was a young Swiss mathematician, Nicholas Fatio de Duillier, briefly a friend enjoying a unique state of intimacy with Newton. He had first lived in England more than two years before, after spending some time with Huygens on his way thither. He was, by his own account, both skilled and inventive in mathematics. In relation to him (as to few, if any, others) Newton wrote: "This rule [a mathematical theorem], or one like it, was communicated to me some while ago by Mr. Fatio."[20]

While in London, in the spring of 1687, Fatio had heard of Newton's incomparable book then in the press, and indeed wrote about it to Huygens. He was in no way less enthusiastic when the book itself came to his hands, and at once contrived to attach himself to Newton after the Royal Society session on 12 June 1689, and so was with him again on 10 July. Fatio, who later proved to be an unstable though ingenious individual, seemed to be determined to assert himself as Newton's dearest friend and champion; his adulation was received by Newton with a warmth never extended to Edmond Halley, David Gregory or any other young man serving Newton's cause. Unfortunately, Newton's would-be champion also set himself to be Leibniz's enemy, initially because he supposed Leibniz insufficiently sensitive to his own genius. As he wrote to Huygens on 18 December 1691 (after access to Newton's private papers):

> It seems to me from everything that I have been able to see up to the present, among which I include documents written many years ago, that Mr Newton is without any question the first inventor of the differential calculus, and that he knew it as well or even more perfectly than Mr Leibniz yet knows it, before the latter had even the notion of it.[21]

This opinion as to Newton's marked priority in the development of certain mathematical techniques was just: Fatio's inference that Leibniz's first inkling of the calculus came from letters written by Newton (which led him to attack Leibniz publicly a few years later) was wholly false, though shared by other Newtonians such as David Gregory.[22]

Newton's warm feeling for Fatio, who bade fair to replace Wickens as a close companion, appears in the earliest surviving letter Newton wrote to him (10 October 1689) wherein he welcomed a proposal to share lodgings in London, undertaking to "bring my books & your letters". Besides mathematics they had in common an interest in chemistry and alchemy.[23] In April 1690 Fatio was recommending a 'servant' (sizar?) to Newton in succession to Humphrey – who would be expected to shave his master – a boy from St Paul's School, of course knowing Latin, ready for the university.[24] Soon after this Fatio made a long visit to his other friend, Christiaan Huygens in the Netherlands, a visit which was to prove troublesome to Newton because Fatio took with him a list of mistakes to be amended in the *Principia* compiled by Newton and himself, which he left with Huygens. After the latter's death (1695) it fell into the hands of friends of Leibniz, who took it to be a list of mistakes found by Huygens and used it against Newton.

Fatio returned to London in the first week of September 1691 and was almost immediately joined by Newton, who stayed for a week.[25] By the end of the year he was firmly entrenched as the chief of Newton's friends. David Gregory – another man eager to edit a new edition of the *Principia* – recorded:

> Mr Fatio designs a new edition of Mr Newtons book in folio wherein among a great many notes and elucidations, in the preface he will explain gravity acting as Mr Newton shews it doth, from the rectilinear motion of [aetherial] particles . . . he says that he hath satisfied Mr Newton, Mr Hugens and Mr Hally in it. Mr Fatio says that he knows the inverted problem of the tangents better than Libnitz. Mr. Newton and Mr Hally laugh at Mr Fatios manner of explaining gravity.

We may guess that Fatio was not taken so seriously by the Newtonian group as he took himself.[26] Nevertheless, Newton revealed his inner thoughts to Fatio, who (being less than perfectly discreet) passed them on to impress his friend Huygens, whom (though a much older man with very distinguished achievements to his credit) Fatio began slightly to patronize soon after settling in London, for example assuring Huygens that Newton would receive 'perfectly well all that you have said to him. So many times have I found him ready to amend his book upon matters of which I spoke to him, that I cannot sufficiently marvel at his facility, particularly as to those matters which you have criticised.'[27] Again, in February 1692, Fatio explained to his friend how Newton believed that the ancient mathematicians like Pythagoras and Plato knew the inverse-square law of gravitation, and all the demonstrations that he had given in the *Principia* of the true "System of the World": though they had made a great mystery of this, so much was evident from extant fragments.[28] Fatio was equally indiscreet in advertising to Huygens – who was in close correspondence with Leibniz – his sense of the originality and superiority of the Newtonian type of calculus; it might easily be guessed that this was Newton's own view.

All this time, as Gregory noted, Fatio confidently expected that Newton would entrust to him the editing of a second edition of the *Principia*, for which Fatio, Newton and Gregory all collected materials. This was not to be; Newton put off the start on the new text till 1710, and its editor then was Roger Cotes.

Fatio expected to work with Newton on this great task in Cambridge, but of Newton's own feelings nothing is known until November 1692, when Fatio suffered an illness to which he reacted in his usual excitable way. He assured Newton that he had burst something inside and might see him no more – he who would live another sixty years! Newton replied with concern, an offer of money and his prayers.[29] A few months later he did press Fatio to join him in Cambridge, as preferable for his health; but (though Newton repeated the offer) Fatio now proved coy and began to talk of returning to Switzerland, where his mother had died and he expected an inheritance on which he might live comfortably there.[30] The remainder of the extant correspondence is largely about medicines, payment for medicines and brass rulers, and alchemy. The old intimacy between the two men weakened and for several years there is no surviving correspondence – which does not prove that they were no longer in touch.

Friendship with Locke

A far more stable friendship begun by Newton at about the same time was with John Locke, who had read the *Principia* in Rotterdam soon after its publication; more than that, he twice took careful notes from the book and wrote an account of it for the *Bibliothèque Universelle* (March 1688).[31] Fatio too became acquainted with Locke, and as late as April 1693, in typical style, wrote of his 'fancy to have us go to Oates [Lady Masham's country house in Essex, where Locke made his home] . . . I think he means well & would have me to go there only that You may be the sooner inclinable to come.'[32] For Locke, Newton had opened the door to a post-Cartesian natural philosophy through which he gladly passed, and he was unstinting in his praise of Newton's work. How Locke, returning to England soon after the

Revolution, first met Newton in person, is unknown; he had of course been a Fellow of the Royal Society since 1668. Besides being a philosopher in the modern sense of the word – with his great works still unpublished in 1689 – he was a physician, a chemist, and in some measure an alchemist. But he was no great mathematician and it was for this reason that in March 1689–90 Newton sent him an easier "Demonstration that the Planets, by their gravity towards the Sun, may move in Ellipses" which, as we have seen, he may have composed in the late summer of 1684.[33] On this point – the comprehension of the *Principia* by those less than expert in geometry – Locke himself wrote:

> the incomparable Mr *Newton* has shewn, how far mathematicks, applied to some Parts of Nature, may upon Principles that Matter of Fact justifies, carry us in the knowledge of some, as I may so call them, particular Provinces of the Incomprehensible Universe . . . And though there are very few that have Mathematicks enough to understand his Demonstrations; yet the most accurate Mathematicians, that have examin'd them, allowing them to be such [as satisfy them], his Book will deserve to be read, and give no small light and pleasure to those, who [being] willing to understand the Motions, Properties and Operations of the great Masses of Matter in this our Solar System, will but carefully mind his conclusions, which may be depended upon as Propositions well proved.[34]

The acquaintance between the two men may have begun before the end of 1689, since Locke was deeply involved in Newton's affairs by mid-February of the following year.[35]

Like the letters to Burnet, and indeed those to Richard Bentley soon to be considered, Newton's correspondence with Locke throws early light on his interests outside science proper – in this case, alchemy and theology. In Locke's opinion, Newton was 'really a very valuable man, not only for his wonderful skill in mathematics, but in divinity too, and his great knowledge of the Scriptures, wherein I know few his equals'.[36] Not surprisingly therefore, he seems to have promoted the realization of the desire Newton had by now half-formed (at least) to abandon the university for London, with the support of some distinguished office. In the absence of all evidence, the historian cannot profitably speculate on Newton's motives in seeking election to Parliament and then, when he was not chosen for the successor to the Convention Parliament, looking discreetly about for some other way to change his life. Poverty cannot have been a motive, nor a desire to marry. We may guess that Newton was tired of teaching (light though his burden was) and perhaps found irksome the narrow mentality of some of his colleagues. The months he spent in London might have taught him new pleasures of friendship and social life, and of the independence he could not enjoy as a don living in college.[37] Perhaps he was reluctant to return to his former intensity of study in mathematics and science – valued even in the university only by a few choice minds – though (as we shall see) he did take up that work again after his long absences from Cambridge.

"Two notable corruptions"

However these things may have been, by the autumn of 1690 Newton was busy preparing, or more likely revising, his first semi-public criticism of the orthodox Trinitarian position: "An historical account of two notable corruptions of Scripture, in a letter to a Friend".[38] Its message was that in two significant places a New Testament text had been falsified in early medieval or even Roman times in order to endorse the doctrine of God as a Trinity of persons: Father, Son and Holy Ghost. In the First Epistle of John, Newton argued, at chapter 5, verse 7, the doctrine of redemption by Jesus is confirmed by 'the Father, the Word and the Holy Ghost; and these three are one' – but the original reading had made the spirit of Christ, the water of his baptism and the blood of the Crucifixion the three testimonies to the truth of redemption. Newton took St Jerome, by tradition the translator of the Vulgate Latin text of the Bible, to be chiefly responsible for this corruption.[39] The second instance Newton brought forward (originally in a separate letter to Locke, the "Friend") was the last verse of chapter 3 of St Paul's First Epistle to Timothy. Not the current reading, "Great is the mystery of godlinesse GOD manifested in the flesh," but the ancient statement, "Great is the mystery of godlinesse WHICH WAS manifested in the flesh," was authentic.[40] The thrust of Newton's preferred version was obvious: it claims only that godliness not God was manifest in Jesus. To justify these purifications, in an anti-Trinitarian sense, Newton invoked innumerable authorities in several ancient languages, texts assembled with great learning, the fruit of nearly twenty years of secret study. The method of his research was both historical and textual: by comparisons, he found that neither the modern Scriptures of the non-Western Churches nor the oldest texts of the Western Church itself contained these Trinitarian readings. Then he showed how words had been altered, or slipped into the text, to give it this unintended sense.

The "Historical Account" was despatched to Locke on 14 November 1690. Besides apologizing to him for the length of the paper – whose content had already been discussed between them – Newton shows that his intention was to have it published abroad, anonymously and in French.[41] A 'Third Letter', also presumably for Locke, of an unknown later date, continues the argument with many other examples of the way in which the texts of the New Testament (as received in the Western Churches) had gradually been modified to support a particular theological formulation of the origins of Christianity.[42]

Sadly, any letters written by Locke to Newton about this massive piece of scholarship are missing; as Newton spent some days with Locke at Oates soon after Christmas 1690 they would have had ample opportunity for discussion. Allusions to theological issues recur frequently in Newton's letters to Locke. The latter's high opinion of Newton's paper appears from the trouble he took it to copy it, and send this copy to the theologian Jean Le Clerc in Amsterdam (without indicating the author's name). Le Clerc was eager to translate and publish it, but then Newton changed his mind. The "Historical Account" remained unprinted until 1754, when the Amsterdam copy was published.[43]

As with all matters that he took seriously, the mass of Newton's surviving manuscripts on theology is very large; the portion of the whole sold by Sotheby's in 1936 was catalogued in forty-three lots estimated to contain one and a quarter million words. A single lot comprised a major incomplete treatise of 850 pages; those of its chapter titles that are recorded show that its later portion was intended to trace the corruption through time of the Western Church.[44] Moreover, Newton's treatment of the history of early Christianity was interwoven with this view of the general history of Antiquity. Biblical prophecy also greatly interested him. His *Observations upon the Prophecies of Daniel and the Apocalypse of St. John* were published in 1733, but besides this he wrote a large tract upon "the language of the prophets" (apparently part of a more extensive work) and many smaller pieces on the same theme. He wrote a complete essay against the Roman Catholic Church and another upon Solomon's temple. He considered the future of the Roman Church in the light of the biblical prophecies. But above all he gave himself to the problem of the true nature of Christ which had so fascinated theologians through the ages.

Newton and Primitive Christianity

Although Newton's embryonic library certainly included theological books, it is impossible to assign a chronology to the preparation of his multitudinous papers. A draft letter of 1674 has theological references on the back – which might be of a later date – and he presumably began quite early a typical notebook with appropriate headings to one or several pages, such as "Christi Passio, Descensus, et Resurrectio" (Christ's Passion, Descent [from the Cross] and Resurrection'). The pages were not always filled up with annotations. Westfall suggests that 'almost the first fruit of Newton's theological study was doubt about the status of Christ and the doctrine of the Trinity.'[45] The notebook shows how Newton began to go deeply into these matters. He may have owned ultimately as many as thirty-one editions of the whole or part of the Bible in English, French, Greek, Latin, Hebrew and Syriac, including of course the invaluable six-volume polyglot Bible of Brian Walton and others (1655–7), one of the great works of English scholarship. He took copious notes upon the Fathers of the Church, not only the most famous such as Athanasius, Augustine, Jerome, the two Gregorys, Origen, Tertullian (all, except Jerome, strongly represented in Newton's mature library)[46] but many, many minor ones. These notes were indexed and re-used. He also studied the modern theologians and historians, Baronius (1537–1607) inevitably, and when Jean Le Clerc (through Locke) referred him to the Oratorian Richard Simon (1638–1712) in relation to the "Historical Account" Newton read Simon too. It is not surprising that Locke was impressed by this formidable learning.

Extrapolating somewhat, one might guess that the notion of God the Creator appealed particularly to both the scientific and the religious facets of Newton's mind. He was perhaps less interested in that great foundation of Judaism and Christianity that inspired Milton, the idea of the Fall and Redemption of mankind. Genesis taught that the universe was created for mankind but also before mankind, and the problems of this creation are therefore prior to those of sin and salvation. Mersenne (not represented in Newton's library) before Newton's time, and his colleague Henry More, had both shown how much there might be to discuss in relation to them. Is it possible that to a mind as logical as Newton's was, and so little anthropocentric, the idea of the single, creative, all-powerful God, "Lord over all, Pantokrator, Universal Ruler," was of weightier significance than that of Jesus, who opened the gates of Heaven to believers?[47] Possibly in Newton's religious upbringing he heard more of God's power and punishment, less of Christ's mercy upon mankind. Perhaps therefore he acquired a bent towards strict mono-theism, or what in the future would be called unitarianism. At any rate, as he read more and more he became increasingly convinced that the trend of Western Christianity through the early ages had been towards error buttressed by fraud.

Newton would surely have relished those pages of Gibbon in which he traced to their Platonic origins the threefold division of the Godhead: the First Cause (God the Father), the *Logos* or reason (God the Son) and the Spirit of the Universe (the Holy Ghost). Then,

> The Christian Revelation, which was consummated under the reign of Nerva, disclosed to the world the amazing secret, that the LOGOS, who was with God from the beginning, and was God, who had made all things, and for whom all things had been made, was incarnate in the person of Jesus of Nazareth; who had been born of a virgin, and suffered death on the cross.[48]

In the eastern half of the late Roman world Platonic mythology mingled in exuberant richness with the religious traditions first of the Jews, then of the Christians. Philosophy raised a hundred questions about the nature of Jesus; at one extreme he was another prophet and teacher, blessed by God and endowed by Him with enormous spiritual power, at the other his human mantle was an illusion, for though born of Mary he was pure divine spirit. Whatever his nature, whence had he come? Had he always been with God the Father, or had he been created as an emanation of the divine essence inspired into a human frame? From a hundred questions sprang a thousand sects and (after Constantine had given the Roman Empire to Christianity) endless struggle by each for the domination of a single Church. One Alexandrian priest, Arius, taught that God was one and simple, Jesus a created though divine Being. By 319 he had been condemned by a synod at Alexandria, a condemnation repeated by the Councils of Nicaea (325) and Constantinople (385). His trium-phant opponent was St Athanasius the Great, also of Alexandria, whom Newton came to regard as the chief villain of early Christian-ity, a scholar who had distorted texts, history and reason with the double object of exalting Trinitarianism and vilifying Arius, the grand heretic.[49] His victory was in the long run Pyrrhic since it led to the split of Christendom.

Just as Newton saw himself as the restorer of truths known to the Pythagoreans, forgotten during the long reign of Aristotelian and

Ptolemaic error, so also he saw himself as the restorer of long-hid truths of religion. The very Greek word which was chosen at Nicaea to express the true faith as seen by Athanasius (*homoousios* = 'of the same nature') was a fraud; according to Newton, the Fathers merely "chose it for it's being opposite to Arius".[50]

Though Athanasius had been dead for over a thousand years, Newton pursued him as though he were a living enemy, tracking down his falsehoods as he would later track down the coiners. He carried his hostility to Athanasius to the point of preparing for publication (as it seems) a thirty-one-page denunciation in "Paradoxical Questions concerning the morals and actions of Athanasius and his followers", of which Brewster printed an outline.[51] This in no way concerned doctrine but the condemnation of Athanasius by the Council of Tyre, especially for the alleged murder of Arsenius, a Melitian priest.

As Newton read history, the corrupt victory of the Trinitarians had led to the evil ascendancy of the Bishop of Rome, for Arianism had always won most followers in the Hellenized portion of the Roman empire. The Reformation had reduced this evil but not corrected the root mistake in belief. In the true sense, the word God "doth always signify the Father from one end of the scriptures to the other"; the Son was divine, not a man with a human soul, but nevertheless subject to the Father and therefore not an equal Person in the Trinity.[52]

That Newton was a devoutly religious man no reader of his scientific works can doubt for a moment. Nor was he a deist; he fully believed in the historical existence of Jesus Christ and in the truth of the narratives of his life and works given in the Gospels. Brewster quoted from an unspecified manuscript the following passage, an expanded credo:

God made and governs the world invisibly . . . And by the same power by which he gave life at first to every species of animals, he is able to revive the dead, and hath revived Jesus Christ our Redeemer, who hath gone into the heavens to receive a kingdom, and prepare a place for us, and is next in dignity to God, and may be worshipped as the Lamb of God, and hath sent the Holy Ghost to comfort us in his absence, and will at length return and reign over us, invisibly to mortals, till he hath raised up and judged all the dead, and then he will give up his kingdom to the Father, and carry the blessed to the place he has prepared for them, and send the rest to other places suitable to their merits.[53]

Newton adds that as God's universe contains many mansions which he governs by agents which pass through the heavens, and since all places we know are full of living creatures, "why should all these immense spaces of the heavens above the clouds be incapable of inhabitants?" Here Newton's religion and his cosmology come close, and it seems likely that some passages concluding his scientific works were borrowed from earlier theological writings.[54] However, it would be a mistake to suppose that Newton's manuscripts are extensively concerned with natural theology (the demonstration of God's existence and purposes from His creation); this is not at all the case.

From the surviving correspondence – very unlikely to have been all that passed between the two men – there is no evidence that Newton ever imparted to Locke another major piece of his theological writing. Theological issues, notably about the occurrence of miracles, did crop up from time to time. Towards the end of Locke's life he sent to Newton for his criticism an essay on the Epistles to the Corinthians, which he and Newton had already discussed at Oates in the autumn of 1702. Newton delayed long in his reply, causing Locke some anxiety, but there was no break in their friendship. Earlier letters, scattered over more than a decade, are concerned rather with business matters of common interest than with deep matters of religion: the treatment of some alchemical recipes left by Robert Boyle (who had died in 1691), the relative values of gold and silver coins – at the time when Locke was a member of the Board of Trade and writing on currency questions, while Newton was at the Mint – and especially, at first, Newton's career.[55] He would not consider the Charterhouse (not, in fact, to become vacant), though he was interested by the Comptroller's place at the Mint (not offered to him). Locke strove to serve Newton's interests. As Newton paused sometimes at Oates on journeys to and from London, so Locke once or twice visited Newton at Cambridge, as he was urged to do in May 1692: "You may lodge conveniently either at the Rose Tavern or Queen's Arms Inn", he was assured.[56] Perhaps colleges then had no guest-chambers?

Newton's Mental Breakdown, 1693

The most intimate and the most terrible letter ever penned by Newton was addressed to Locke from the Bull Inn, Shoreditch, on 16 September 1693. In earlier letters there is some sign of a neurotic concern on Newton's part about his future prospects, and (significantly) few of Newton's letters from the immediately preceding months survive, apart from those from Fatio:[57]

> Sir,
> Being of opinion that you endeavoured to embroil me with woemen & by other means I was so much affected with it as that when one told me you were sickly and would not live I answered twere better you were dead. I desire you to forgive this uncharitableness. For I am now satisfied that what you have done is just & I beg your pardon for my having hard thoughts of you for it & for representing that you struck at the root of morality in a principle you laid down in your book of Ideas & designed to pursue in another book & that I took you for a Hobbist. I beg your pardon also for saying or thinking that there was a designe to sell me an office, or to embroile me, I am
> your most humble & most
unfortunate Servant
Is. Newton[58]

No one knows what Newton was doing in Shoreditch, well out of his usual beat, and it is obvious that he was still very disturbed, though beginning to recover his usual sense of things. A few days before Samuel Pepys had received an equally unaccountable letter:

> Sir,
> Some time after Mr. Millington had delivered your message, he pressed me to see you the next time I went to London. I was averse; but upon his pressing consented, before I considered what I did, for I am extremely troubled at the embroilement I am in, and have neither ate nor slept well this twelve month, nor have my former consistency of mind. I never designed to get anything by your interest, nor by King James's favour, but am now sensible that I must withdraw from your acquaintance, and see neither you nor the rest of my friends any more, if I may but leave them quietly. I beg your pardon for saying I would see you again.[59]

According to John Millington, a Fellow of Pepys's college, Magdalene, the story concerning himself was pure delusion. Since it occurred in both letters, the idea that he was trying to win himself an office by unworthy means preyed heavily on Newton's mind. It may be too that both letters reflect some crazy recognition in Newton's mind that both his friends enjoyed female companionship outside the bonds of marriage. Pepys and Newton are not known to have been friends before – indeed, this letter is the first occasion of Pepys's name appearing in Newton's correspondence. (At the end of the following November Pepys received word that Newton had been in town, but had left again; he took the opportunity to write proposing a question about odds in throwing dice, perhaps as a way of discovering his condition, leading to a little correspondence between them.)

Both Pepys and Locke responded compassionately to these distraught letters. The former sent his 'nephew Jackson' to Millington to enquire after Newton; then wrote directly to Millington, revealing his receipt of a letter which had put Pepys 'into a great disorder . . . from the concernment I have for him [Newton], lest it should arise from that which of all mankind I should least dread from him and most lament for, – I mean a discomposure in head, or mind, or both.' Millington replied – but not until 30 September – that he had just met Newton in Huntingdon, and that 'before I had time to ask him any question' Newton had confessed to writing Pepys 'a very odd letter, at which he was much concerned' while suffering from 'a distemper that much seized his head, and that kept him awake for above five nights together'. Newton was now well again, but very much ashamed that he had insulted one 'for whom he had so great an honour'.[60] As for Locke, he wrote an understanding and affectionate letter to Newton after taking thought (or perhaps making quiet enquiries) beginning: 'I have ben ever since I first knew you so intirely & sincerely your friend & thought you so much mine that I could not have believed what you tell me of yourself had I had it from anybody else.' He begged Newton to impart any criticisms of his *Essay concerning Human Understanding* and to arrange a meeting between them.[61] Newton replied with a fuller version of the explanation given to Millington: a habit of dozing by the fire during the winter of 1692–3 had got him out of a regular way of sleeping and during the summer following an "epidemical distemper" had further worsened his health "so that when I wrote to you I had not slept an hour a night for a fortnight together & for 5 nights together not a wink." This fits in with Humphrey's reminiscence of Newton's ill manner of sleeping. Newton closed this curiously curt note on a formal level: "I remember I wrote to you but what I said of your book I remember not. If you please to send me a transcript of that passage I will give you an account of it if I can."[62] It is possible that Locke was hurt by this cold acknowledgement, for a long gap in the (known) correspondence followed, ended by Newton's purely factual piece of information about money.[63]

Newton's own account of his breakdown seems inadequate, concealing deeper causes. Many of these have been proposed: overwork, frustration (both in his career and in his private investigations) – for he had left both *Opticks* and a new mathematical essay unfinished – perhaps distress caused by another fire and destruction of his manuscripts, the collapse of his friendship with Fatio de Duillier and their plans for a life of collaborative studies.[64] Other scholars suggest that Newton had been poisoned by the vapours of heavy metals during his long alchemical experimentation. Analysis of hair supposed to have been Newton's is said to confirm this diagnosis.[65] Against this last explanation it may be urged that the characteristic signs of mercury poisoning (salivation, hatters' shakes) were not reported by Newton, nor did he complain of the colic caused by lead. He lived for thirty-four years in full mental and physical vigour after his illness, and therefore suffered no permanent damage to his brain and nervous system. That his (post-mortem) hair so long after his exposure to metal fumes should still contain arsenic etc. derived from them is incredible. Moreover, the hair tested showed ten times the normal quantity of gold; however creditable to a would-be alchemist this might be, it suggests something odd about the sample. The conclusion of Dr R. W. Ditchburn that Newton suffered from a normal episode of depressive illness seems reasonable.[66]

It is also possible that Newton's breakdown had its origin in his religious beliefs. The post-Revolution settlement had reiterated penalties against Arians, which some years later were to be invoked against Newton's friend William Whiston. If it is to be believed that Newton in the mid-1670s was fearful of taking holy orders because of the oaths imposed, how much more in the 1690s must he have dreaded the probability that any preferment from the Lucasian professorship would require him to swear equally unpalatable oaths. For this very reason Robert Boyle had always declined any office, even that of President of the Royal Society. A choice between honest, if private, adherence to his true convictions and an entry into the national world of affairs might almost certainly lie before him. Men have taken their own lives for less. Newton in the early 1690s may already have laid this difficulty before such intimate friends as Locke, who certainly knew what Newton's private opinions were. The difficulty may have impeded Newton's appointment to a post in the service of the State before 1696; the breakdown might have been the price paid for a decision not to place scruple before preferment, which permitted his move to the Mint in 1696. It should be added, however, that never in his life was Newton – or Boyle or Locke – out of communion with the Church of England.

After his illness, Newton lived many years, published *Opticks* in four editions and *Principia* in two; he managed with high efficiency a major organ of the State and effected a complete reform of the British currency. He was an energetic and forceful President of the Royal Society for twenty-two years. There was thus no basis for the story that got about, which Brewster felt it necessary to confute, that illness ruined Newton's intellect so that he turned to biblical studies in his dotage. That Newton had been disturbed in his mind was retailed to Huygens, and by him passed on to Leibniz and so to all Leibniz's circle.[67] But Leibniz never made the mistake of supposing Newton to be feeble-minded. The breakdown did Newton harm, obviously, but the move to London three years later, the chance to take on great responsibilities and start a new life, brought him confidence and authority which he enjoyed to the full. One has only to read his letters to Roger Cotes about the second edition of the *Principia* (1710–13) or to follow his unrelenting hostility to Leibniz to realize what a powerful man Newton remained, in both personality and intellect, until he was well over seventy years old, how swift was his 'taking a point' and how great his power of work in dealing with it.

Newton advises Bentley, 1692–3

Marginally related to Newton's thoughts about religion at this time were his letters to Richard Bentley (from 1700 Master of Trinity College, Cambridge) about God and natural philosophy. Robert Boyle had died on 30 December 1691. Newton attended his funeral in London, and a few days later met Samuel Pepys at his home in Clapham, together with John Evelyn and a former Trinity scholar, Thomas Gale (now High Master of St Paul's School) to choose under the terms of Boyle's will the first preacher to deliver a series of sermons on natural theology. They chose yet another Trinity man, Bentley, then a rising divine and the best British classical scholar of his age. (In 1697 Bentley would attempt to organize a philosophical club comprising Newton, Wren, Locke, Evelyn and himself.) He now planned, and in the final pair of his 'Boyle Lectures' executed, a demonstration of the reality of God's power from the 'Origins and Frame of the World', drawing upon Newton's *Principia*.[68] The Scots mathematician John Craige supplied him with an overpowering bibliography to master before tackling Newton's book.[69] Newton himself gave Bentley a shorter list from which he particularly noted

Descartes's *Geometria*, Bartholin's introduction to it, Mercator's *Institutionum astronomicorum* and Huygens's *Horologium oscillatorium*.[70] How far Bentley went into these authorities is not known, nor when or what he wrote to Newton at the end of 1692, after delivering his sermons, before preparing them for publication. Newton's reply opens with a marvellous paragraph:

> When I wrote my treatise about our Systeme I had an eye upon such Principles as might work with considering men for the beleife of a Deity & nothing can rejoyce me more than to find it usefull for that purpose. But if I have done the publick any service this way 'tis due to nothing but industry & a patient thought.[71]

Presumably Bentley asked Newton: can there be a cosmogony without a Creator? Newton answered, NO; gravity could make uniformly diffuse matter into discrete large bodies, but that some should be opaque while others became hot and lucid "like the Sun, . . . I do not think explicable by mere natural causes but am forced to ascribe it to the counsel & contrivance of a voluntary Agent." Similarly with the formation of the solar system: the Cartesian evolutionary hypothesis must be false. The planetary motions must have been "imprest by an intelligent Agent":

> To make this systeme therefore with all its motions, required a Cause which understood & compared together the quantities of Matter in the several bodies of the Sun & Planets & the gravitating powers resulting from thence, the several distances of the primary Planets from the Sun & secondary ones from Saturn Jupiter & the earth, & the velocities with which these Planets could revolve at those distances about those quantities of matter in the central bodies. And to compare & adjust all these things together in so great a variety of bodies argues that cause to be not blind and fortuitous, but very well skilled in Mechanicks & Geometry.[72]

Bentley replied with further difficulties, to which Newton responded on 17 January 1693. In this letter he touches on an example of how a mathematical principle may illuminate philosophy: "The generality of mankind consider infinities in no other way then definitely, & in this sense they say all infinites are equal, though they would speak more truly if they should say they are neither equal nor unequal nor have any certain difference or proportion one to another." With such a conception of infinities, attempts to compare them produce error: "There is therefore another way of considering infinites used by

Mathematicians, and that is under certain definite restrictions & limitations whereby infinites are determined to have certain differences or proportions to one another."[73] He also mentioned the notion attributed (inaccurately) by Galileo to Plato, that if all the planets fell from one certain point in space towards the Sun, their appropriate orbital speeds would be reached at the due distances from the Sun. (He returned to this fancy in the 'Classical Scholia'.)[74] This letter closes with a famous sentence on which many have pondered: "You sometimes speak of gravity as essential & inherent to matter: pray do not ascribe that notion to me, for the cause of gravity is what I do not pretend to know, & therefore would take more time to consider of it."[75]

On 18 February 1693 Bentley at last told Newton that his eight sermons were in the press, and asked Newton to confirm positions taken by Bentley in the seventh. Some of them relate to the size of the universe (as then understood), on which Newton corrected Bentley; he again corrected his expressions with respect to infinities. But in general he was very satisfied with Bentley's arguments, and in particular agreed with him that

> Tis unconceivable that inanimate brute matter should (without the mediation of something else which is not material) operate upon & affect other matter without mutual contact; . . . That gravity should be innate inherent & essential to matter so that one body may act upon another at a distance through a vacuum without the mediation of anything else by and through which their action or force may be conveyed from one to another is to me so great an absurdity that I believe no man who has in philosophical matters any competent faculty of thinking can ever fall into it.[76]

The agent causing gravitation must fill space and act according to fixed laws, but Newton would not tell his readers whether it was of a physical or spiritual nature.

Nevertheless, this is one of several pieces of evidence indicating that at any rate during the post-*Principia* years gravitation was thought by Newton to be best accounted for by a direct action of the divine power, rather than by some mechanism created once for all at the beginning of the world. For at this time he was unable to imagine that gravity could be transmitted from one particle of matter to another by a physical process, across the vast empty or near-empty spaces between them. As we shall see, he changed his mind later.

10

Fluxions and Fury, 1677–1712

Leibniz and Newton

The link between Britain and G. W. Leibniz, now firmly stationed at Hanover though often travelling about Europe in the interests of his master, was decisively broken by the sudden death of Henry Oldenburg early in September 1677. A last letter from him to Leibniz had advised the philosopher to expect no speedy response to his recent mathematical letters from either Collins or Newton; the silence in fact was to last for fifteen years.[1] Whether or not Newton would have been willing to carry on the correspondence with Leibniz, he lacked easy means to do so. Both Collins and Newton studied Leibniz's letters, however. We have no opinion at this stage about Leibniz's mathematical discoveries from Newton himself, but it would seem that he, Collins and John Wallis agreed that Leibniz had (like James Gregory) made an independent discovery of the 'method of series' based on the work of earlier mathematicians which he had developed in his own way, and to which he had added an improved 'method of tangents'. Collins imagined (mistakenly) that Leibniz had gleaned something from Barrow's lectures, and suggested that the Royal Society might print Leibniz's letters to Oldenburg. He never hinted that Leibniz owed anything to Newton.[2]

As previous chapters have shown, Newton did little to promote *his* calculus in the twenty years after 1671. Leibniz too gave few signs of his slumbering mathematical genius for eight years after leaving Paris. He produced a couple of papers reporting the fruits of his work there with Huygens, notably the series

$$\frac{\pi}{4} = 1 - \tfrac{1}{3} + \tfrac{1}{5} - \tfrac{1}{7} + \ldots$$

It is perhaps a little conspicuous that Leibniz did not think fit to mention earlier cognate work by Mercator, Gregory and Newton, the unpublished part of which had been made available to him by Collins. Then, in 1684 and 1686, he printed in the *Acta Eruditorum* – the newly founded German learned journal, edited by Otto Mencke, with which Leibniz was long associated – two papers on the differential and integral calculus.[3] The first of these, very short, very compressed, very hard to follow, caused John Craige to complain of its opacity in his *Methodus figurarum . . . quadraturas determinandi* ('A method of determining the quadratures of figures', 1685). Craige, who had learned from Newton in Cambridge, adopted and praised Leibniz's new methods also. In response to his complaint, Leibniz prepared the second paper to augment and clarify what he had written before.[4]

Between them, the major and the minor mathematician were bringing the methods of the calculus into the open. Newton added a little more in the *Principia*. At the beginning of Book II, Section II, the second Lemma is not the least of the puzzles in the volume, for Newton introduces yet another way of treating fluxions (though neither this word nor the thing were used in the book) and explains simple differentiation (as we call it) without ever employing such an algebraic method in the volume. Perhaps the Lemma was written as a peg on which to hang the historical statements in the Scholium following it.[5] In it Newton names *moments* the momentary increases or decreases of changing quantities which are "indeterminate or variable, increasing or decreasing as it were in a perpetual movement or flux"; the *moment*, however, is not a finite particle of quantity (as a differential was to Leibniz), for as soon as a *moment* becomes finite it ceases to be such. It is only the first nascent principle of a finite quantity. In algebraic practice (Newton continues more clearly) the *moment* of $A^{n/m}$ is $a.n/m.A^{\frac{n-m}{m}}$, where a is the *moment* of A. In Leibnizian formulation we write to the same effect dA as the differential of A, and the differential of $A^{n/m} = n/m.A^{\frac{n-m}{m}}dA$. The parallelism is obvious. In the algorithm, though not at all in its justification, what Newton has called the *moment* is the equivalent of Leibniz's differential (as he had explained it in his letter of 1677). Upon this follows Newton's Scholium:

> When, in letters exchanged between myself and that most skilled geometer G. W. Leibniz ten years ago, I indicated that I possessed a method of determining maxima and minima, of drawing tangents and performing similar operations which served for irrational terms just as well as for rational ones, and concealed the same method in transposed letters [which, when correctly arranged] expressed this sentence: "Given any equation involving flowing quantities, to find the fluxions, and vice-versa"; that famous person replied that he too had come across a method of this kind, and imparted his method to me, which hardly differed from mine except in words and notation.[6]

Since no reaction by Newton to Leibniz's 1677 letter at the time of its reception is recorded, this is the first statement by either party to the calculus dispute of the similarity between their respective discoveries, and of their effectively close simultaneity in time. The Scholium was obviously a response to Leibniz's first calculus paper in the *Acta Eruditorum* for 1684, whose content cannot have taken Newton by surprise. As far back as 1677 John Wallis, in writing to Collins, had advised that 'Mr Newton should perfect his notions, & print them suddenly'; instead it was Leibniz who had done so.[7] The points made in the Scholium are acceptable except for the implication (immediately after the 'decoded' lines about fluxions) that Leibniz, in saying that "he too had come across a method of this kind" (Newton's words) referred to the *method of fluxions*. Since even now (in 1687) the purport of this method was wholly concealed from Leibniz, he could not possibly have acknowledged that Newton in his *Second Letter* had dealt with that topic, nor that Newton had been the first to put it on paper. Leibniz never at any time made such acknowledgements.

As a manifesto of priority the fluxions Scholium was weak. Supposing the string of letters in the *Second Letter* of 1676 to have meant what Newton now revealed 'in clear', Leibniz could hardly be expected to balance this gnomic utterance against the clearly outlined foundations of his own method in his 1677 letter, still less against the formal expositions he had recently printed. The difficulties in these last had not obstructed Craige nor (more significantly) the brothers Jakob and Johann Bernoulli. The Scholium lacked conviction unless rapidly backed up by a detailed account of what fluxions could do. This Newton failed to furnish.[8]

Mathematics at Cambridge after 1690

Whiteside has painted a dark picture of Newton's last years in Cambridge, whither he returned 'gloomily' in the winter of 1690, entering a period of months in which he wrote no letters. 'He must have felt the deep sadness of realizing that whatever else it was still in him to do would not stand comparison with' the *Principia*, about whose 'bright-burning candle' he could only 'flit moth-like . . . now and then fanning its flame to a yet purer blaze.' Although Whiteside insists upon the high quality and importance of Newton's proposed multivolume 'Geometry' (see p. 254) he also recognizes in Newton's mathematical writings of the early 1690s 'a slow but accelerating decrease in his elasticity to absorb fresh findings and his hitherto matchless capacity to attack novel problems and evolve new techniques of solution'.[9] For the first time in Newton's experience he was defeated by a problem, the gravitational theory of the Moon's motion, a problem only to be solved (still using Newton's principles) by his successors. Also in these years – scholars believe – Newton successfully completed the text of his second great work, *Opticks*, but here too his success was flawed; he was unable to complete it as planned and its Third Book was broken off short.

In the summer of 1691 Newton was brought back to pure mathematics and the quadrature problem in particular by David Gregory, whom he then first met in London in July. At once and at Gregory's request Newton recommended him to the electors of the Savilian professorship of astronomy at Oxford, as one skilled in analysis and geometry and understanding astronomy very well. Apparently Newton preferred him to Halley and Caswell, both also candidates.[10] Gregory now began to cultivate Newton's friendship, writing out for him an account of the Scottish universities, then reporting to him on a visit to Flamsteed (27 August), and finally acquainting him with a plan to republish 'his' quadrature series (see p. 149).[11] This republication was to be in the form of a Latin letter to Newton (*sapientissime!*), to be followed, Gregory hoped, by Newton's answer. Newton took his time about his response, also of course in Latin, which he prepared (in draft) as a general account of the development of such quadrature series, with just attention to Newton's own priority, leading into an extract from the *Second Letter* to Leibniz of October 1676. But probably Gregory received no written response at all. After his Oxford success, his plan to republish was dropped.[12]

Once again Newton was brought against the fact that he had to hand no convenient account of his method of quadrature (or integration) outside his early papers except that contained in his October 1676 letter to Leibniz: in his draft response to Gregory he did mention the 1671 tract on fluxions. This he no longer thought of printing as it stood. Therefore now, late in 1691, much as seven years before, he took the opening paragraphs of that draft reply as the start of a longer tract, ultimately to be known as *De quadratura curvarum* ('On the quadrature of curves'). The first draft of a brief text, extending to thirteen propositions, is incomplete.[13] Its chief purpose is to set out a process whereby "all curves whose equations are of three terms are either geometrically squared, or compared with conics, or transmuted into other curves of the four simplest forms which conics and the cubic parabola assume."[14]

Almost at once Newton threw this draft aside and began another which (after the first few lines) ran a very different course. This time, at Proposition IV, Problem I, Newton disclosed the technique concealed by the "anagram" of the letter to Leibniz: "Given an equation involving any number of fluent quantities, to find the fluxions, and *vice-versa*." Into a far clearer exposition than that in the *Principia* he introduced his henceforth standard notation, denoting $\dot{x}$ as the fluxion of x (etc.). Remembering that the fluxion is a rate of change, in a very small time of flow o (but Newton does not here say it is a time) x becomes $(x + o\dot{x})$, etc.; this is what he had already called the *moment* of x, the exact equivalent of a Leibnizian differential.[15] We may follow the process by finding the fluxion of a very simple equation, $x^2 - y^2 + 1 = 0$. According to the rule stated by Newton, its fluxion is $2\dot{x}x - 2\dot{y}y$. To check this by Newton's route, we write down the *moment* of the equation, which is $(x+o\dot{x})^2 - (y+o\dot{y})^2 + 1 = 0$, or $(x^2+2o\dot{x}x+o^2\dot{x}^2) - (y^2+2o\dot{y}y+o^2\dot{y}^2) = 0$. Subtracting the original equation from this and omitting powers of the fluxions $\dot{x}$ and $\dot{y}$ as negligible leaves $2o\dot{x}x - 2o\dot{y}y = 0$.[16] Then dividing both sides by the time o gives the fluxion of the expression as obtained by the rule, which may be rewritten $\dot{x}x = \dot{y}y$. After a few more algebraic steps, Newton is able (in Proposition 9) to address the problems of integration – deriving fluents (quantities) from their given fluxions – which with illustrative applications of the techniques involved take up the rest of the (not quite finished) text, much longer than the first draft.[17] Newton shows (with examples) how the fluxions of higher orders ($\ddot{x}$, the fluxion of a fluxion, etc.) may be handled and how infinite converging series may be used to find the required fluents. At Proposition 13 he begins to solve problems in mechanics by the fluxional method: an interesting point here is Newton's use of higher-order fluxions in defining changes in the curvature of curves, denoted by multiple dots over the letters.[18] This second state of *De quadratura curvarum* ends with some problems having "some connection with the physical world", that is, problems in the theory of gravitation. The first of these is solved by geometry alone, in the style of the *Principia*, but the second was to be solved by a fluxional analysis which Newton did not complete: it is in fact the inverse problem of central forces.[19]

Though he left the concluding page of this draft as a sketch, we need not doubt that the working out of a fluxional solution was well within Newton's powers, or that (had he so chosen) he could have set out the mathematical structure of mechanics in the algebraic form, thereby considerably anticipating the achievements of the Leibnizian school. Perhaps the best proof of this is his swift solution of the Bernoulli challenge problem – the elucidation of the brachistochrone – in 1697 (see p. 263).[20] But Newton's second draft of *De quadratura* was, like so much else, only to be printed in recent years. In 1693 he incorporated a shortened version of it into his projected large 'Geometry', adding to the compressed draft some fresh material of which the most significant were the tables of integrals lifted, almost unchanged, from *De methodis serierum* (1671). From this third version of the original tract the treatment of problems in mechanics was excised. Finally, standing on its own again but with a new introduction, this last version was published by Newton with *Opticks* in 1704.[21]

Analysis and Synthesis in Mathematics

It may seem strange to us that Newton should thus, with an original work of his own composition, have concluded a planned study of the geometry of the Ancients, but such a procedure appeared less unusual in Newton's day. Gregory even recorded without comment Newton's intention to attach the dual treatise to the revised edition of the *Principia* which also occupied him at this time (1694).[22] To many mathematicians of the seventeenth century – but among these Descartes is not to be numbered – what posterity was to regard as original discovery appeared rather to be the recovery of what the Ancients had once known. Others besides Newton believed (with some subsequent justification) that Greek mathematicians had built up but never systematically expressed procedures of analysis to which were owing their elegant synthetic demonstrations in geometry. This lost ancient analysis was thought to prefigure that recently discovered by the moderns in all but notation (as Newton said). It is a curious paradox that the same kind of erroneous attribution of a prior analytical process was made in relation to Newton's own *Principia*; however, Christiaan Huygens, whose synthetic demonstrations were much admired by Newton, did practise a prior method of algebraic analysis, subsequently suppressed. In more than one place Newton draws a distinction between mathematical analysis as a method of investigation, and geometrical synthesis as the method of proof *par excellence*. Accordingly, he deplored the influence of Descartes and his successors in treating analysis as the only significant branch of the subject or, in the words of Henry Pemberton, editor of the third edition of the *Principia*, would 'censure the handling [of] geometrical subjects by algebraic calculations' praising Barrow, Huygens and Sluse among his contemporaries 'for not being influenced by the false [Cartesian] taste, which then began to prevail'.[23]

This 'Geometrical Analysis of the Ancients', later simply the first of 'Three Books on Geometry' was Newton's tribute to his Greek predecessors whose work had been so imperfectly transmitted to modern times. Adding his various drafts together they fill some 300 pages in the *Mathematical Papers*. Newton did not treat the available material – mainly writings of Euclid, Apollonius and Pappus – in a strictly historical way: rather he presented a view of what he took to be the nature of Greek theoretical geometry, mingling his own with ancient mathematical procedures. Running through all is Newton's sense that algebra is for bunglers in analysis:

> he who uses [algebraic] analysis when a question is solvable without it uses a circuitous route . . . the geometer, however, is better trained and comes upon conclusions that are often simpler – and gained with less effort – than [those of] a man who strays further from the direct path. For almost all problems have a natural way of being solved, and its discoverer will attain the solution with no trouble, whereas one who deviates from the [geometric] path will do violence to a problem . . . Whence it happens, I think, that the ancients, whose aim was composition [or synthesis] frequently arrived at simpler conclusions than the moderns, who are more devoted to algebra. [Innumerable problems] are by algebra, following the conventional method, brought only with extreme difficulty to an equation, and innumerable ones . . . cannot be so reduced at all, and yet their solution if anyone should go about it the right way, is easy enough [by geometry].[24]

Upon such claims as these, however, the modern editor observes that Newton's 'appeal to the relative ease and neatness of a geometrical construction found without analysis is essentially spurious'.[25]

If some may regard the drafts of this classical geometrical disquisition as of rather dry academic interest, since they bore no fruit, they do at least show us two things. One is that Newton's mind would not allow him to expend scholarship on mathematics as he did upon history and theology. In the end he had to return to writing about mathematics that was done after his own fashion. Secondly, we may note that all this drafting went on, apparently, in the dreadful year 1693 and into 1694 without any sign of Newton's mental breakdown. Obviously we cannot be sure that there was no gap in Newton's preoccupation with his 'Geometry'; we do, however, have the silent testimony of David Gregory that Newton seemed as usual when he visited Cambridge early in May 1694 and went through this material.

This visit inaugurated a friendship between Gregory and Newton lasting until the former's death in 1708, at the age of 49, when his place was to some extent taken by his pupil, John Keill. His post at Oxford (from 1692) enabled Gregory to visit London often, and to record Newton's conversation. Unfortunately, no complete edition of Gregory's *Memoranda* exists, but the extracts available in print throw light upon Newton's thought and activities at various times. Gregory's abortive plan to edit a second edition of the *Principia* has already been mentioned. Further, as a result of his inspection of Newton's papers in 1694 he seems to have drawn up a plan for a general treatise on calculus, which also came to nothing. His *Elements of Physical and Geometrical Astronomy* (1715; in Latin, 1702) was an important exposition of Newtonianism, also printing for the first time Newton's ideas about the Pythagoreans as his precursors from the draft 'Classical Scholia', and his theory of the Moon. In July 1707 Gregory was sent by Newton to take charge of the Scottish recoinage at the Edinburgh Mint, a business of several months. He seems to have exerted no obvious formative influence upon Newton's life, unlike Halley.

The Cubic Curves

In the year following Gregory's visit to Cambridge, Newton's last full year in Cambridge, he composed another mathematical treatise, also codifying his studies over many years and also to be published with *Opticks* in 1704. This *Enumeratio Linearum Tertii Ordinis* ('An Enumeration of Lines of the Third Order'), a classification and study of the cubic curves, is very much a geometer's masterpiece: made after the example of Omar Khayyam, had Newton but known it. In Newton's taxonomy there are four equations describing these curves, and nine principal "cases" of them divided into sixteen genera and seventy-two species – six short, in fact, as was discovered in the early eighteenth century. The result is rich and subtle, but not perhaps stimulating to the imagination of the non-specialist: 'When faced for the first time with this boiled-down puree of Newton's lifetime of discovery in the pure and analytical geometry of curves, one is filled with admiration for the range and deftness of its techniques and minimally demonstrated arguments, and also less than confident in one's capacity to digest its drily bottled intricacies.' Hence Newton's work was laid aside by those contemporaries who tasted 'its supra-sweetness, and its insights were not to be absorbed into the common store of mathematical knowledge for many years.'[26]

Although Newton had made several studies of the cubic curves in earlier years, it is not known why or when he decided to organize his investigation into a taxonomic form, first sketched by him a quarter of a century earlier. Perhaps David Gregory had stimulated Newton to the task in the course of his visit to Cambridge, as Whiteside suggests; if so, he did not see fruit of his suggestion until July 1698. He then vainly begged Newton to be allowed to lay this tract before the public. Gregory, it seems, was little qualified to plumb the depths of Newton's dry text; nor, later, was Roger Cotes, editor of the second Principia (1713), who proposed to edit a large collection of Newton's mathematical papers as vainly as Gregory before. In fact, the work made little impression upon its own time, either at home or upon such continental mathematicians as Leibniz and Johann Bernoulli.[27]

The Calculus Dispute begins

It was in 1699 that the question of priority in the discovery of calculus, as between Newton and Leibniz, first became an open wrangle. The dispute had been latent throughout the last decade of the century, indeed, since Newton had printed the fluxions Scholium in the Principia (1687). After that Newton remained silent on the subject for a number of years, during which his elder Oxford colleague John Wallis served as his shield-bearer. Recall that in 1685 Wallis's English Algebra contained matter extracted from Newton's First Letter to

Leibniz (June 1676); six or seven years later he began to prepare a revised edition (in Latin) of this book, for a three-volume collected edition of his mathematical writings. Already sensitive to Newton's neglect of his own accomplishments in the advancement of mathematics, he sought fresh material from him in order to display them to better advantage in the forthcoming Latin edition. A copy of the *Second Letter* was also in Wallis's files, received from Collins long before, but he could not have used this without Newton's consent. Wallis apparently – for all these letters are lost – requested an elucidation of this *Second Letter*, especially of its 'anagrams'. The upshot of all this was that in August and September 1692 Newton sent Wallis a long text on the method of fluxions (and its use in problems of tangents, maxima and minima and the quadratures of curves), virtually lifted out of *De quadratura curvarum*. This Wallis printed in his Latin *Algebra*, as part of the second volume of his *Opera*, 1693.[28]

At last a terse account of Newton's calculus was before the world, nine years after Leibniz, eight years after Craige. When Leibniz read Wallis's pages after their publication, noting that Newton's calculus agreed with his own and expressing gratification at seeing the sense of the 'anagrams' sent to him eighteen years before, he also expressed disappointment that Newton's process of quadrature led to infinite series, a method known to himself long before. He had hoped for something new.[29]

As for Leibniz's friend and champion, Johann Bernoulli, Wallis's account prompted in him the kind of ill-natured, mischievous reflection that he was often to transmit to Leibniz later: Newton's method differed in nothing from the differential calculus, save that it called a differential a fluxion, a sum [an integral] a fluent. The operations were the same, 'so that I do not know whether or not Newton, having seen your calculus, did not thereupon fabricate his own method, Particularly as I see that you had communicated your calculus to him [in 1677] before he had published his own method.'[30] However unjust and imperceptive Bernoulli's contempt for Newton's mathematical achievements in his letters to Leibniz – of which a great deal more might be quoted – it must be said that he did not wholly blind Leibniz to Newton's genius.

This genius had become known to Leibniz in stages. He had first come across Newton as an experimenter upon light. We may imagine the Huygens's judgement of Newton's success in this field would have become known to Leibniz in the early 1670s. Then, from the

Oldenburg/Collins letters, his visits to London and Newton's *Two Letters* to himself he would have seen what Newton could do in mathematics. Finally, when the *Principia* came to his hands, Leibniz read Newton's work as a natural philosopher and applied mathematician. What he found there astonished and appalled him.

Leibniz's reaction to the *Principia* – again, not different in principle from that of Huygens – was that the book's quantitative analysis of physical forces was 'outstanding' but that the physical principle Newton seemed to posit, gravitational attraction between masses, was philosophically indefensible. With typical intellectual energy Leibniz at once threw himself into the task of composing a wholly different, neo-Cartesian account of orbital forces, in due course published (in accordance with his custom) in the *Acta Eruditorum* of Leipzig; the paper, entitled 'Tentamen de motuum coelestium causis' ('An essay upon the causes of the celestial motions') appeared in February 1689. Leibniz supposd the planets to be borne around the Sun by a 'harmonic' vortex of aether, split (like Ptolemy's spheres) into layers so that each planet might have its proper speed.[31] While revolving the planet was also made to oscillate to and fro upon its radius vector to the Sun by the opposite action of two forces: a centrifugal force, arising from its revolution, and a gravitation towards the Sun, which Leibniz supposed to be effected by a second aether. The oscillation caused the planet to describe a Keplerian ellipse, as Leibniz demonstrated with a good deal of geometrical complexity. Even so, his demonstration yielded only half the correct value of the centrifugal force, as Newton discovered and used to good effect against Leibniz, who only learned of his mistake in 1704 from the French mathematician, Pierre Varignon.[32]

We may see the 'Tentamen' as an attempt to confute the reasoning of the *Principia* with its concept of real, abstract forces, or at least to make it redundant, by developing an aetherial celestial mechanics of equal usefulness. Fair enough: but Leibniz also sought to enhance his independence of, and indeed priority over Newton, by affirming that when he wrote the paper his knowledge of the *Principia* was drawn solely from the review of the book in the *Acta* of the year before, and that he had entertained the ideas now expressed for some years. Only very recently has an able young scholar proved by study of Leibniz's own manuscripts, including notes he made upon the *Principia*, that Leibniz's assertions were untrue.[33] He had carefully studied the book in Vienna, long before he reached Italy, where he wrote the 'Tentamen'.

Newton suspected, from his first knowledge of the 'Tentamen', but this was perhaps as late as 1710, that Leibniz had lied and that its aetherial physics was much more a 'mirror-image' of his own theory than Leibniz wished to admit.[34] He always regarded Leibnizian physics as contorted, redundant and geometrically unsound, opinions applying to Leibniz's treatment of resisted motion and of light, as well as his celestial mechanics.

No doubt Newton felt little distrust of Leibniz when he received, probably by the hand of a visitor, a letter written by him on 7 March 1693.[35] Leibniz had news out of England from Huygens and other friends; he was aware of (though he had not yet seen) Newton's contributions to Wallis's *Opera mathematica*. He paid compliments to Newton on his series investigations and the *Principia*. He asked about Newton's reaction to Huygens's book *Traité de la Lumière* (1690),[36] and assured Newton that he expected some great mathematical triumph from him, suggesting (with a naivety perhaps more apparent than real) that it might lie in clarifying the relations of tangents and rectifications to the quadratures of curves. And finally he remarked, 'above all things I would wish that, being so complete a geometer as you are, you continue (as you have begun) to treat Nature in the mathematical way, in which field you have done work of tremendous value which few can rival.' Newton did not soon respond to this overture,[37] but when he did it was in the softest terms. He valued very highly his friendship with Leibniz, "one of the leading geometers of this century", and begged his censure if he deserved it on any point, since "I value friends more than mathematical discoveries." He gave Leibniz a general solution in fluxional terms of the rectification problem, called Huygens's comments upon his own work "most ingenious" and admitted that the heavens might contain an excessively subtle matter; nevertheless firmly rejecting a mechanical, vortical aether of the Cartesian type: "since Nature is very simple, I have judged that all other causes than [gravitation as described by myself] should be rejected and that the heavens should be deprived of all matter so far as may be, lest the motions of the planets and comets should be impeded or rendered irregular." However, a mechanical explanation of gravity not founded upon a dense material aether would be welcome to him.[38]

That was the end of this brief exchange, but Fatio de Duillier, responding to an indirect message from Leibniz, confirmed in late March 1694 Newton's resistance to the various objections raised by Huygens in his *Traité* and annexed *Discours*. Newton was still (Fatio wrote) undecided whether gravitation was 'inherent in matter due to a direct Law of the Creator' or was the effect of some 'mechanical cause such as I have discovered'. Incidentally, in this letter Fatio still sees himself as Newton's prime spokesman, giving the opinions of 'Mr Newton and me', and hints at nothing of that abrupt breach between the two men that some scholars attribute to 1693.[39]

While Newton avoided letter-writing (as he confided to Leibniz) and was unforthcoming, Leibniz wrote to the Royal Society in 1694 urging that Newton be pressed to print his improved second edition of the *Principia* (on which all Europe knew him to be engaged) and three years later he addressed John Wallis to the same effect, also expressing some eagerness to understand exactly the likenesses and differences between their two methods of 'infinitesimal analysis'.[40]

He could not have sought a warmer ally. Wallis had turned again to Newtonian fluxions, circulating on the Continent 'with great applause, by the name of *Leibniz's Calculus Differentialis*' in his first volume (published second, 1695) of collected works in Latin, and would do so again in the third (1699), where he printed the 1676 letters to Leibniz in full. His appeals to Newton to defend his own reputation and that of the English nation fell upon deaf ears, though they may have moved Newton's spirit against Leibniz (unreasonably) as a man who was profiting from Newton's reticence. He would release neither the *Principia* nor the *Opticks* nor any of his mathematical treatises, invoking (after 1696) the excuse that all his time must go to the King's business.

Old Wallis had some inkling of the truth, that the Leibnizian school of mathematicians was moving swiftly ahead of the Newtonians, who scarcely constituted a school, and to which recruits could hardly be drawn. The Bernoulli brothers were not only able mathematicians themselves but great proselytizers of Leibniz's calculus, which they and their pupils extended rapidly within the science of mechanics. Mathematical physics, expressed by Newton in 'classical' geometry, was by 1720 transmuted and developed in Leibnizian language – except in Britain. The Bernoullis trained distinguished students from Switzerland and Germany, one of whom (Jakob Hermann) carried Leibniz's influence to Padua; Johann Bernoulli naturalized it in France. He had great success with young men associated there with the neo-Cartesian philosopher Nicole Malebranche, who were (like their master) interested in both pure and applied mathematics. One of these, Pierre Varignon, later developed a direct link with Newton also. So too did Charles Reyneau, but the most conspicuous in this group at first was the Marquis de L'Hôpital, who engaged Johann Bernoulli to be his master in calculus and even bought the right to publish Bernoulli's lessons under his own name, as *Analyse des infiniment petits* (1698), the first textbook on calculus.

Justifiably, the *Principia* brought Newton a great reputation as a brilliant mathematician handling the problem of astronomy and physics with inventiveness and rigour, just as his optical papers in the *Philosophical Transactions*, nearly twenty years before, had brought him fame as an experimenter. But in both optics and applied mathematics Newton's theories, his ideas of broad compass, had failed to win widespread support outside Britain. His anti-Cartesianism inevitably made him appear a bizarre, unsound thinker to those, that is the majority, who remained attached to the principles of simple Cartesian materialism. By men like Huygens, Leibniz and Malebranche Newton was admired and respected, but not imitated. Only very slowly in the new century did opinion on the Continent begin to allow that there might be merit in Newton's revolutionary principles of physical thinking, as well as in his mathematical and experimental methodology.

The view taken by European *savants* of Newton as a natural philosopher was intimately entwined with their view of him as the inventor (or plagiarist) of the calculus. As his stock rose on the one account so it rose also on the other, and Leibniz's fell correspondingly on both counts. Moreover, he had the misfortune to die eleven years before Newton. In 1699, when the Académie Royale des Sciences of Paris was reorganized by Louis XIV, Newton was elected a Foreign Associate only after Leibniz, and was indeed last on the list. Twenty years later he was unquestionably the intellectual giant of all Europe, and scarcely anyone now whispered that perhaps he had taken the calculus from Leibniz.

Fatio accuses, Bernoulli challenges

The calculus question came into the open in 1699, when Fatio de Duillier printed the accusation of near-plagiarism which (in the opposite direction) Johann Bernoulli had only hinted in a private letter. There can be little doubt that he did so partly out of personal pique, partly out of a genuine (if slightly absurd) devotion to Newton's interests, and partly (once again) to prove himself the best of Newton's friends. The odds are that both Newton and Leibniz would have let the question of priority in discovery slumber quietly if their 'best friends' would have permitted it. Of course Fatio had learnt from Newton and his files of the 1666 and 1671 tracts, the correspondence with Leibniz of 1676–7, and perhaps heard Newton opine that Leibniz had been unhandsome in forgetting Collins and his study of unpublished British mathematics in those far-off years. But it is inconceivable that Newton set Fatio to attacking Leibniz. The immediate occasion of Fatio's onslaught was the challenge problem (the brachistochrone) – or rather pair of problems, for there was a second annexed to this – put before the finest mathematicians of the world by Johann Bernoulli late in 1696 (privately) and in January 1697 by a printed fly-sheet. On the Continent, besides the problem-setter himself, only Leibniz and Jakob Bernoulli solved the problems. In England John Wallis (now eighty years old) and David Gregory were defeated by them. So were all the French mathematicians. The problems came to Newton's hands on 29 January 1697, as he arrived home about 4 p.m. after a hard day of Mint business at the Tower. He did not sleep till it was mastered, some twelve hours later. His anonymous answers, without demonstrations, were rapidly inserted in the *Philosophical Transactions*. Bernoulli recorded that as soon as he read them he recognized the author as Newton, 'as the lion by his claw'.[41]

The likelihood is that both Bernoulli and Leibniz – certainly the latter – had expected a solution from Newton. The challenge had not been intended to 'expose' him, though Leibniz incautiously remarked that only those who were 'masters of our calculus' could succeed with it. Fatio, taking this to be an insult to Newton and himself (for he believed that he too could resolve the problem) in 1699 published 'A Double Geometrical Investigation into the Line of Quickest Descent'. Into this pamphlet he introduced an assertion of his own independence as an innvator in infinitesimal analysis, continuing

> Yet I recognize that Newton was the first and by many years the most senior inventor of the calculus . . . as to whether Leibniz, the second inventor, borrowed anything from him, I prefer to let those judge who have seen Newton's letters and other manuscript papers, not myself. Neither the silence of the more modest Newton nor the eager zeal of Leibniz in ubiquitously attributing the invention of this calculus to himself will impose on any who have perused those documents which I have myself examined.[42]

This was clear and brutal, proclaiming that Leibniz had no just claim upon the fame that he had arrogated to himself.

Though the 'Investigation' did nothing to enhance Fatio's reputation as a mathematician, neither (despite Leibniz's real anger) did it provoke immediate hostilities. Leibniz wrote that learned men should not quarrel like fishwives; he was very busy grandly bringing his scheme for a Berlin Academy to fruition and assuming office as its first president, while Johann Bernoulli was serving his turn as Rector Magnificus of the University of Groningen. In a public reply to Fatio (May 1700) Leibniz took the line that Newton would not countenance such rubbish as Fatio's, that only Newton and himself were original masters of the calculus, and that the test of this was success in solving problems of maxima and minima, such as Newton had already demonstrated in 1687. The piece, by its generosity to Newton – short of conceding him an absolute priority – tended to make Fatio look like a fool and a trouble-maker. Nevertheless, Leibniz was made more sensitive in his distrust of the British.[43]

Again the potential quarrel had been smoothed over. Peace was a little disturbed by the publication in 1703 of a book by George Cheyne, a Scots physician and mathematician recently settled in London. (The combination was not unusual among the Newtonians; other examples of it are Archibald Pitcairne and Henry Pemberton.) Cheyne had already published a little book giving a 'Newtonian' explanation of fevers. Newton and his friends disliked Cheyne, about whom David Gregory recorded every spiteful tale in his journal. Hence his next book on *The Inverse Method of Fluxions* was not well received by them though Johann Bernoulli (while recognizing its many mistakes) called it 'a most remarkable little book stuffed with clever discoveries'. Yet it was a book filled with praise of Newton and studied silence about Leibnizian mathematics, even while (it seems) making use of its results. Cheyne even wrote that everything published about methods of quadrature during the previous twenty-four years 'relating to these methods [of Newton], *or to other not dissimilar methods*, is only a repetition or an easy corollary of what Newton long ago communicated to his friends or the public' (my emphasis). The implication as regards Leibniz and his followers is obvious, for Cheyne's oddly chosen starting-date (1680 at latest) was well before Leibniz's first paper on the differential method.[44]

Bernoulli retorted that this claim made all others 'Newton's apes, uselessly retracing his steps of long before' and Leibniz denied that the new series method he had published in 1693 was previously known to Newton; more, if Newton had been at times first in mathematical discovery, so had Leibniz also: 'Certainly, I have found no indication that the differential calculus, *or an equivalent to it* was known to him before it was known to me' (my emphasis). This was a flat assertion of his own priority in knowledge as well as in publication.[45]

Opticks and its Appendages

According to Gregory's journal, Newton now began to perceive that publication by such friends as Cheyne might cause him more trouble than bringing out his own writings might do, so he published the two mathematical treatises, already discussed, that accompanied *Opticks* in 1704. Fortuitously, Hooke's death in the previous year had cleared the impediment obstructing its printing for so many years. There was no triumph in this first belated appearance as a pure mathematician (which, as Newton had feared, brought further embarrassments to him). Fourteen or more years of intense work by Leibniz and his followers had brought them level with, or even ahead of, Newton's concealed achievements of long before. In Whiteside's words: 'Newton's historical importance as author of the "De Quadratura Curvarum" is the minimal one of a lone genius who was able, somewhat uselessly in the long view, to duplicate the combined expertise and output of his contemporaries in the field of calculus.'[46] His fluxional calculus was therefore to be without long-term effect upon the history of mathematics, though his achievement in this as in other fields was long an inspiration to others. And before *De Quadratura* was published the modern science of mechanics had been set in a Leibnizian analytical mould by the continental mathematicians. Not surprisingly, Leibniz and Bernoulli, besides judging the *Enumeratio* remarkable but dull, found the *De Quadratura* empty of novelty. The former's long, anonymous review of *Opticks* in the *Acta Eruditorum*, however, gave Newton mortal offence. Into the printed text of *De Quadratura* he had introduced prefatory sentences on the origin of his methods: how, by "naming these velocities of the motions or increments [of quantities] *fluxions*, and by naming the generated quantities *fluents*, I gradually in the years 1665 and 1666 hit upon the method of fluxions, which I have here employed in the quadrature of curves." To this roughly justifiable statement Leibniz responded with a flat negative in his review:

> instead of the Leibnizian differences Mr Newton employs, and always has employed, *fluxions, which are almost the same as the increments of the fluents generated in the least equal portions of time*. He has made elegant use

of these both in his *Principia mathematica* and in other publications since, just as Honoré Fabri in his *Synopsis Geometrica* substituted the advance of movements for the method of Cavalieri [italics in original].[47]

The intrinsically implausible implication of the earlier part of this passage – that after mastering Leibniz's 1684 paper Newton had devised the method of fluxions supposedly used to prepare the *Principia* published three years later – would hardly have been strengthened had Leibniz mentioned his 1677 letter to Newton, which he did not; the real sting of the passage is in the carefully worked allusion to the two Italian mathematicians. Fabri had near-plagiarized his method from the earlier, highly innovative work of Cavalieri. There was some analogy, too, between Fabri's method and Newton's. It is possible that Leibniz's rhetoric was more wounding than he had intended. His review was generally approving of Newton's work, and if he referred readers to Craige and Cheyne for further light upon it, can he be blamed in view of Newton's long silence? When another mathematical text of Newton's came out – *Arithmetica universalis* in Whiston's edition (1707) – Leibniz wrote in the *Acta Eruditorum* that things were to be found in this little book missing from large tomes on algebraic analysis.

Again, there was more psychological pressure, upon Newton this time, but no explosion. For this, it seems, a new voice was required. It came from John Keill, another Edinburgh graduate and pupil of David Gregory, who had followed his master to Oxford after 1694. Keill had there given courses of lectures on physics and astronomy illustrated by experimental demonstrations, thus founding a tradition of the ocular teaching of 'Newtonianism' that throve in the next century. Nothing indicates that this man – a generation younger than Newton – was ever his close friend. But he was an ardent and combative partisan, with whom Newton later co-operated closely in his onslaughts upon Leibniz. Like Fatio eleven years before, Keill introduced into a paper on mechanics in the *Philosophical Transactions* quite irrelevant aspersions upon Leibniz.[48] His results followed, Keill wrote, 'from the nowadays highly celebrated arithmetic of fluxions, which Mr Newton beyond any shadow of doubt first discovered, as anyone reading his letters published by Wallis will readily ascertain, and yet the same arithmetic was afterwards published by Mr Leibniz in the *Acta Eruditorum* having changed the name and the symbolism.'

Before considering Leibniz's response to this ruthless assault, we must recognize that his opinion of Newton's concept of immaterial force had become more hostile during the first decade of the new century – that is, he now began to feel that the foolish 'English philosophy' should be refuted in the *Acta*. Its reviewers now attacked John Freind's *Chymical Lectures* (1709) and George Cheyne's *Philosophical Principles of Natural Religion* (1705) for teaching Newtonian ideas of chemical and gravitational attraction, and so reverting to the 'occult qualities, such as sympathy and antipathy were in the schools of philosophy'. To the Leibnizians, Newtonian force mechanics represented a reversion to the empty philosophy prevailing before Bacon, Galileo and Descartes transformed it. In 1710 this point at issue between them and the English became confused with the quite distinct issue of the discovery of the calculus. Accordingly, when Leibniz protested to the Secretary of the Royal Society (of which he had been a Fellow since 1673) against Keill's offensive statements, and Hans Sloane consulted the Society's President, Sir Isaac Newton, about what answer to make, and Newton in turn sought an explanation from Keill, the unrepentant author was able to show Newton how his philosophy and his originality had both been attacked in the *Acta*. Newton now began to believe that he had better grounds for complaint against the anonymous *Acta* writers than Leibniz had against Keill.[49]

Newton, his champion and the Royal Society were now all agreed that Keill should formally present on paper his case against Leibniz, which he did by setting out the evidence for Newton's mastery of calculus years before 1684, going back indeed to Newton's letter about his advances in mathematics written to Collins on 10 December 1672 and to *De Analysi* in 1669. (These materials, like all Collins's papers, were by this time in the hands of another mathematician, William Jones, who was soon to publish a collection of some of Newton's early mathematical writings.)[50] Suddenly the matters in dispute had acquired a new historical dimension, bearing on Newton's early discoveries.

Keill's letter of justification was sent to Leibniz at the end of May 1711, with a curt note from Sloane, having been approved by the Royal Society on the 24th.[51] A second demand for an apology from Leibniz, reaching London in January 1712, failed to mollify Newton and his friends; though dignified, it made no admission of Newton's fundamental priority in the ideas of calculus. At this point the President himself emerged from behind the scenes, pointing out in a speech to the Royal Society in February or March 1712 that he was not the first aggressor though he was the "first author" of the calculus.[52] He refrained from the suggestion that Leibniz had taken the differential calculus from himself. He, or a friend, proposed that the Society create a committee to report on the dispute; it ultimately consisted of eleven members, mostly Newtonian mathematicians (inevitably), but including William Burnet, a friend to Leibniz.

The Commercium Epistolicum, *1712*

This committee, appointed on 6 March, reported on 24 April. Such speed in assembling and interpreting a considerable mass of documents was only possible because Newton did (or had already done) the committee's work, and further hastened its labour by drafting its report in all essentials, including the decisive judgement, "For which reasons we reckon Mr Newton the first inventor". Someone else continued the sentence with '. . . and are of opinion that Mr Keill in asserting the same has been noways injurious to Mr Leibnitz.' The Royal Society, accepting the report *nem. con.*, appointed Halley, Jones and John Machin to be editors of its publication with full documentation. Again, Newton saved their trouble. The *Commercium Epistolicum* (or, translated in full, 'The Correspondence of Mr John Collins and others concerning the promotion of Analysis') printed at the end of that year had occupied Newton throughout the summer of 1712.[53]

Even allowing for the different practices of a different century it seems odd that a group of respectable men, including the Ambassador of Brandenburg-Prussia, should have put their names to a report speaking so strongly for one party to a dispute, in whose writing that same party had had (to say the least) a preponderant role. Leibniz, the other party, knew nothing of what was going on. And yet Newton once retorted to Leibniz that no man should be a witness in his own cause! It is true that the documents printed in the *Commercium Epistolicum* were genuine and accurately quoted; but the documents were not allowed to speak for themselves and Leibniz was invited to submit none. Newton was a dangerous man to offend. He would not have subscribed to Whiteside's calm conclusion that, in the long perspective of subsequent history, 'the priority in time of creation of his fluxional method which Newton indubitably had must seem of minimal significance.'[54]

Though the *Commercium Epistolicum* was not printed in a large edition, Newton saw to it that its 250 or so copies were freely distributed over Europe, not forgetting the Italian mathematicians. In the nineteenth century, when the manner of its composition was first discovered, indignation was voiced against its 'atrocious unfairness' to Leibniz. In its omission of any early evidence from Leibniz's side – which, indeed, he never made available – the tract was unfair; but it was true that Leibniz had seen Gregory's and Newton's unpublished material in 1676 (less early than Newton imagined). It was unfair also in inferring that because Leibniz had pretty full knowledge of Newton's methods of quadrature, he was therefore equally familiar with the method of fluxions. With justice, Leibniz retorted that he had never heard the name before Wallis's account appeared. In fact, he had no access to the 1666 tract in 1676, nor would he have found the name 'fluxion' in it. *Commercium Epistolicum* was again unfair in supposing that Leibniz's delay in answering Newton's *Second Letter* was caused by his concocting the differential method; and Newton's contention (often to be repeated) that fluxions were conceptually superior to differentials in being rigorously geometric is at least dubious.[55]

The *Commercium Epistolicum* in the eyes of the Leibnizians marked the British as enemies of truth and progressive mathematics. Everything coming from their pens must be suspect. The decade after 1712 was thus that of their harshest opposition to all Newtonian influence, causing Newton himself considerable trouble and distress. Meanwhile, French and Italian mathematicians began to find good in Newton's writings, largely due to the effect of *Opticks* (1704). To this we shall turn in the next chapter.

Flamsteed and the Moon

Meanwhile, something must be added here about the other great quarrel in Newton's later life. As we have seen, Newton had appealed to the Astronomer (or *Mathematicus*, as he preferred) Royal, John Flamsteed, for data on the planetary orbits in the course of writing the *Principia*. Thereafter for some years there is little evidence of their relationship, which was surely closer than this silence might indicate. A rare surviving letter shows Newton introducing David Gregory to Flamsteed, after failing to bring him to Greenwich in person (10 August 1691). Newton indicated continuing interest in the motions of Jupiter and Saturn, and further study of optics.[56]

The reply to this letter from his 'affectionate friend and Brother' already lays bare the differences of policy and temperament that were to divide the uneasy allies over the years.[57] Flamsteed was a perfectionist, slowly accumulating positional observations that would be reduced and published in his own good time. 'Would it be wisely done of me to cease my designed Observations of the Constellations that yet remaine to be taken, ' he asked rhetorically, in order to print those he already had, 'to gaine a little present reputation?' But Newton wanted data now, not in twenty years' time. Flamsteed loathed Halley and despised his catalogue of the southern stars; soon he would feel much the same about Newton's other friend, Gregory. In fact Flamsteed was an even greater paranoid than Newton. Another difficulty that arose was that Flamsteed considered his observations made at the Royal Observatory to be his private property, because he had bought instruments and hired assistants out of his own pocket.

In 1694, as part of the programme for improvement of the *Principia*, Newton took up the gravitational mechanics of the Moon's motions (despite his illness of the year before). Seemingly, he had already made some progress before paying a visit to Flamsteed at Greenwich, along with Gregory, on 1 September 1694. He sought a larger stock of more precise lunar observations. According to his own story, Flamsteed allowed Newton to copy some of these under the curious restriction that any theory derived from them should be imparted to Flamsteed alone. A contemporary letter displays greater cordiality to Newton.[58]

Returning from what may have been a longish visit to London, Newton began to press his requirements upon Flamsteed, assured now that it should be possible to predict the Moon's position in the sky to within two or three minutes of arc. He needed positions just before and just after the Moon's quadratures, and begged again for positions of the outer planets too. He asked Flamsteed to be sure to measure the atmospheric temperature and pressure, with a view to correction for atmospheric refraction. When Flamsteed sent more positions Newton was sharp about some errors in transcription – and would not glance at some magnetic research by Flamsteed's friend Caswell.[59] So it went on. Flamsteed was annoyed and jealous because Newton discussed his investigation with Halley, Newton was irritated with Flamsteed for meddling with theoretical questions when all Newton wanted was data. Time and again he urged Flamsteed simply to supply him with specified raw materials, while the astronomer desired to be treated as an equal partner in a mathematical-philosophical voyage of discovery. When Newton sent him a table of refractions he could not rest till he had (as he thought) fathomed its basis. While Flamsteed thought in terms of fidding with the numbers, Newton was after a generall notion of all the equations on which [the Moon's] motions depend & considering how . . . to determin them. For the vulgar way of approaching by degrees is bungling & tedious . . . Sir – [he under-lined this sentence] – if you can but have a little patience with me till I have satisfied my self about these things & made the Theory fit to be communicated without danger of error I do intend that you shall be the first man to whom I will communicate it.[60]

It was Flamsteed's turn to ask Newton to be patient when, after the latter had particularly requested positions taken during the next lunation, he pleaded the necessity of performing his parish duties. He made a great fuss because he supposed Newton to be offering him money for his work, and all the while of course Flamsteed was by no means content to be pinned down to his lunar task. Not that Newton allowed him to forget the planets: "I intend to determin the Orb of Saturn within a few days & then I'le send you the result."[61]

Newton did manage to calculate a table of atmospheric refractions (which he allowed Halley to publish), and sent further tables to Flamsteed for the computation of lunar parallax, latitude and longi-tude. By this time (April 1695) he had decided that there was too much random uncertainty in Flamsteed's earlier lunar positions to permit very precise orbital calculations, and therefore set himself to await better ones. "I reccon it will prove a work of about three or four months" to establish the lunar theory "& when I have done it once I would have done with it forever." He made an interesting remark to Flamsteed about the intensity of his working habit: "When I set myself wholly to calculations . . . I can endure them & go through them well enough. But when I am about other things, (as at present) I can neither fix to them with patience nor do them without errors. Which makes me let the Moons Theory alone at present."[62]

During the winter and spring both Newton and Flamsteed, not to say Pepys, were concerned in the affairs of the Christ's Hospital mathematical school, where Edward Paget was replaced by Samuel Newton (no relation), a practical rather than a learned teacher, to Flamsteed's disgust. Isaac Newton thought his namesake the best man available. He also took the practical line that Latin was of little use to boys intended for the sea, and that the time allotted to mathematical teaching should be increased.[63]

In the summer of 1695 relations between Newton and Flamsteed virtually collapsed, despite the latter's expressions of willingness to continue their 'collaboration'. At the end of June Newton wrote curtly, even rudely, that Flamsteed should provide raw lunar observations for 1692, "or else let me know plainly that I must be content to lose all the time & pains I have hitherto taken about the Moons Theory & about the Table of Refractions". Flamsteed protested that illness (bladder-stone) had hampered his work; 'Let the world Judg whether Mr Newton had any cause to complaine of want of observations.' He was deeply hurt by tales running about London (Halley's fault, of course!) that the new *Principia* was obstructed by want of observations and showered more data upon Newton in July and August. His only reward was a bitter sentence from Newton (9 July 1695):

> After I had helped you where you stuck in the three great works, that of the Theory of Jupiters Satellites, that of your Catalogue of the fixt stars & that of calculating the Moons places from Observations, & in all these things freely communicated to you what was perfect in it's kinds (so far as I could make it) & of more value than many Observations & what (in one of them) cost me above two months hard labour which I should never have undertaken but upon your account, & which I told you I undertook that I might have something to return you for the Observations you then gave me hopes of, & yet when I had done saw no prospect of obteining them or of getting your Synopses [of observations] rectified, I despaired of compassing the Moons Theory, & had thoughts of giving it over as a thing impracticable.

Newton did, however, thank Flamsteed for the offer of further observations.[64] Whether, as half-promised, he went back to the lunar theory at this time seems doubtful, for (as he told Flamsteed on 14 September) a great new astronomical interest had presented itself. Then he wrote to him no more for three years. He was not softened by Flamsteed's avowal that 'the agreement of your [theory of comets] with them [observations] demonstrates its truth & confirmes the Theory of Gravity at the same time' – of which, it must be confessed, Flamsteed had once been doubtful.[65]

In fact, the Moon emerged victorious from Newton's long siege, hence the frustration vented upon Flamsteed, a man who might have tried the patience of an archangel. Years later Newton told John Machin that it was the only problem that had made his head ache. He had been unable to contrive a logical dynamical theory of the Moon in the first *Principia*, and though in the 1690s (influenced by Flamsteed) he modified his lunar ellipse to the form devised by Jeremiah Horrocks about 1640, he still could not closely approximate to the complex phenomena without arbitrary adjustments. Newton in the final version of his lunar theory – in which the precise numbers altered from time to time – quantified the fixed parameters of the solar and lunar motions and showed how annual equations varying the latter were to be computed from the Sun's annual equation arising from "the excentricity of his Orb" (Newton still using the old expressions). Altogether seven inequalities had to be reckoned with, some changing with the reciprocal of the cube of the Earth's distance from the Sun, all based on an arbitrary Horrocksian model in which the centre of the Moon's elliptical orbit (constant in the length of its diameter, varying in its eccentricity) rode an epicycle upon a deferent circle whose centre is the Earth. Clearly this was far from being a dynamical theory of the Moon, yet difficulties remained: in particular Newton could obtain only about half the observed speed of rotation of the lunar apsides. He told Flamsteed that the table sent him in April 1695 "never errs above 10 or 12 minutes, & so is twice as exact as your printed Tables" (which were also based upon Horrocks's model).[66] Despite a great deal of computation and fine-tuning of parameters, this was about the limit of Newton's success, still very far from that practical solution of the problem of the longitude which Hooke had proposed to Newton so long before. The final non-dynamical scheme for the Moon's motion was complete by 1699 – just when is uncertain – since in February 1700 Newton's paper on the theory of the Moon was copied by David Gregory, to be printed in his *Elements of Physical and Geometrical Astronomy* in 1702 (in Latin translation). The same scheme, with modified numbers, was to appear in the second edition of the *Principia* (1713).[67]

Newton had dismissed Flamsteed with the words: "Mr Halley was with me about a designe of determining the Orbs of some Comets for me." To Halley's success in fixing the orbit of the comet of 1683 by Newton's theory, and confirming this orbit precisely by observations, I shall return later; meanwhile, the tragic story of the irreconcilable hostility between Newton and Flamsteed may be taken from the moment in January 1699 when Newton wrote a very angry letter to Flamsteed, because the astronomer had mentioned Newton's lunar researches in a letter addressed to John Wallis and intended for the press.[68] Newton did not wish to be thus "publickly brought upon the stage" so as to cause the world to expect from him what it might never receive:

> I do not love to be printed upon every occasion much less to be dunned & teezed by foreigners about Mathematical things or to be thought by our own people to be trifling away my time about them when I should be about the Kings business . . . You may let the world know if you

please how well you are stored with observations of all sorts & what calculations you have made towards rectifying the Theories of the heavenly motions: But there may be cases wherein your friends should not be published without their leave.

There could hardly be a clearer example of Newton's paranoiac fury. Naturally, Flamsteed was deeply offended and moralized in his reply on the forgiveness of pride in others, more to the point noting to himself: 'Was Mr Newton a trifler when he read Mathematicks for a sallery at Cambridge. Surely the Astronomy is of some good use tho his place [as Warden of the Mint] be more beneficiall.' He was also badly upset by tales of Halley going about boasting that *he* had supplied the data upon which Newton had founded a near-perfect theory of the Moon.[69]

In the summer of 1700 Flamsteed wrote to Newton a formal letter (never printed and perhaps never sent) about his observations of the Earth's 'parallax' (really a phenomenon not yet discovered, the aberration of light) and the dreadful discovery of the slow subsidence of the wall carrying his mural quadrant, causing an increasing error in his measure of star positions (and so of all other observations referred to these). A calculated correction was required to all his data dependent upon the mural quadrant. Presumably this letter was addressed to Newton out of a sense of duty, for Flamsteed had written to a friend not long before that Newton was a man not to be trusted, prejudiced against himself and a betrayer of confidences.[70]

Flamsteed's Great Book

Meanwhile, the notion that Flamsteed was 'so perversely wicked that he will neither publish nor communicat his observations' seems to have been spread abroad and was already shared by both Newton and Halley.[71] By 1704, however, the astronomer himself claimed that the first part of his *Historia Coelestis Britannica* was ready for the printer, and the second and third could be made ready while the first was being printed. This was gross over-optimism. The whole work, Flamsteed estimated, would be of some 1500 pages with accurate and large maps of the constellations. He meant to print not only the ancient star lists of Ptolemy and others, but observations made by the 'northern astronomers', William Crabtree and William Gascoigne, whom he revered as his immediate predecessors. Queen Anne's consort, Prince George of Denmark, Lord High Admiral, undertook to pay for the printing. Court circles appointed Newton, Wren and other Fellows of the Royal Society a committee to pass Flamsteed's text for printing, which within a month was done for the first two parts and the two star catalogues, despite errors introduced by Flamsteed's copyists, who were still busy transcribing the material that he had said was prepared for the printer.[72] Flamsteed's (short) distance from London, his ill-health, his long visits to his living in Sussex, all delayed progress, but an agreement between the committee of referees, Flamsteed and the printer (Awnsham Churchill) was at last signed on 17 November 1705. As Flamsteed seems to have signed the agreement with unspoken reservations about the practicality of fulfilling some of its terms, future trouble was certain. Flamsteed was a man incapable of sticking to businesslike rules, Newton was eminently a man of business. Dissensions about 'errors' in the hundred pages of copy submitted by Flamsteed by the end of 1705 led the astronomer to claim that Newton was holding back the Prince's remuneration for his own trouble and expense. Newton's 'temper wants to be cried up and flattered,' he wrote, but 'I have allways hated all such low practices.' Nevertheless, he ate Newton's dinners. Aided by further visits of Newton to Greenwich – probably his closest approach to the sea – Churchill did manage to print off some pages of proof copy, and by April 1708 the referees had paid Flamsteed £125 of the Prince's money. The whole of volume I was with the printer, and it was agreed that he should at once deliver the copy for volume II and the star catalogues. But he did not.[73]

Newton for the time was quiet, leaving Flamsteed in possession of the field, but he had in his hands a useful weapon, a partial star catalogue deposited by Flamsteed in 1706 (it lacked six of the northern constellations, and much else). At last, in the late winter of 1711, with the Prince long dead, he moved through friends at Court to put pressure on Flamsteed. The Queen's physician, Dr John Arbuthnot (a poetic wit, a friend of Swift and of Newton, something of a mathematician) begged Flamsteed to finish his book at the Queen's command, particularly by supplying a perfect catalogue of the fixed stars. Finding Flamsteed less than co-operative with Arbuthnot, Newton now secretly placed Halley in charge of the second volume of *Historia Coelestis [Britannica]*, in order to correct and complete Flamsteed's imperfect materials already in Newton's hands. By April 1711 some proof pages of Halley's volume had been given to Flamsteed by Arbuthnot, to the astronomer's rage and contempt: 'Make my case your own,' he wrote, '& tell me Ingeniously [ingenuously], & sincerely were you in my circumstances, and had been at all my labour, charge & trouble, would you like to have your Labours surreptitiously forced out by your hands, convey'd into the hands of your declared profligate Enemys, printed without your consent, and spoyled as mine are in the impression?'[74]

He withdrew altogether, talking of making his own edition. But Newton was far from outdone. Some time before, at the end of the year 1710, he had secured an unprecedented royal warrant appointing the President and a committee of Fellows of the Royal Society as Visitors to the Royal Greenwich Observatory with powers to require from the Astronomer Royal a yearly record of his observations and the making of specified observations from time to time, as well as the right to inspect, repair or replace the instruments there. Newton had thus at a stroke transformed the Astronomer Royal from being an independent gentleman following his own whims into a civil servant supervised by the Royal Society. It was a move any modern administrator might envy, carried out with quiet simplicity, logical, and leaving Flamsteed no choice but to resign or comply. Seemingly, these powers were first exercised on 30 May 1711 when the astronomer was ordered to observe expected eclipses of the Sun and Moon in that year, especially the solar eclipse of 4 July.[75] In subsequent years he was regularly reminded to submit his records.

By mid-1711 Halley's *Historia Coelestis*, the companion volume to that already published, was finished enough for Halley to send a proof copy to Flamsteed, with a civil letter regretting the lapse of their former friendship and offering (with tongue in cheek?) to have mistakes amended. Flamsteed, finding 'more faults in, and greater, than I imagined the impudent editor either could, or durst have committed', did nothing. The book appeared in 1712. After changes at Court in 1714, Flamsteed's friends were able to acquire for him the 300 remaining undistributed copies (the edition being, presumably, 400 again) which he sacrificed to the truth by fire. Six years after his death in 1719, Flamsteed's wife and devoted assistants published the three volumes of *Historia Coelestis Britannica* which he had by then partially prepared.[76]

The completion of the publication begun so long before was by no means the end of the matter. Newton determined to examine the state of the Greenwich instruments, sending for Flamsteed to meet a Royal Society committee on 26 October 1711 (the committee, besides Newton, consisted of the two physicians, Hans Sloane and Richard Mead: perhaps others prudently absented themselves). According to Flamsteed's story, the President was much deflated on being told that the instruments were either given to Flamsteed personally or were bought by him. The Crown had contributed none. Flamsteed kept his temper, but Newton (so he said) flared up on being told that he had robbed Flamsteed of the fruits of his labours, 'cald me all the ill names Puppy etc that he could think of'. Clearly Flamsteed's sanctimony was more irritating than his anger. After this unseemly row he came across Halley, accepted a dish of coffee from him, '& told him still calmly of the Villainy of his conduct'![77]

Apart from the fireworks sparking from a clash of temperaments and methodologies, what may be said of the consequences of this long quarrel? Newton started from acceptance of Flamsteed as the finest observer of the time, deferring to him on all astronomical matters until his patience broke. Flamsteed, a man with tremendous confidence in himself beneath a misleading diffidence of expression, had never accepted Newton as the finest natural philsopher of the time. The astronomer entered their relationship, therefore, with the feeling that he did not need Newton's lunar theory and that the world could well wait until his own (founded upon unimpugnable observations) was ready for it. Just as Newton had objected to being Hooke's mathematical drudge, so Flamsteed resented being Newton's observational drudge but, since he could hardly refuse co-operation openly, he played a canny part, readily furnishing observations that Newton had not asked for while passing over those that he wanted. This may well not have been conscious malice. That, in consequence, Flamsteed destroyed Newton's chances of working out a complete and fully satisfactory theory of the Moon in 1694 (as older writers claimed) is doubtful; possibly Newton was shorter of new ideas and techniques than he was of raw material.[78] But only a very close and expert examination of the computations through which Newton went could decide the point. For Newton himself, however, it was clear that Flamsteed must be the scapegoat for his failure: but for his laziness, disobligingness, incompetence – wherever the weakness might lie – the lunar theory might, in Newton's eyes, have made in his hands the decisive advance that for ever eluded him. Thus, when it came to printing Flamsteed's observations the two men started with intense prejudice against each other: Flamsteed was intent upon not being cheated of the due acknowledgement of his observational skill and devotion (as he felt he had been cheated by Newton with respect to the lunar theory). Newton doubted Flamsteed's goodwill and competence. One could not spend Crown funds without great prudence! Once self-assured that it was his duty to browbeat Flamsteed, to prise his observations out of him by force and cunning, Newton lost the finer scruples of ordinary conduct. As the Mint procured the punishment of coiners of false money, so the Royal Society must correct falsity or incompetence in science. Newton's high moral crusade was, inevitably, in vain. Halley's, as some now say 'pirated', edition of the observations was never of any significance. Quite possibly Flamsteed would have published his great record of observations – among them, obviously unknown to himself, the first record of Neptune, the second major addition to the solar system as described by astronomers – within his lifetime if Newton had left him alone. In this dispute both men were unwitting enemies to the advancement of science.

11

Opticks, or a Treatise of Light, 1687–1704

On the first of March, 1704, David Gregory noted in his diary that Newton had been 'provoked by Dr Cheyns book to publish his Quadratures, and with it, his Light & Colours, &c.' And just one month later Newton indeed signed and dated his "Advertisement" to this volume.[1] Some sixteen months had passed since he had promised 'Mr [Francis] Robarts, Mr Fatio, Capt. Hally & me to publish his Quadratures, his Treatise of Light, and his treatise of the curves of the 2^d Genre [*sic*]'. In the interval Robert Hooke had died, thus clearing the way for the publication of *Opticks*, a book which Newton had sworn to keep to himself so long as Hooke lived.[2] *Opticks*, like the pair of mathematical treatises that was to appear with it, had been long anticipated by Newton's friends. In 1694 Gregory had examined its three Books – hence it was substantially complete then, though Newton was not yet 'fully satisfied about a certain kind of colours and the way of producing it' – and summarized it in his diary of his visit to Cambridge (5 to 7 May). Newton meant to publish it after leaving the university, in English as it was written, or translated into Latin if he remained at Cambridge.[3] By April 1695 John Wallis at Oxford knew of it (through Flamsteed's protégé Caswell) as 'a Treatise about Light,

Refraction and Colours' already completed. 'Tis pitty it was not out long since. If it be in English (as I hear it is) let it, however, come out as it is; & let those who desire to read it, learn English.' Thus spoke this fervid Englishman! Through successive letters Wallis continued to prod Newton, but he would not budge.[4]

Wallis accused Newton of thirty years of silence; by the time *Opticks* was issued this was more than true, though Newton himself claimed only that "Part of the ensuing Discourse about Light was written at the Desire of some Gentlemen of the Royal-Society, in the Year 1675 . . . and the rest was added about twelve years after to complete the Theory." The rest, of course, was the fundamental Book I, derived from the 'New Theory' of 1672, the optical lectures and that whole course of prismatic research which went back to 1664. Presumably the first, Latin draft of *Opticks* was written at this time (1687–8), in natural continuation of Newton's Latin prose for the *Principia*; this draft extends about two-thirds of the way through the printed Book I, whose English text is more or less a translation of this draft, as far as it goes.[5] Of the composition of *Opticks* during the quiet interval between the publication of the *Principia* and the Francis Alban affair there is otherwise no trace. Nor is it precisely known when Newton returned to it. It is generally supposed that, with the end of his involvement in public life and his gloomy return to Cambridge, Newton in the early 1690s proceeded as far with its main text as he would ever go, but no firm chronological footing can be given to his statement that "the third Book, and the last Proposition of the Second . . . were since [the year 1687] put together out of scatter'd Papers."[6]

The comments of both Gregory and Wallis indicate that Newton had already before Gregory's visit in 1694 abandoned his experiments on diffraction, never to resume them again. As he wrote himself: "The Subject of the Third Book [that is, diffraction] I have also left imperfect, not having tried all the Experiments which I intended when I was about these matters, nor repeated some of those which I did try." Thus only a portion of Part I of Book III was ever printed, and what was to be in Part II is uncertain, presumably a theory to explain the experimental observations of Part I. What form such a Newtonian theory of diffraction might have taken is only hinted by the Scholium to Proposition 96 in Book I of the *Principia* and the early Queries upon which Newton embarked after finishing his observations on diffraction. Newton took the phenomena as he saw them to suggest that light-rays are inflected at a distance from bodies by some force in them, and that in some complex fashion involving a sinuous "motion like that of an Eel" the colours were caused to separate, revealing themselves as diffraction bands. Such an hypothesis had the advantage of confirming that in all interactions of light with matter (reflection, refraction, diffraction) the coloured constituents of white rays moved differentially, along divergent paths. Nature uses ever to be consonant to herself.[7]

Yet the sad truth is that in (at least) two areas of optical experimentation – that is, double refraction and diffraction – though Newton put up a bold front of hypothesis to hide weaknesses inherent in his basic ideas of light, those ideas which he had formed early in his career were inadequate to cope with the new phenomena of which he learned relatively late. Newton first read of diffraction effects in May 1672, when he received from John Collins the present of Honoré Fabri's *Dialogi physici [sex]* (1669).[8] Fabri gave an account of Francesco Maria Grimaldi's original discovery of diffraction fringes, published in 1665; and we know that Newton read it in Fabri because he said so, on 18 March 1675. On this day, during a long stay in London, when Newton attended for the second or third time a meeting of the Royal Society (following that at which he had been belatedly admitted to the fellowship and signed the charter book), he heard Hooke give a lecture on several new properties of light, unknown to the writers on optics. Among these, Hooke had observed diffraction effects. Newton at the finish teased him by remarking that these phenomena were not new, having been first described by an Italian author. And he told this story in his "Hypothesis explaining the Properties of Light discoursed of in my severall Papers", sent to the Royal Society at the end of 1675.[9] But for long after this Newton himself saw no diffraction effects, and was no more aware that diffraction could produce *colours* than Hooke had been. In the *Principia* he treated it simply as an anomalous bending of light-rays into the shadow. When he wrote Book III of *Opticks*, however, after making some exact experiments, Newton was fully aware of the significance of the separation of colours in diffraction fringes, and had satisfied himself (Observation 11) that violet light was less susceptible to "inflection" round a hair than was red light.[10] We may possibly infer, then, that these last optical experiments of Newton's were made between 1687 and 1694 (when Book III, seen by Gregory, was already abandoned). There is nothing to say more precisely in what year; I am myself a little inclined to doubt that Newton ever dusted off his prisms and lenses after his year in high politics was over. At any rate, it was this kind of colour that (as Gregory put it) Newton was not fully satisfied about, and that caused *Opticks* to remain an incomplete book.

Double refraction was not treated at all in the main text of *Opticks*, appearing almost as an afterthought in Queries 17 to 20 added to the Latin *Optice* (see p. 350) of 1706.[11] Newton *could* have learned of the extraordinary refractive properties of Icelandic spar (calcite) long before, in 1671, from Henry Oldenburg's lengthy summary of Erasmus Bartholin's book in *Philosophical Transactions* number 67.[12] But nothing indicates Newton's attention to these bizarre phenomena until late in life, in fact until he talked about double refraction before the Royal Society on the occasion of his first meeting with Christiaan Huygens on 12 June 1689, when Huygens described his still unprinted *Traité de la Lumière*. Again, nothing as yet affixes a date to Newton's own experiments upon Icelandic spar.[13] It is just possible that Newton suppressed Bartholin's discovery because he considered it to deal with phenomena too far out of the usual course of Nature, until he heard Huygens accounting for it by the wave theory of light; this would have put him upon his mettle and might be a weak indication that the main text of *Opticks*, including Book III, was complete by the summer of 1689. At any rate, the factual account given in Query 17/25 does indicate first-hand experience of calcite by Newton at some time, from which he had reached the primary conclusion that double refraction is caused by an "original property of the Rays". He did not mean that the rays are of two kinds, one diverging one way in calcite, one the other, for Newton insisted that what happens in the crystal depends upon the orientation of the incident ray to it. "Every Ray of Light has . . . two opposite Sides, originally endued with a Property on which the unusual Refraction depends, and the other two opposite Sides not endued with that Property."[14] Newton's notion has been much praised as an anticipation of double refraction interpreted in terms of the polarization of wave-trains, but Newton's was *not* a wave hypothesis, nor did he conceive of the first refraction as imposing order in place of a former disorder. The notion seems to involve much more of the familiar lock-and-key simile, or indeed of Descartes's idea of left-hand and right-hand screwed particles. If so, it looked back rather to the past than forward to the future.

Book IV of Opticks

Book III, Part II, was not the only planned section of *Opticks* never to achieve print. Among Newton's manuscripts there is the beginning of a "Fourth book [of Opticks] concerning the nature of light & the power of bodies to refract & reflect it."[15] If it had ever been completed it would have been a most elaborate statement of Newton's particle theory of light, worked out in full dynamic detail. Here, as on numerous other occasions, Newton endeavoured to set out his conviction that "if Nature be most simple & fully consonant to her self she observes the same method in regulating the motions of smaller bodies which she doth in regulating those of the greater."[16] This principle of uniformity justified the extension of force dynamics from the heavenly bodies (for which Newton regarded the universality of gravitational action to be amply demonstrated) through the medium-sized corpuscles involved in the ordinary phenomena of physics and chemistry to the most minute particles of matter, among which he reckoned the particles of light. In a suppressed conclusion to the *Principia* Newton had already examined the way in light may be created by fermentation and heat:

[By the] motion of fermentation bodies can expel certain particles, which thereupon by their repulsive forces are caused to recede from each other violently; if they are denser, they constitute vapours, exhalations and air; if on the other hand they are small they are transformed into light. These last [particles] undoubtedly adhere more strongly [to their matrix] since bodies do not shine save by a vehement heat. After they are separated they recede from bodies more violently, then in passing through other bodies sometimes they are attracted towards them, sometimes repelled; and by attraction they are certainly refracted, and sometimes reflected as I have explained above [in *Principia* Book I, Section XIV]; by repulsion they are always reflected.[17]

Newton then went on to explain how the surface texture of bodies determined the quality of reflection, its brightness, coloration and so forth. *Opticks* too would have expressed identical ideas if Newton had not on second thoughts excised such passages from the main text, only – much later – to rewrite them again for the Queries.[18]

More specifically, in this proposed "Book IV" Newton worked out geometrically the dynamics of the sine law of refraction by a different method from that already printed in the *Principia*, and of the hypothesis (already found in his earliest notes) that the colour characteristic of light-particles is their velocity: "The most refrangible rays [that is, particles in Newton's language] are swiftest. For the light of Jupiters satellites is red at their immersion." Or so Newton guessed! For when he asked Flamsteed about colours at the immersion or emergence of the satellites, the astronomer denied that any could be seen. Hence these drafts can be placed by the related letters before February 1692. After an intermediate recasting into three propositions, likewise rejected, Newton divided all these materials into fragments inserted, 'sometimes in sentences verbally identical to sections of the draft', into Book II, Part III of *Opticks* and the Queries.[19] This Part is much concerned with the "Fits of easy Reflexion and easy Transmission" transformation of the aetherial hypothesis put forward in 1675, which also figured tentatively in "Book IV".

It seems likely, then, that the greater part of Books I and II of *Opticks* was prepared in a form close to that printed before Newton was despatched to Westminster, but a remark to Huygens (June 1689) indicates that Newton did not then regard his book as complete. Returning to it in 1691, he drafted the abortive "Book IV" and "Conclusion" to the work, abandoned these and transferred part of the material to the extended Book II, Part III, and wrote Book III (unless, indeed, this had been written before 1689).

The Structure of Opticks

Roughly speaking, the twin poles of the book are the study of prismatic colours in Book I and the study of the colours of thin plates and surfaces in Book II. Prismatic colours are both objects of study in themselves and a means towards the study of the coloration of light passed through lenses; the key discovery is that announced in 1672 that light consists of rays differently refrangible, so that by a single, simple refraction these rays must necessarily be caused to diverge. As regards the distribution of colours within the spectrum Newton was ambiguous and contradictory. On the one hand he taught that any point on the axis of the spectrum could be considered as the centre of an image of the Sun's disk, the unique base of a cone of refracted light: we may "conceive that there are other intermediate Circles without Number, which innumerable other intermediate sort of Rays would successively paint upon the Wall".[20] Now colour and refrangibility are in one-to-one correspondence (Proposition II, Theorem II in Book I, Part II) – this is absolutely fundamental; therefore every point of light along the axis of the spectrum, having a unique refrangibility, must also have a unique colour. But some closely adjacent points of light in the spectrum differ more from each other in colour than do others, equally separated, and in any case even the best eye cannot clearly distinguish every shade, so that bands of colour are seen. Yet

Newton never made it clear that this is a subjective effect. On the contrary, in the very next proposition he begins to treat the bands as though they have a real existence, and, as is well known, by a bizarre analogy placed their successive divisions one from another in the proportions of the musical octave. Now Newton held it as a principle that each pure coloured ray observes Snel's Law of refraction (according to its own degree of refrangibility) and that therefore a proportionality of the sines may be stated for every ray from that defining the visible limit of the red to that defining the limit of the violet; thus constant proportions may also be stated for the divisions between the bands, as Newton denoted them. But he failed to point out that between the limits of the spectrum an infinite number of such proportions may be stated. He let it appear that the change, say between red and orange, is discontinuous.[21]

A cognate ambiguity infected Newton's attitude to the relationship between dispersion and refrangibility. According to Newton's 'atomic' conception of light, these properties (together with colour, our sensation of another physical property) should be inherent in the light-particle, consequently inviolable; but he was also aware that he had no experimental proof of this supposition.[22] Unused drafts suggest that Newton cited the (in)famous Experiment 8 of Book I, Part II to prove that a beam refracted by a composite prism of glass and water can be white, contrary to the (factually false) opposite contention drawn from this experiment in the printed text. There he settled for the message that refraction is loosely proportional to the density of the transparent medium, and that the dispersion is calculable from the refraction. On this basis the improvement of lenses to achromatism seemed, as he alleged, a desperate hope.[23]

The argument to assert the 'atomicity' of light – the physical uniqueness of each ray (particle) – is more fully and more cogently illustrated by experiments in *Opticks* than in Newton's previously published papers, notably the divergence of reflected coloured light through the prism with which the book opens is new. The experimental derivation of rays of pure, homogeneous colour was greatly refined. Newton silently admitted that his early claim for the simple experiment with two parallel prisms, that it was "crucial" for his theory, was too sweeping; in *Opticks* this experiment is quoted merely to show "that the Light which being most refracted in the first Prism did go to the blue end of the Image, was again more refracted in the second Prism than the Light which went to the red end of that Image" without any claim that a pure homogeneous ray was thus produced. That claim is postponed to Propositions 4 and 5, and especially

Experiment 12 (of Book I, Part I), to be stated again strongly in Proposition 2 of Book I, Part II.[24]

Continuity of Optical Effects

A new theoretical point in Book I, absent from the previously published material though of major significance for subsequent sections of *Opticks*, is the concept that physical continuity is concealed behind the apparent antithesis of reflection and refraction. Book I, Part I, Experiment 9 teaches that reflexivity and refrangibility are directly related. As with refraction, no process of reflection can alter the colour of light (Book I, Part I, Experiment 6). Colours are separated from each other by reflection as well as by refraction, that is, when the surface reflects some colours and absorbs others; indeed, this is the explanation of the colours displayed by "natural bodies" just as refraction explains the colours of raindrops, prisms and rainbows (Book I, Part II, Propositions 9 and 10). At an interface between two different mediums, the reflexivity is greatest when the difference between the refractive indices of the two mediums is greatest and so is the refraction (Book II, Part III, Proposition 1). Like refraction, reflection is effected by a force in bodies acting upon light, not by the bouncing of light-particles off the particles of matter (ibid., Propositions 8 and 9). An additional physical factor is required to account for partial reflection and refraction, and the fate of a ray on meeting an interface (ibid., Propositions 12 to 14). Finally the principle of continuity is squarely put in Query 4: "Do not the Rays of Light which fall upon Bodies, and are reflected or refracted, begin to bend before they arrive at the Bodies; and are they not reflected, refracted, and inflected [diffracted] by one and the same Principle, acting variously in various Circumstances?"[25] To which it may be added that this leading question looks straight back to the dynamical justification of such continuity, already provided in *Principia*, Book I, Section XIV.

The wider context of this principle is of course that assurance of the simplicity and consistency of Nature that Newton frequently reiterated. Nothing is more disparate and confusing than the phenomena of light and colours. How, for example, do the blue of the sky, of woad, of Boyle's shining fish, of the cornflower, of the rainbow fit into one coherent explanation? Newton resolves the problem by stating that the blue sensation is our perception of a particular species of light-particles having particular physical properties, after which he explains (in part geometrically, in part dynamically) how the 'blue' particle arrives at the eye by processes of selection from the multiplicity of particles of all kinds which is white light. 'Blue' particles, all individual 'coloured' particles, are elements of the divine creation in Newton's view; like the atoms of matter their character cannot be altered by any optical, physical or chemical process. They can only be made to appear by the selective processes of dispersion or the action of the "fits" or else by the absorption of all (or most) of the rest of the whole multiplicity. The fundamental unity of colour theory, therefore, is found by Newton to spring from the concept of light which made – or rather tried to make – optics a special branch of dynamics.

But there were grave difficulties, even ignoring such anomalies as double refraction. Since Newton could only see colours with his eyes, how was he to reconcile positive statements about the nature of colour with ignorance of the eye's process of perception, a problem he did not attack? How can we trust this physiological instrument when (unlike the prism) it fails to distinguish a pure green, orange or purple from a compound formed from two or more pure colours? Thought along these lines raises many questions which Newton might have pursued had he reflected upon his proclivity for making musical harmonies reappear in optics, but he did not. He also found difficulties in explaining why pigments do not correspond to the 'physical reality' of colours in Nature, and precisely how (even in the case of prismatic colours) this reality is to be distinguished from physiological illusion.[26]

There were graver and more obvious difficulties too which the simple particle theory of light adopted by Newton in his earliest days of optical experiment could not resolve. In the 1670s Newton had necessarily followed Hooke in recognition of the periodicity of the (interference) phenomena of thin plates, and had measured the intervals concerned with consummate skill. The experimental investigation reported by Newton to the Royal Society in December 1676, and repeated in Book II of *Opticks*, pays the highest tribute to his imagination, dexterity and intelligence. His resource then had been to account for periodicity by a subsidiary aetherial hypothesis of wave motion. The waves were not themselves light, but (in certain circumstances) determinants of the motion of its particles. Newton's hypothesis postulated an universal aether, filling both matter and space, of which he allowed traces to survive in the main text of *Opticks* even though he seemed (during the 1680s) to have abandoned aether and the 'full' universe for force dynamics and an 'empty' universe.

For example, in Observation 11 of Book II, Part I, a white spot is formed between two wet glass plates which is filled not by air but by a "subtiler Medium, which could recede through the Glasses at the creeping of the Water".[27] However, in the maturer development of the hypothesis of waves into the *Opticks* theory of "fits", Newton chose to impose the "fits" as periodic dispositions upon the rays (or stream of particles), causing them to be more easily transmitted, or more easily reflected, at the next interface. (A more modern term to invoke here instead of "fits" (that is, alternations) might be modulation – the rays are modulated by the "fits".) Thus in the main text of *Opticks* the aether as the vehicle of waves is done away with. Many ambiguities remain, nevertheless. The length of the "fit" must vary with each colour and with each refraction (it is of the order of 10^{-5} inches). What kind of force is it in transparent bodies that maintains these distinctions over many hundreds or thousands of cycles?[28] In detail, Newton's fundamental idea of particle dynamics broke down altogether in dealing with this theory, which the more one probes it, the more it seems to vanish as mere camouflage of the (accurate) experimental measurements stated by Newton in Part I. What is Newton really saying about the measured periodic recurrence of interference rings other than that they recur at these intervals because something causes them to recur at these intervals – a tautology? As often as Newton tries to characterize the "fits" he repeats what he has said under the "Observations".

In Book III, Part I, Newton could not even think of so much disguise for his observations of diffraction.

It must have been with some relief that in the second English edition of *Opticks* (1717) Newton was able to return to a modified version of his original (1675) notion of aetherial waves modulating the light-rays. This came about because Newton had (as we shall see later) reneged upon that confidence in the 'empty' universe that he had upheld for more than twenty years. Accordingly, while leaving the main text of the book and the earlier Queries unchanged save for the addition of a qualifying phrase here and there, in the new Queries numbered 17 to 23 inserted into the series at this time (so that the former Queries 17 to 23 of the 1706 *Optice* now closed the set as numbers 25 to 31) Newton developed aetherial mechanisms for light and gravity. He had formerly ended Query 16 with the suggestion that vision is a vibratory *physiological* process in the eye; now in Query 17 he reinterpreted this process as involving *physical* vibrations in an aetherial medium. Such vibrations, originating from the first interaction of light and matter, are (probably) propagated from the point of Incidence to great distances? And do they not overtake the Rays of Light, and by overtaking them successively, do they not put them into the Fits of easy Reflexion and easy Transmission described above? For if the Rays endeavour to recede from the densest part of the Vibration, they may be alternately accelerated and retarded by the Vibrations overtaking them.[29]

True, Newton was careful to emphasize (Query 21) that this newly minted aether was very different from the coarse, space-filling medium of the Cartesians. The mass of matter thus reinserted into the formerly vacuous space of the *Principia* by Newton was thus small in the extreme. Yet, however small, this new aether wholly destroyed Newton's former dynamic concept of light; it was a reversion to the ideas of 1675 and rejection of the dream of an optics of motion and force.

Thus, in the end Hooke won a victory over Newton in the realm of optical hypothesis. *His* discovery of the periodicity of interference rings became more determinative of Newton's final optical speculations than Newton's own discovery of dispersion. With the 1717 renunciation of dynamic explanations depending upon the motions of light-particles there was little left for particles to do except explain why light beams travel in straight lines. But Huygens's principle, published in his *Traité de la Lumière* (1690), already showed how this seeming flaw in the wave theory of light could be amended.

Opticks in its final form (the fourth edition, 1730, is the one reprinted in modern texts)[30] became a strange work. Not only had its author failed to carry through his original design for the book, but in the full tale of Queries he had negated it! The Queries themselves became incoherent in 1717 because (whereas the group added in 1706 had voiced many of those far-reaching speculations often penned by Newton before, but always deleted upon second thoughts, speculations about "Powers, Virtues and Forces" among others, consistent with the dynamic philosophy embraced by Newton from about 1680 to 1710) that group now followed the latest set reviving a revised aetherial mechanism. Yet, what is no less strange, Newton had not at all been induced by any fresh considerations of *optical* science to make this transformation in his ideas – surreptitiously effected, as it were, since he made so few alterations to the main text of the book that to this day it carries (as indeed do the late editions of the *Principia*) a double, contradictory message.

No wonder that some corpuscularian Newtonians like Robert Smith, who must have found Newton's change of hypothesis puzzling, declined to have any truck with "fits" or wave motion of any kind. They were content to disregard Book II, Part III of *Opticks* and the Queries following as speculations that experimental, geometrical optics could well do without.

Most early eighteenth-century readers were undoubtedly more struck by the power of Newton's methodology to lead him to profound conclusions and the uniformity inherent in his analysis of phenomena than by these contradictions. Both the aether and Newtonian forces figured in the contemporary language of science, hence to such readers it did not seem inconsistent that Newton should appeal to both systems of physics, given Newton's insistence that *his* aether was markedly distinct from that of the Cartesians and could not at all impede the revolutions of the heavenly bodies. We may feel that after 1717 the now aged Newton was both keeping and eating his cake; they felt that he was making good eclectic use of available principles. And presumably most readers were content to gain from *Opticks* an extraordinarily wide (and quite unprecedented) scientific comprehension of colour, without concerning themselves too minutely with ultimate (and necessarily hypothetical) questions about the physical reality of light. Since colours had never been much discussed by either mathematicians or philosophers before Boyle's book of 1664 – for even the pre- and post-Cartesian geometrical analysis of the rainbow had scarcely examined its colours – little being said beyond the specious distinction between 'real' colours intrinsic to the concept of something (as grass must be green and blood red) and 'fantastic' or ephemeral colours (as when red light through coloured glass tinges salt with its own hue), Newton's book was wonderfully original and wonderfully informative. Granted that the eye is a perfect detector, Newton demonstrated that in its physical essence colour is quite unlike paint; it is not something applied by Nature with a brush, but absolutely part of the frame of things, for *colour* and *light* are inseparable concepts. Colours are not shown to us by light, but are light. This was an extraordinary step from the subjective to the objective.[31] Newton could not of course have said that 'redness' is the label people normally give to the perception of a certain range of frequencies in the electromagnetic spectrum, but the idea would immediately have been clear to him and in a reply to Hooke he expressed its essence.[32] Though as aware as anyone of the immense complexity and power to deceive of colour effects, he asserted convincingly that it is possible to penetrate with the aid of experiment and geometry to a true and constant physical theory, just as one can in mechanics and might hope to do in chemistry.

While Book I, with the prism as its experimental device, set out in qualitative detail the outline of this approach, Book II (where the thin plate becomes the experimental tool) established in quantitative elaboration a second-order effect, interference in our terms, which is more ordinarily productive of natural colours than is simple refraction.[33] The strategy is like that of the *Principia*, with the difference that in the older book Newton's theoretical concept (attractive force) is developed mathematically in Book I, whereas in *Opticks* the theoretical concept (the atomicity of the light-ray) is developed experimentally. In both books the second portion is devoted to exemplification of the theory in the world of experience (one might compare Newton's placing of the explanation of the rainbow in Book I, Part II of *Opticks* with that of the experiments on air resistance in *Principia*, Book II). Even more than the *Principia* – where the chief rhetorical pointers to it are in the rules of reasoning and the preface – *Opticks* directs attention to the barely accessible microcosmic realm of corpuscles and particles as the ultimate seat of Nature's action, where Newton was able to achieve the triumph of establishing by optical means the order of size of the largest surface particles. What the reader would not find in *Opticks* were geometrical or physiological optics, methods of ray-tracing through multiple lenses, procedures for designing optical instruments, or metaphysics. For these he would have to turn to other works like those of Malebranche or Molyneux. Even Roemer's decisive proof of the finite velocity of light is barely mentioned.[34]

Comparison of Opticks *and* Principia

Although *Principia* and *Opticks* have the same axiomatic structure, displaying ontological similarities and to a lesser extent methodological ones also, the latter book appealed far more to readers. In Newton's lifetime only a few men of great mathematical capacities could duly value the *Principia* but almost any reader could appreciate and admire *Opticks*, especially the imaginative, far-reaching and devout speculations of its Queries. (To some these were virtual certainties because Newton had conceived them.) *Principia* was mathematical and written in Latin; *Opticks* was experimental and written in English: these were for British readers strong points in its favour. Newton himself alleged that he had described his experiments so fully in order that "A novice [might] the more easily try them."[35] If celestial mechanics were the grander science, optics was the more spectacular. Newton's work on colour was presented to the London public in the early eighteenth century by such scientific lecturers as Francis Hauksbee and J. T. Desaguliers. Artistically, as in Roubiliac's famous statue of Newton in Trinity College chapel, he was generally identified by his prism and the spectrum it cast.

In the influence of *Opticks* there was certainly some difference between Britain and the Continent. To Newton's great satisfaction, the Parisians did indeed take charge of a fine translation of the book in 1722.[36] This edition confirmed the recognition by the French, growing since 1706, of Newton's pre-eminence in this field. But generally the continentals judged Newton as a mathematician and his *Principia* was for them a far more important book than *Opticks*, if also (in its philosophical implications) a more problematic one; however, influential continental mathematicians also found great difficulty with the corpuscular hypothesis of light. In Britain, the development of the mathematical sciences was less notable than that of the experimental sciences – especially electricity and chemistry. The *Principia* was a matter for national pride, and was indeed rapidly published in a vernacular translation by Andrew Motte soon after Newton's death, but there were few to follow its example or to repair its weaknesses. To the British the experimental method of *Opticks* proved far more suggestive, and though little was added by them to the technical and theoretical aspects of Newton's achievement, the inspiration derived from its Queries in particular directed many new investigations. Light treated as a physical entity in the Newtonian manner offered many new speculative possibilities, some of which could be followed along the experimental path. As that awkward Newtonian George Cheyne emphasized in 1715:

> Those who desire full Satisfaction in this wonderful Appearance of Nature must go to that admirable Treatise of *Opticks*, written by Sir *Isaac Newton*; that great Person having before shown how far *Numbers* and *Geometry* would go in *Natural Philosophy*, has now manifested to the World to what surprizing Heights, even vulgar Experiments duly managed and carefully examined in such Hands may advance it.[37]

Cheyne makes no scruple to assert that 'Light is a Body, or a material substance' and shows how light considered in this way is comprehensible and capable of contributing in several diverse ways to the working of Nature.

In this broad sense *Opticks*, like the *Principia* in a different way, was one of the great formative books of the eighteenth century. For a time, too, it had a major effect on Newton's personal life, for it was *Opticks* that brought the French to London and rendered Newton a great international figure.

12

Life in London, 1696–1718

Warden of the Mint, 1696

It is a fair guess that Newton was never fully content with his professorial and collegiate life in Cambridge after the university had sent him to Parliament in 1689. Indeed, it may be that his eye began to rove as early as 1687, in the period of lassitude that must inevitably have followed the intense and exciting task of bringing the *Principia* to a conclusion in the summer of that year. By that time the acquaintance or friendship between Samuel Pepys PRS and Isaac Newton must have been established; besides the Royal Society, Christ's Hospital mathematical school had brought the two men together. Pepys was then at the height of his career and of his influence. At the peak of his psychological crisis in 1693 Newton wrote to Pepys: "I never designed to get anything by your interest, or King James favour," an obvious allusion to days before the Revolution, after which Pepys's influence was nil.[1] If we may read this anxious denial as evidence that there really was some such talk or manoeuvre (as with Newton's parallel denial to Locke), it must have occurred in 1687 or 1688.

Similarly and in the same disturbed frame of mind, Newton wrote to Locke: "I beg your pardon also for saying or thinking that there was a designe to sell me an office."[2] In this case, letters do hint at

Locke's enlisting the help of the Earl of Monmouth (a favourite of William III's and recently First Lord of the Treasury) who had been a patron of Locke himself.[3] Again, Locke seems to have thought of Newton as a possible Comptroller of the Royal Mint and also as Master of the Charterhouse, a post depreciated by Newton as too poor and confining him to the City, away from the power and society of Westminster.[4] But no one would have thought of making Newton a prebendary or a bishop! His reliance upon the friendly patronage of Pepys, Locke and Montague to start him upon a new official career evidently irked Newton, wounding his pride and provoking not only the morbid sensitivity of his 'disturbed' letters of 1693, but the charge (to Locke in 1692) that Charles Montagu was a false friend because of an "old grudge", and a violent disclaimer to Edmond Halley some three years later that he had any intention of accepting a post at the Mint, written at the very moment when he must have been deciding to become its Warden.[5] Rumours had circulated as early as the autumn of 1695 that there was to be such an appointment for Newton – one reason why he was rough with the unoffending Halley – and the matter must have been settled between Newton and his success- ful young friend, who had become Chancellor of the Exchequer and President of the Royal Society, before Montagu informed Newton officially on 19 March 1696 that the wardenship was his: 'the Office is most proper for you 'tis the Chief Officer in the Mint, 'tis worth five or six hundred pounds per An, and has not too much bus'nesse to require more attendance then you may spare.'[6]

Letters patent appointing Newton were issued on 13 April 1696; a week later he left Cambridge and was at work early in May. Like many other government offices, the wardenship of the Mint was what the Warden made it; it could be a sinecure; exercised by deputy, or a place of authority (as with Newton). For many years the Mint had been managed by its Comptroller, James Hoare, who had recently died. The Warden, as Montagu said, was nominally the first officer of the Mint, but the chief executive authority had tended to fall into the hands of the Master and Worker, and was certainly there when Newton transferred to that post. Presumably he could, as Warden, have continued to live in Cambridge, but this was not his object; on the contrary he became at once a non-resident Fellow and an inactive professor. He appointed William Whiston his official deputy as Lucasian professor (with full salary) less than a year before he resigned from the post (10 December 1701), soon after his second election to Parliament. That any mathematical lectures were given in the five (or more) previous years is unlikely. Whiston had given up a living at Lowestoft to take Newton's place at Cambridge, having formerly been Bishop Moore's chaplain at Norwich, so he was certainly not teaching at Cambridge before 1700. His much-read *New Theory of the Earth* had been published in 1696. Newton continued to receive his dividends as a Fellow of Trinity until his resignation (also just before the end of 1701) and indeed remained on the buttery books after it.[7]

Presumably Newton retained his rooms in Trinity for some time after his removal, taking lodgings in Westminster as Montagu had suggested. He must have left Cambridge with little regret, being a wholly unsentimental man; his last years had been neither happy nor productive, though far from idle. His period of ill health (a year, he told Pepys) had been followed by his long duel with the Moon, increasing annoyance with Flamsteed and ultimate defeat. From his illness in 1693 to his departure two and a half years later Flamsteed was by far Newton's most frequent correspondent and there is almost nothing of personal interest in these or in his few other letters. To this time, however, belongs a trivial anecdote (like the stories of repeated accidental fires, the dog Diamond, and Newton's cat, more likely to be false than true). On 7 May 1694, finding some scholars of St John's hanging about a house opposite to that college, reputedly haunted, he berated them with the words: "O ye fools, will you never have any wit? Know you not that all such things are mere cheats and impostures? Fie, fie, go home for shame!"[8] The effect of this is not recorded.

Hardly less strange is the last 'scientific' document penned by Newton in Cambridge, his note in March 1696 of the visit from an anonymous Londoner (see p. 196). Here he carefully recorded how the stranger, after a process of digestion lasting nine months, had produced a "menstruum" which "dissolves and volatilizes all metals & gold dissolved & volatilized may be digested with it to the end". Newton, who possessed all the writings mentioned in this memorandum, made a number of scholarly cross-references indicating the connections of the stranger's narrative with the standard literature. Perhaps this was a fitting farewell to Cambridge and his own well-used furnaces, for though certainly in touch with adepts such as William Yarworth after his move to London, he never set up a laboratory there.[9] As often in alchemical tales, a mysterious unknown plays a major role.

Newton measures Heat

But, in this context, it may be noted that from a late stage in his laboratory work Newton derived a fundamental paper on thermometry, published anonymously in 1701.[10] Some dated notes on relative temperatures show that Newton's investigations were incomplete on 10 March 1693; one may readily accept J. A. Ruffner's guess that they originated in Newton's studies of the fusion of metals in pre-*Principia* days.[11] It is clear that material in the 1701 paper was worked out after that date, and very probably after 1694 when Gregory visited Newton in Cambridge (further proof that Newton was by no means incapacitated by his illness from doing what he had done before). Newton's exceptional ingenuity and manual skill are once more apparent in this investigation. He made a glass thermometer filled with linseed oil, reaching to higher temperatures than the alcohol thermometers then usual. He fixed two points, the freezing point of water at 0°, blood heat at 12°. Supposing the expansion of the oil by volume to be proportional to the change in temperature, he marked other points as shown, including those of some readily fusible alloys. Temperatures far above those of boiling water were clearly not measured with lineed oil. Instead, Newton employed an arbitrary law of cooling that has since borne his name: the temperature of a hot body cools to that of its surroundings in geometric proportion, as the time increases in arithmetic proportion.[12] Accordingly time measurements could be substituted for temperature measurements. Having found the cooling-rate of a particular block of iron within the range of his thermometer, Newton could extrapolate to the cooling-rate at much higher temperatures and so, by measuring the time taken by the block to cool from (say) dull red heat to boiling-water heat, he could estimate the number to be assigned to red-heat on his scale. Then,

TABLE 12.1 The scale of heat

°N	°C	Effect
0	0	Freezing point of water
6	20?	Summer heat
12	37	Blood heat
34	100	Boiling point of water
72	232	Melting point of tin
96	327	Melting point of lead
192	6–800?	Red coal fire

noting how long a time was taken from red-heat to the solidifying points of metals, their melting-points were established.

As the table shows, 1°N is roughly equal to 3°C, the multiplier increasing as the temperature rises because of insufficiencies in Newton's law of which he could not be aware. Of course it was obvious that a large body (like a comet) must cool more slowly than a small one, and that different materials might cool at different rates. Taking such factors into account, the law of cooling had some cosmological interest for Newton and his successors, apart from its interest in physics and chemistry.

A longer and more speculative essay "On the nature of acids" has long been associated with the "Scale of Heats" and was already in existence when Newton made the dated note on temperatures just mentioned. A text of "On the ntaure of acids" was given to Archibald Pitcairne in the course of a two-day visit to Newton at Cambridge in March 1692, just a year before. It was first published in 1710, and foreshadows Query 31 of *Opticks* (1706; chapter 14).[13]

The Second Principia

Far more important than these chemical essays was Newton's continued improvement of the *Principia* during his last years in Cambridge and in the early London period. Fatio de Duillier's correspondence (especially with Huygens) throws light on Newton's work of revision during the early 1690s, when Fatio hoped to become editor of a second edition of the book and to foist certain superior ideas of his own upon Newton; for Newton's last years at Cambridge David Gregory is the best authority.[14] A vast deal of material prepared by Newton for the new book exists, in the form of rewriting entered in printed copies of the *Principia* (1687), interleaved pages, and separate drafts, but no one has undertaken the thankless task of systematically collating and ordering all these fragments, many of them rejected when the text of the second edition was settled in 1709–13. It would perhaps not be easy to determine what belongs to the earlier phase of revision, what to Newton's final years in Cambridge, and what to the seventeen London years. For example, Fatio early made Newton aware of the mistake in Book II, Proposition 37 (on the flow of water from a tank) but the treatment (now

Proposition 36) was finally quite reconstructed, with Cotes's assistance, in the final consideration of 1710–11. Perhaps the most serious of all corrections to the book (in Book II, Proposition 10) was made so late that it involved reprinting some pages.

One proposed addition to the *Principia* that we need not greatly regret was the series of Scholia on the wisdom of the Ancients (both mythical and historical), especially in relation to astronomy and the theory of gravitation, intended for Book III. As already mentioned, a connected version of Newton's manuscript sketches was later published by David Gregory as his own in 1702, in the preface to his *Astronomiae . . . elementa*.[15] Newton's confidence in the extent and profundity of ancient natural philosophy was by no means unusual in the seventeenth century; only his attribution to the Pythagoreans of such recent and controversial concepts as gravitation would have seemed arguable.

In the first edition of the *Principia* there were many other more pertinent technical or mathematical weaknesses demanding Newton's corrective hand, and indeed receiving his attention during these years or in the course of his final review of the text, in conjunction with Roger Cotes, during the years 1709 to 1713. If we may trust Gregory's memoranda, the work of revision went on continuously through Newton's last years in Cambridge and into his London period. So Gregory noted in 1698 that Newton 'will determine the orbit of Saturn, but not those of the other planets for lack of observations.' He would reconstruct the theories of the precession of the equinoxes and of comets. As for the Moon, 'Its theory would not be completed because of Flamsteed's bad temper nor will his name be mentioned; the theory will be finished to four minutes of arc, which he would have perfected to two, if Flamsteed had supplied the observations.'[16] There was no love lost between the two astronomers, the observer regarding the mathematician as little better than a charlatan.

By 1702 Gregory thought Newton ready to publish his revised text, as again in 1704 and 1706; in that year Newton showed Gregory 'a copy of his *Princ. Math. Phil. Nat.* interleaved and corrected for the Press. It is entirely finished as farr as Sect. VII. Lib.II. pag. 137 . . . He letts alone his Doctrine of Comets as it was . . . Most of the corrections made in the book printed at Hamburg are in his copy word for word.'[17] The last sentence is of interest as an indication of the persistence of many early amendments in the corrected copy of 1706.[18]

Halley on Comets

But yet more interesting is Gregory's allusion to the treatment of cometary motion in the *Principia*.[19] One of the major additions to Book III in the second edition was Halley's calculation of the orbits of the comets of 1664–5, 1682 and 1683, extending the final Proposition 42 of Book III by some seven pages. To find the origin of this extra material we must go back to Halley's letters to Newton of 1695. About the end of August Halley again visited Newton, discussing with him a plan to compute the orbits of comets more accurately than before, with a view to confirming Newton's parabolic (gravitational) orbit more fully. As the latter told Flamsteed: "He has since determined the Orb of the Comet of 1683 by my Theory & finds by an exact calculus that it answers all your Observations & his own to a minute." On 13 September Newton made a note of positions of the great comet of 1680–1 (about whose path he had formerly had debates with Flamsteed and which Newton had taken as an example in the *Principia*, 1687), but he was away from Cambridge for most of this month, partly in Lincolnshire. On his return he found that Halley had pursued his aim to ease Newton 'of as much of the drudging part of your work as I can' to such good effect that he had worked out the comets of 1644–5, 1680–1 and 1683. Furthermore he announced a considerable philosophical discovery in his habitually modest way:

> I find certain indication of an Elliptick Orb in that Comet [of 1680–1] and am satisfied that it will be very difficult to hitt it exactly by a Parabolick . . . I must entreat you to procure for me of Mr Flamsteed what he has observed of the Comett of 1682 particularly in the month of September, for I am more and more confirmed that we have seen that Comett now three times, since the Yeare 1531, he will not deny it you, though I know he will me.[20]

This last was 'Halley's Comet', whose orbit and period of return he set before the Royal Society on 3 June 1696. In further letters to Newton in October Halley added much detail about his inquiry into these three comets, confirming Newton's analysis of cometary motion and proving that the one of 1680–1, as well as that of 1682, had an elliptical orbit. This Newton accepted, writing to Halley "I can never thank you sufficiently for this assistance, & wish it in my way to serve you as much" and adding, in reply to a question from Halley: "How far a Comets motion may be disturbed by [Jupiter] and [Saturn] cannot be affirmed without knowing the Orb of the Comet & times of its transit through the orbs [of the two planets]." He thought the orbital period of the comet might by this gravitational perturbation be lengthened or shortened up to a year, or even more.[21]

It is notable that Newton wrote to Halley as though certain of being understood, and on terms nearer to equality than he extended to most men. With this letter he sent Halley "a Box of brass Rulers & beam compasses", probably the very same that Newton had once bought from Fatio.[22] At this point the burst of correspondence about comets ends, and one cannot say what further work Newton may have done on their theory before his move to London. Halley too, it seems, was distracted from astronomy by his naval work.[23] But there is enough to make it clear that he moulded an idea that was merely potential in the first *Principia* into a demonstrated concept. From observations of a comet as it passes through perihelion within the solar system it was impossible, then, to determine confidently whether its orbit was parabolical or elliptical; but Halley showed that sufficient indication existed of the characters of some comets in the past to permit the conclusion that these were identical bodies with others seen recently. Conceptually – though the great test had to be postponed to 1759 – this was a major new achievement for Newtonian mechanics.

The seven pages added just before the new General Scholium closing the second edition of the *Principia* were wholly based upon Halley's computations, lightly revised by Newton, relating to the comets already mentioned, of which the one of 1664–5 was treated at greatest length:

> By these examples [Newton wrote] it is well enough shown, that the motions of comets are no less accurately represented by our Theory, than the motions of planets use to be by their theories. And further-more the orbits of comets can be enumerated by this Theory, and the periodic time of a comet revolving in any orbit can be known, and finally the *latera transversa* and altitudes of the aphelion of the elliptical orbits found out.[24]

Newton then disclosed the near identity of the parameters of the orbit of the 1607 comet with those of the 1682 comet, to which therefore a period of seventy-five years might be assigned, to be confirmed if "the comet should indeed return in the future in this orbit, after 75 years". Newton may be thought a little grudging in his public acknowledgement of Halley's extension of Newton's cometary theory to include an elliptical orbit (though this was infinitely more than

Cotes received for his labours on the book); nor is it clear whether or not Newton now supposed *all* cometary orbits to be ellipses.

London and Catherine Barton

Newton's first home in London was in Jermyn Street, where he settled in the summer of 1696 and stayed for more than ten years (see Appendix C). All this part of Westminster was newly built, the houses in Jermyn Street being about thirty years old when Newton moved there, and less grand than those in Pall Mall or St James's Square, where some of the nobility lived. Newton was close to his parish church, St. James's Piccadilly (built by Wren in 1684) where his protégé Samuel Clarke was rector from 1709 to 1714.[25] That Newton was concerned with the affairs of this church from soon after his settlement in Jermyn Street may be gathered from some pages in his hand headed "The Accounts of [Ambrose] Warren relating to the Tabernacle neare Golden Square". The date is 1700. The Tabernacle and its school for poor boys was a dependency of St. James's founded by Archbishop Tenison a little after the principal church; Newton was to remain a trustee (and evidently a benefactor) until he died.[26]

Obviously Newton had now to set up a household of his own, a new experience. To manage it for him he gave a home to his niece Catherine Barton. The sickness and death of her father Robert in the autumn of 1693 had called Newton away from Cambridge for a time and may have contributed to his own mental disturbance immediately thereafter. He had a particular affection for his youngest sibling, Hannah Smith, who became Catherine's mother, and her three children; the family lived in the rectory at Brigstock, Northamptonshire, whence Hannah had appealed to her half-brother for help and comfort when she was 'overwhelmed in sorrow'.[27] Catherine may well have come to London, aged about seventeen, before the end of 1696; certainly she was living with Newton when he solved the brachistochrone problem (29 January 1697).[28]

According to her husband, Catherine lived with her uncle for twenty years before her marriage in 1717, which fits perfectly with her arrival in London about 1697.[29] Of her first years there with Newton nothing is known; that he directed her education is mere supposition. The only solid fact is that she was attacked by smallpox in the summer of 1700, apparently while staying with friends in Oxfordshire, where she certainly made her recovery, and whither

Newton wrote her a letter of local London gossip.[30] We may guess
that she was fortunate in that the disease left her face unmarked –
Newton asked particularly about this – for within a few years she was
as noted for her beauty as her wit.

Little other trace of Catherine Barton before her marriage remains
in Newton's papers. Signs of his care for her family do exist. At some
point after her father's death – perhaps as early as 1693 but before
1711 – Newton bought an annuity of one hundred pounds for two
lives, his own and then after his death the lives of Robert Barton the
younger, Catherine, and her sister Margaret (later Mrs Pilkington).[31]
Since Newton had a more than adequate income of his own, one may
assume that he presented this sum annually to those named during
his lifetime. When Lieutenant-Colonel Robert Barton – whose steps in
promotion must surely have been bought by Newton – was drowned
in the St Lawrence River in the course of an expedition against French
Canada in 1711 ("Eight transport ships with about eight hundred men
on board were cast away by striking upon rocks & the rest escaped
narrowly," Newton wrote to the unfortunate man's father-in-law), he
was active in securing a pension for the widow. According to
Brewster, Robert Barton left three children to whom Newton gave 'an
estate in the parish of Baydon in Wiltshire' a short time before his
death.[32]

The great enigma of Catherine Barton's life, in which Newton was
certainly concerned, emerged in 1706. In that year his friend and
patron Charles Montagu, now Lord Halifax, made a codicil to his
will leaving to Catherine Barton all his jewels and £3,000 'as a small
token of the great Love and Affection I have long had for her'. There
was also an annuity for her benefit taken out in Newton's name, of
which he must have been aware. In explanation of this extraordinary
bequest Halifax's anonymous biographer (1715) wrote that Halifax,
then a childless widower resolved not to remarry, 'cast his Eye upon
the [sister] of one Colonel Barton and Neice to the famous Sir Isaac
Newton, to be Super-intendant of his domestic Affairs'. However,
this bequest was nullified by an even more extravagant codicil
(February 1712) making Newton a token present of £100 (the same
sum was left to others for mourning symbols) 'as a Mark of the great
Honour and Esteem I have for so Great a Man' and bequeathing to
Catherine the sum of £5,000, 'my Manour of Apscourt in the County
of Surrey together with all the Rents, Profits, and Advantages
thereunto belonging' and the rangership and lodge of Bushey Park
with contents and furniture.[33] These rich gifts, Halifax now wrote,
were made because of the 'sincere Love, Affection and Esteem I have

long had for her Person, and as a small Recompense for the Pleasure and Happiness I have had in her Conversation'. No evidence survives to show whether Catherine actually received anything of this legacy; she certainly lost Bushey Park, which was in the gift of the Crown. What Newton's papers do abundantly reveal is his involvement, after Halifax's death in 1715, in the implementation of his will to Catherine's benefit. They reveal him strongly intent upon defending Catherine's claim to a substantial bequest from the Halifax estate.[34]

In September 1710, after some months in the village of Chelsea (which he clearly did not care for) Newton moved into a tall, narrow house in St Martin's Street, south of Leicester Fields (now Square) which he occupied for most of the remainder of his life. There he was (for the next few years) a near neighbour of Jonathan Swift, no friend of Newton's, the Royal Society or modern science. In his *Journal to Stella* (September 1710 to June 1713) Swift never mentioned his great neighbour the Master of the Mint, though he dined from time to time with Halifax. Yet he was a close friend of 'Mrs Barton' with whom he 'dined today alone at her lodgings' soon after the start of the (extant) *Journal*. On this occasion she passed on to him a bit of juicy scandal. On another she related a story to prove the shortage of virgins in London. He refers to her brother's death, calling Robert Barton a coxcomb. Catherine obviously was at this time, and had been for some period, an intimate and admired friend of that formidable satirist. She was a leading figure in Swift's smart, slightly raffish world and (it is said) a toast of the Kit-Kat Club. Yet her name is never linked by Swift with Halifax's, and this great lover of scandal and scurrility wrote none about her in this private record, rather defending her probity and honour (see Appendix D).

It was not so with many others. The strange content of Halifax's published will was soon known everywhere. More than ten years later Voltaire in London picked up the tale – with Catherine now ten years married – 'In my youth,' he wrote in his *Dictionnaire philosophique* long afterwards (1757):

> I believed that Newton's fortune was founded on his supreme merit. I had imagined that the Court and the City of London had named him Master of the Royal Mint by acclamation. Not at all! Isaac Newton had a very charming niece, Mrs Conduitt; she made a conquest of the Lord Treasurer, Halifax. The Infinitesimal calculus and gravitation would have done nothing for him without a pretty niece.[35]

The bite of this joke is not lessened by its total anachronism. Again, soon after the publication of the biography of Halifax, Flamsteed told his friend Abraham Sharp of Mrs Barton's good fortune, underlining the words *excellent conversation* 'quoted' from Halifax's will. He continued: 'Sir I. N. loses his support in him [Halifax]. & having been in with Ld Oxford Bollingbrock & Dr Arbuthnet is not now looked on as he was formerly.'[36] This hint of damage to Newton in the reign of George I by his former high Toryism was misconceived, but it was true that the new monarch brought Flamsteed friends at Court.

What is one to make of Halifax's generosity, which so amused London's lampoonists and tale-bearers?[37] The intimate and affectionate relation to Halifax of the gay, handsome, young niece of Sir Isaac Newton is beyond doubt. That she was not his wife is equally certain: when wedded to John Conduitt she assigned herself as a spinster. Was she like Pepys's Mrs Skinner, a *maîtresse en titre*, Halifax's acknowledged hostess and concubine? This, despite the gossip, seems equally impossible. No contemporary evidence supports such a view of the situation. Emphatically, Newton felt no cause for shame in his family relation to Halifax, no more than in their friendship.

Four days after Halifax's death Isaac Newton wrote to the man whom he respected as the head of his clan an apology for not calling upon Sir John Newton to take leave before Sir John's departure for Lincolnshire: "The concern I am in for the loss of my Lord Halifax & the circumstances in which I stand related to his family will not suffer me to go abroad till his funeral is over."[38] He could hardly have been more respectful had Halifax been a close and revered relative. What sane man would have presumed in his chief a knowledge of a discreditable connection, while alleging it as a reason for failing in a polite obligation?

Again, there is the problem of Catherine's apparent separation from Newton, though up in Lincolnshire it was supposed that they were in daily converse. If Swift dined alone in St Martin's Street, conveniently at hand, why was Newton never present in his own house? What was the reason for the secrecy about a situation which both Newton and Swift evidently considered respectable and unworthy of remark, secrecy that give the lampoonists their opening for humour? Probably no viable hypothesis about Catherine's situation can ever be framed without new evidence, but I do not accept Westfall's dichotomy that if not a wife, then necessarily a mistress. History does not know enough about Halifax to impose such a choice.

The Royal Society

On 30 November 1703 Isaac Newton was elected President of the Royal Society, the first untitled President since Samuel Pepys (1684–6). Soon after his move to London Newton had been chosen for the Society's council, but because of Mint business he attended no meetings nor any meeting of the Society itself until 1699, when he was present twice. On one of these occasions he described a form of reflecting sextant, which Hooke at once claimed to be his own earlier invention.[39] In any case the business of the meetings at this period, arranged by the Secretary, Hans Sloane, held little of interest for a mathematician or a physicist.

While we may guess Newton to have been Sloane's candidate, the two previous noble Presidents having almost completely dissociated themselves from its affairs, his was not a unanimous choice in 1703; indeed through all the first half of his life-long presidency there was strong opposition to him and his policies. Some existed from the first – perhaps among the naturalists, who recalled how Newton's friend Halley had declined to correspond with Antoni van Leeuwenhoek – more was created by his powerful leadership within the Society, which some judged domination. This he exercised almost at once after his election as President by (for example) bringing in Francis Hauksbee (FRS 1705) as, effectively, a new 'Curator of Experiments', a man active for a decade and the author of experiments having a strong effect on Newton's ideas.

Hooke was not an opponent only because death had removed him on 3 March 1703. Other Fellows, whom we may reckon Hooke's friends and partisans rather than Newton's, such as Richard Waller, Hooke's biographer and publisher of his posthumous works, were still influential in the Society. Other Fellows, seeing President and Secretary as one, particularly resented the high-handed (as they saw it) management of the Society's business by Hans Sloane and his casual attitude to its meetings. The geologist, John Woodward, stood out for his criticisms of the Secretary; in 1706 the Council ordered the Fellows to be admonished (when Woodward was present) that a Fellow casting reflections upon his colleagues made himself liable to ejection. In 1710 Woodward was so suspended from his fellowship for his constant abuse of Sloane. Woodward counter-claimed that Sloane had made faces at him during meetings! Yet he was not without supporters, including the second Secretary, John Harris (author of *Lexicon Technicum*, 1704).[40]

Even more serious dissensions grew up over the meeting-place of the Royal Society. Hooke's death removed the Society's claim to consideration at Gresham College, other than long usage. He had managed to prevent the City companies responsible for the institution rebuilding it, as they were eager to do; but in 1705 the livery companies informed the Society that they could not be accommodated in the reconstruction. The Society explored a variety of expedients: in 1705 Newton drafted a petition (refused) for a grant of land in Westminster, whereon the Society might build.[41] For five more years it maintained its insecure tenure within Gresham's crumbling walls, many Fellows (even Council members), Woodward chief among them, seeing no reason for there ever being a change. Newton, however, clung to his conviction that the Society's meetings and property – its library and repository – should be properly secured by ownership or a proper lease of a house of its own. At last, early in September 1710, he told the Council that a house belonging to a deceased Fellow, Dr Edward Browne, was for sale "and being in the middle of the town and out of noise, might be a proper place to be purchased by the Society for their meetings." It was off the Strand, in Crane Court, between the City and Westminster, and the Society was to occupy it for two generations. In fact, two dwellings were bought for £1450. Although the Society is recorded as having invested funds of £1050 in 1697, the purchase caused embarrassment since the Society could only put £550 down in cash. Moreover, £300 extra had to be found for Wren's gallery, added to contain the repository. Newton gave £100 towards the deficit, others contributed similarly, and £900 were borrowed from the Society's wealthy 'operator' and Librarian, Henry Hunt. At his death a few years later Newton and others again delved into their coffers to clear the Society's debt to his estate.

Finance was always a problem to Newton and the other officers, partly because investments made in the past were not always productive, mainly because, ever since the foundation of the Society half a century before, many Fellows had failed to pay their subscriptions, its preponderant source of income. Newton revived the practice of imposing a contractual obligation on Fellows to pay, but there was still no effective sanction to bring the money in – putting a Fellow out was no remedy. However, thanks to abnormal contributions the Society had a positive balance by 1716, and fresh investments were bought in the next year. From 1718 to the end of Newton's presidency there was a favourable balance each year, approaching £400. By 1727 the assets of the Society, besides the two houses and their contents, amounted to £2100.[42]

Sloane and Newton were both strong, wilful characters and his recorded comments indicate that the President by no means always applauded the Secretary. In 1713 Sloane resigned, after holding his office for twenty years, to be replaced by Edmond Halley; he would return after Newton's death to occupy the presidential chair himself and continue his predecessor's sound administration. The historian of the Society's business, Sir Henry Lyons, judged the first decade of Newton's term of office to 'constitute a period of exceptional importance in [its] history since [it] was a turning-point in its fortunes and its efficiency . . . the years of difficulty and steady decline were left behind, and a steady progress was setting in . . . From this time onwards both the number of Fellows and the financial resources of the Society continued to increase.'[43] Not everyone, then or since, has taken quite so rosy a view of the Newtonian administration – Flamsteed for one. For him the first decade of Newton's presidency was a time of oppression and annoyance. This did not end with the printing of Halley's *Historia Coelestis*, for Newton continued to 'hinder me all he can . . . lately he was for makeing me New Instruments I want not.' As chief of the Royal Society's Visitors Newton involved the Officers of the Ordnance in an examination of Flamsteed's equipment at Greenwich, but this move Flamsteed blocked by proving it was all his own property (February 1714). After Flamsteed had used his new influence to obtain the undistributed copies of 'Halleys Copy of my spoyled Catalogue' for his bonfire, Newton seems to have left him alone. Flamsteed's last known letter to Newton is a demand for the return of papers put into his hands years before.[44]

Foreign Savants

One factor in the increase of the number of Fellows of the Royal Society during Newton's presidency was the more frequent election of foreigners. The Society had always been generous in such elections. In the ten years 1699 to 1708 twenty-three foreign Fellows were elected, but in the twenty years 1709 to 1728 no fewer than ninety-six, an annual average more than twice as great. Partly this was due to the return of peace and the frequent arrival of learned or socially distinguished voyagers in London, partly to Newton's wish to make friends, for such foreigners were often proposed by the President. As we may infer from his presentations of his books, Newton was quite aware of the possibility of building up a Newtonian party on the Continent, and the years after 1715 saw visits to England of French and Italian scholars (and others) who were not unwilling to be converted to Newtonianism, especially (at first) experimental Newtonianism.

Quite a few of these foreign Fellows figure in the last volumes of Newton's *Correspondence*, among them the Duc d'Aumont and that great Pooh Bah of French academic life, the Abbé Bignon, along with his colleague (Newton's first biographer) Bernard le Bovier de Fontenelle. One of Tsar Peter's efficient ministers, Prince Menshikov, wrote from St Petersburg to Newton requesting election (and it was done). Another visitor to London (and FRS) Pierre Rémond de Monmort, sent Newton fifty bottles of champagne after his return to France in thanks for Newton's courtesies to him. Newton replied with an ornament for Madame chosen by Mrs Barton; he was by no means unschooled in social graces. Monmort was an important intermediary in the calculus dispute between Newton and Leibniz, linked in this with a better-known Italian traveller (also FRS), the Abbé Antonio-Schinella Conti, who had some correspondence with Newton on this topic in 1716, and more with Leibniz.

Although Newton was elected a Foreign Associate of the reorganized Académie Royale des Sciences in 1699, he had no contact with the body before 1713. He may have been unaware of an obligation to report his scientific investigations and send copies of his publications. Being now about the King's business he paid no attention to the Académie. Newton's dignity as President of the Royal Society and his need to vindicate himself against Leibniz forced him out of this self-imposed silence. If he (reasonably) did not distribute *Opticks* abroad, it was otherwise with *Optice*, soon being read in France and Italy. Ahead of Newton, Hans Sloane sent a copy of *Opticks* to his friend Etienne-François Geoffroy in 1705, of which the latter read a French summary to the Académie des Sciences at ten sessions between August 1706 and June 1707. His brother, Pierre Claude-Joseph Geoffroy, Monmort and the Chevalier de Louville came to London in 1715 to observe an eclipse: all were elected FRS and became personally acquainted with Newton. Louville was (in a veiled way) one of the first authors to expound Newtonian gravitation in France; all three were fully convinced by Desaguliers's demonstration of Newton's optical experiments. With *Optice, Commercium Epistolicum* and the Amsterdam *Principia* circulating on the Continent Newton had established himself as a natural philosopher of the highest standing, and his kindly reception of foreigners in London encouraged others to seek his acquaintance.[45]

Catherine Barton was no mean social asset to her uncle, now in his seventies. Like a gallant and susceptible Frenchman, Monmort was particularly ardent in admiration of Mrs Barton's intelligence and beauty. Word of her charms had reached him in France and when they met he found in her besides much beauty a most delicate and witty manner. The dry testimony of Newton's household inventory confirms his hospitality: his modest wealth is expressed in forty plates but only one decanter (so that wines must have been poured from their bottles directly); 370 ounces of silver plate (including two chamber-pots); nine table-cloths with six and a half dozen napkins but only one coffee-pot; and besides six loose spoons and forks a dozen place-settings in a shagreen case. Newton's dinner-parties cannot have been large since his dining-room numbered only eight 'India back chairs with seats and cases' which were presumably placed around the 'one oval table'. The room must have been a fair size since it also held 'two walnut tree card tables', 'two old tea tables' and a 'gilt leather screen', perhaps to reduce the draught from the door. It was warmed by 'one iron stove compleat' and lighted by four (silver?) sconces; eleven pictures and 'a figure cutt in ivory of Sir Isaac in a glass frame' enlivened the walls. Since Newton – at least in earlier days – had only a halting reading knowledge of French, conversation with his foreign guests must have been difficult, for even in Latin our peculiar English pronunciation would have defeated strangers to it.[46]

Newton as President

Richard S. Westfall has already made the point that Newton was not only the first scientific President for many years but also the most dutiful. He was rarely absent from the chair and he concerned himself with all the business of the Society, not only mathematical and physical matters. Soon after he had pushed through the acquisition of Crane Court against considerable opposition he had the council make standing orders for the good conduct of ordinary meetings. These more or less formalized the decent orderliness which Samuel Sorbière had remarked in the very early years of the Society: no chattering, all Fellows to speak to the chair only, and so on. Stukeley wrote of Newton's presiding with 'singular prudence, with a grace and dignity – conscious of what was due to so noble an Institution – what was expected from his own character . . . Indeed, his presence created a natural awe in the assembly.' Like a good modern chairman, Newton started off the discussion of a paper himself 'with a just commendation, where it might be improv'd, where any experiment might be better directed'.[47]

Further, and unlike any President since Brouncker, Newton presented (optical) experiments himself, in 1704 and again in 1708, or (more often) proposed experiments to be shown to the Society by Francis Hauksbee the elder and his successor, John Theophilus Desaguliers. Both curators *de facto* repeated old experiments – going back to Boyle's investigations with the air-pump – including such experiments directly relevant to Newton's work as those on the air resistance to the free fall of heavy bodies. Desaguliers gave demonstrative confirmations of Newton's optical discoveries; Hauksbee explored for him the phenomena of capillarity and surface tension, including the famous 'oil of oranges' experiment cited in *Opticks*. A former Curator of Experiments now back in London, Denis Papin, induced Newton to make (in 1708) his only allusion to the steam-pump and there and then invent the notion of reaction – or jet – propulsion.[48] In 1714 Newton talked about chemical reactions mentioned in Query 31 of *Opticks*, and in the same year Desaguliers repeated Newton's study of heat and cooling (at Newton's request). It was Newton who set Colonel Molesworth's new sand-glass before the Society and commented upon its imperfections; and on another occasion he spoke of earlier work of his own on burning-mirrors. While Newton was obviously more personally concerned with the physical sciences, the President seems often to have been interested in biological or antiquarian topics too; once he gave an account of a death by lightning-strike in Grantham that had occurred thirty years before. Historians agree that Newton gave a firm push to the return to 'experimental discourses' at meetings, further encouraged by Sir Godfrey Copley's legacy of 1709.[49]

Of all the experimental work in these years, that causing most revision of Newton's own natural-philosophical speculations was carried out by the elder Hauksbee (d.1713), the virtual founder of a whole new branch of experimental physics, of great importance in London (and everywhere else) during the eighteenth century. The French astronomer Jean Picard had first reported a strange glow in the 'vacuum' of a mercury barometer if the instrument were (perilously) shaken; that great anti-Newtonian Johann Bernoulli returned to this experiment in 1697. Hauksbee took it up in 1705, demonstrating before the end of that year that glass rubbed *in vacuo* would shine brightly. Before another year had passed he had inverted the arrangement: an exhausted glass globe rapidly whirled on a spindle and rubbing against the hand produced a brilliant glow; and (as Newton wrote),

> if at the same time a piece of white Paper or white Cloth, or the end of ones Finger be held at the distance of about a quarter of an Inch or half an Inch from that part of the Glass where it is most in motion, the electrick Vapour which is excited by the friction of the Glass against the Hand, will by dashing against the white Paper, Cloth or Finger, be put into such an agitation as to emit Light, and make the white Paper, Cloth or Finger, appear lucid like a Glow-worm.[50]

From these luminous beginnings came the origins of triboelectricity, which grew to so much splendour.

Newton followed Hauksbee's investigation carefully and may have had a hand in developing its line. The experiments besides being quoted in *Opticks* (1717) re-emerge in the final sentence of the *Principia*. They had a decisive effect in shifting Newton from the firm vacuist position of the first editions of both his great books to the specialized aetherism of their later editions – hinted in the *Principia* (1713), justified with some elaboration in *Opticks* (1717). 'There can be no doubt [writes Henry Guerlac] that Newton, having followed Hauksbee's experiments with keen interest, identified the electrical effluvium or electrical spirit with a kind of subtile, penetrating, elastic and highly active matter pervading all bodies, and was willing to give it the name of "aether".'[51] Equally, thermometric experiments by Desaguliers confirmed Newton in this new idea.[52] We may well believe that his thinking about these two sets of experiments in 1716–18 and consequential modifications to his books was the last real thinking in natural philosophy that Newton did.

Newtonianism in the Royal Society

Under Newton's presidency the fellowship of the Royal Society and the composition of the council annually elected inevitably came to reflect his authority and his pre-eminence in mathematics and physical science. Of the men active in Newton's defence and the exposition of his ideas during the early eighteenth century a few had been elected Fellows long before – Joseph Raphson in 1689, David Gregory in 1692, Richard Bentley in 1695, Abraham de Moivre in 1697. Among those still fairly new to the Society when Newton took office were John Keill, George Cheyne and the physician Richard Mead. After Hauksbee (1705) there was a gap in the election of Newton's supporters until John Machin came in (1710), followed by Roger Cotes, John Freind and another physician, James Keill (brother of John), all in 1711. John Colson, editor and translator of *The Method of Fluxions*, Desaguliers, and the Netherlander W. J. 'sGravesande, one of the great exponents of Newtonian physics, had all been elected by 1715, then followed William Stukeley, Stephen Hales the physiologist, James Jurin, Robert Smith (later Master of Trinity), Colin Maclaurin and Henry Pemberton, editor of the third and last *Principia* (1720). By this time every significant figure in the Royal Society was a Newtonian.

As we have seen, the Society had decisively aligned itself with its President in 1712 against its foreign Fellow, Leibniz. The *Commercium Epistolicum*, a carefully edited selection of documents going back half a century, was Newton's historicist answer to Leibniz's claim to the first discovery of the new infinitesimal analysis. In Newton's eyes the only valid rejoinder to it would be a similar production of documents by Leibniz showing that his technique was at least older than 1672; this impossible task Leibniz never attempted, though he did begin to sketch his own autobiography of the calculus.

The *Commercium* broke Newton's silence on the fluxions–calculus issue that had lasted since 1687. The year before had seen William Jones's publication of mathematical treatises by Newton, the next year would witness the second *Principia* containing in its editor's preface Roger Cotes's strong (if rash) reply to Leibniz's assertions of the absurdity of universal gravitation. In 1714 this book was pirated at Amsterdam. Next came Joseph Raphson's *History of Fluxions* (1715), while 1717 saw not only the third edition of *Opticks* but Samuel Clarke's publication of his exchanges with Leibniz on the merits of Newton's philosophy. Since by the end of the decade the first of the flood of Newtonian popularizations had begun to appear, it is evident that with the Royal Society behind them the Newtonians were strongly responding to Leibniz and other neo-Cartesians.

The *Commercium*, though advocating a case, contained no falsehoods that Leibniz was ever able to demonstrate; its fault is that of omission of Leibniz's side of the story. Newton only cheated in a rather unimportant way in making out that *On the Quadrature of*

Curves was an early work and thus its dot notation was too. The book quoted Barrow's letter of 20 July 1669 introducing Newton to Collins and gave the text of *De Analysi*. On this at one point (where he had mentioned moments of areas) Newton commented: "N.B. There is here described the method by fluents and their moments. These moments were afterwards called differences by Mr Leibniz: hence the name differential method." Similar comments, rubbing salt in the wound, indicate that Newton's *Second Letter* to Leibniz of October 1676 had "explained the methods of infinite series and fluxions at the same time", that Leibniz had had no competence in geometry before 1674, and had had no inkling of quadrature by infinite series until those of Gregory and Newton were sent him. According to Newton's chronology, a plausible reconstruction, Leibniz had not formulated his differential calculus before his return to Germany in the autumn of 1676, but in truth he had it over a year before. This chronology, a curious example of Newton as a historian, was based on errors of date and a priori prejudice; for example, Newton always believed that Leibniz had stolen the Newton–Sluse rule for finding tangents from his own letter to Collins of December 1672, whereas in fact Leibniz had found it in the *Philosophical Transactions* for 1673. In Newton's eyes, Leibniz's claim to have been a first discoverer, where on occasion he had been the second (but independent) ruled him out as a cheat who had truly discovered nothing.

The little book made it clear that Newton had been a very fertile and ingenious mathematician long before Leibniz began in that field. Leibniz and his friends did not seriously challenge this truth; they argued rather that what Newton had invented was *not the differential calculus* (trivially true) and then shifted the quarrel to other grounds: Newton had displayed so much mathematical ignorance that he could not possibly have been master of a genuine infinitesimal calculus. This line of attack was much developed by Johann Bernoulli, who strove to extend it by making all Newton's followers look ridiculous by their failure to resolve the problems he posed to the world. Meanwhile, Leibniz himself sought to demolish Newton's natural philosophy, especially his concept of universal gravitation, though this had nothing to do with the original issues in dispute. Leibniz still believed, as when he wrote the *Tentamen* in 1689, that the planetary revolutions were caused by an aetherial circulation, gravity by an aetherial pressure.

Bernoulli published criticisms of the theoretical mechanics of the *Principia* in 1713 and 1714, making great play with a real error (in Book II, Proposition 10) unperceived by Newton and his friends. In

Newton's general analysis of the effect of the resistance of air or water on the motion of bodies through them, he had made an error producing a result false by a factor of 3 to 2. Bernoulli had informed Leibniz of this in 1710. In September 1712 his nephew, Nikolaus Bernoulli, paid a visit to London, where he was recommended to Newton's friend, Abraham de Moivre, for some years a correspondent of Johann's. De Moivre introduced Nikolaus to both Halley and Newton, who invited the two men to dine with him twice. Nikolaus produced as his own his uncle's correction of Newton's false result, admitting that no one had yet found a fault in the argument that yielded it. 'Two or three days later when I [that is, de Moivre] went to [Newton's] house, he told me that the objection was valid and that he had corrected the result, which now proved to agree with your nephew's calculation. Thereupon he added that he intended to see your nephew and thank him, and begged me to bring him to his house, which I did.'[53] Newton had, indeed, devoted immense labour to putting right his proposition; to discover where he had gone wrong and correct the mistake he filled fifty sheets of paper, and he proved the correctness of his (and the Bernoullis') calculations five times over. It was thus Newton himself who actually found the mistake, which involved some partial reprinting of the second *Principia*, then in the press. Nikolaus later produced an ingenious argument to the effect that Newton's error arose from a misconception in the use of series, thereby convicting him of the error of not knowing how to derive a second-order differential; in fact, Newton had used a purely geometrical argument in which calculus had no place. Subsequently in the printed text he omitted any allusion to the Bernoullis.[54]

Newton's correspondence of this first fortnight of October 1712 gives a rich picture of his many and intense occupations. The new edition of the *Principia* was advancing with Cotes at Cambridge; Newton was writing some additional pages on comets, sent to Cotes on 14 October. Then there was Proposition 10 to rewrite. At the same time he was working out the exchange rates to be used in paying the British troops occupying Dunkirk, in French money. He was also trying to secure a widow's pension for Katherine (Greenwood) Barton, his niece by marriage. And he was attending to agricultural matters at Woolsthorpe, writing to his bailiff, Henry Ingle, about the pasturing of animals on common and fallow. Newton was rarely idle, but he did find time to attend to the affairs of his poorer relations.

When he heard of the *Commercium Epistolicum* and other doings of the English Leibniz was furious against the British who, he wrote to

Bernoulli, made a habit of appropriating German triumphs to themselves: it had been so with Boyle and Guericke's air-pump and now with Newton. Leibniz (like Newton) was a very busy man, too busy to defend himself properly; and also a devious one, as when he praised himself in anonymous reviews. He now attacked the *Commercium* in a Latin fly-sheet ("Charta volans"), anonymously of course, quoting Bernoulli's letters to himself about these matters without naming him. No more innuendos; here came the straight accusation of theft:

> it will be gathered that when Newton took to himself the honour due to another of the analytical discovery of differential calculus, first discovered by Leibniz in numbers and then transferred (after having contrived the analysis of infinitesimals) to Geometry, because Newton was not content with the fame of advancing [geometry] synthetically or directly by infinitely small quantities (or as they were formerly but less correctly called, the indivisibles of geometry,) he was too much influenced by flatterers ignorant of the earlier course of events and by a desire for renown; having undeservedly obtained a partial share in this, through the kindness of a stranger [that is, Leibniz], he longed to have deserved the whole – a sign of a mind neither fair nor honest. Of this Hooke too has complained, in relation to the hypothesis of the planets, and Flamsteed because of the use of his observations.[55]

This fly-sheet came to Newton's hands in the autumn of 1713. But it may have been its reprinting in continental journals, with added 'Remarks' by Leibniz openly, that made him feel a reply to be necessary.

John Keill had before this, in the early summer of 1713, contributed to the *Journal Littéraire de la Haye* a Newtonian history of the calculus, not dissimilar to a letter which he had composed for Leibniz in May 1711, in refutation of Leibniz's complaints against himself.[56] Definitely now established as Newton's champion, Keill opined that 'Mr Leibnits should be used a little smartly and all his Plagiary and Blunders showed at large.' Newton did actually draft an answer to the "Charta volans" intended for the *Journal Littéraire* but suppressed it in favour of an 'Answer' by Keill that came out in July/August 1714. Newton's draft contains some points of technical mathematical interest and the sharp remark that "Mr Leibnitz confines his Method to the symbols dx & dy, so that if you take away his symbols you take away the characteristick of his method." Keill had also written to Newton: 'I think I never saw any thing writ with so much impudence falshood and slander.' Newton agreed, and was not at all reluctant to urge Keill to his task of reproof.[57]

From this point onwards, so far as mathematical matters were discussed, the calculus quarrel passed from the hands of the principals to secondary figures: Keill, Brook Taylor (a brilliant young mathematician, the friend of Rémond de Monmort), de Moivre, des Maizeaux, the Bernoullis, Monmort, Conti and still others, some of whom attempted maladroitly to effect a reconciliation of the principals. Accusation was met by rebuttal and counter-charge. The British now brought up Leibniz's mistakes, particularly in his *Tentamen* of 1689. Johann Bernoulli charged that Newton had given an incompetent proof of the inverse theorem of central force, and continued to throw challenge problems at the British mathematicians (which, indeed, they were less well able to solve than he was).[58] To their credit, French and Italian mathematicians stayed clear of the wrangling, as did some of the younger Germans like Jakob Hermann, though they followed the methods of Leibniz rather than those of Newton. Their sense of the merits of Newtonian mathematical physics increased steadily between 1710 and 1720. On the English side, Newton continued to be deeply involved, though always behind the scenes; yet he had refused to give Roger Cotes hints for the editor's preface to the second *Principia* (1713), indeed would not read his draft, "for I find I shall be examined about it".[59]

Newton's last major contribution to the dispute was "An Account of the Book entitled *Commercium Epistolicum*" published in the *Philosophical Transactions* for February 1715. This fifty-page article was anonymous, but drafts prove it to have come from Newton's pen, a fact surely known to Edmond Halley, the editor of the *Transactions*, and to all intimate members of Newton's circle. It was a skilfully prepared piece of advocacy against Leibniz, founded upon true facts, saying nothing new and winning few new adherents. Newton contrasted the firm evidence for his own achievement before 1669 with Leibniz's mathematical inadequacies as late as 1677. Leibniz's series were nothing after what Newton and Gregory had done. How could Leibniz have written in March 1677 of the need for an 'analytical table of tangents' if he was already aware of how generally the calculus handled problems of tangency? In fact, Leibniz's method of tangents was Barrow's "exactly, excepting that he has changed the letters a and e of Dr Barrow into dx and dy". Discussing the differences between fluxions and differentials Newton argued strongly for the conceptual superiority and greater utility of the former; notably, he wrote, "By the help of the new analysis Mr Newton found out most of the propositions in his *Principia Philosophiae*; but because the Ancients for making things certain admitted nothing into geometry before it was demonstrated synthetically, he demonstrated the propositions synthetically, that the system of the heavens might be founded upon good geometry." Newton's claim here is categorical; did he in 1715 believe it to be true? Many have repeated his statement, but we must believe it to be false (see chapter 8). All evidence agrees that he composed the book as we have it. Finally, Newton rebutted Leibniz's claim that the neo-Cartesian mechanical philosophy was superior to Newton's philosophy of forces: "We are not to fill this philosophy [of Nature] with opinions which cannot be proved by phenomena." Imaginary mechanical causes, however plausible, were not good enough, nor should "Experimental Philosophy be exploded as miraculous and absurd because it asserts nothing more than can be proved by Experiments, and we cannot yet prove by experiments that all the phenomena in Nature can be solved by mere mechanical causes."[60]

The "Account", duly translated into French in the *Journal Littéraire de la Haye*, was duly diminished by Leibniz as twice-cooked cabbage. He claimed nothing of Newton's: the position simply was that since 1684 the learned world had recognized Leibniz as the inventor of the differential calculus, and only afterwards had the English taken to thinking otherwise. Newton continued to spend long hours on the dispute, even after Leibniz's death in November 1716, correcting the mistakes of those who sought to encourage a compromise, denouncing the subterfuges, prevarications and irrelevancies of Johann Bernoulli, and maintaining that he was not responsible for the words of those who upheld his own case. When in 1719 Bernoulli made a half-hearted attempt at self-exculptation and *rapprochement*, Newton treated him with cold, formal politeness.[61] The last details of this deplorable and thoroughly egocentric affair may be left to the specialist.

Clarke and Leibniz

A work of permanent interest to historians of metaphysics did appear in these last throes of the quarrel, Samuel Clarke's publication of the letters about Newton's philosophy that had been exchanged between himself and Leibniz, a series cut short by the philosopher's death.[62]

Leibniz's opposition to the non-mathematical features of Newtonian mechanics went back (as we have seen) to his first reading of the *Principia* (probably in Vienna) soon after its publication. After reiteration in the *Acta Eruditorum* over the years it received a formal elaboration in his *Essais de Théodicée* (1710). In Britain the new Princess of Wales, a philosophically inclined lady in correspondence with Leibniz since her upbringing in Germany, sought Samuel Clarke's reaction to Leibniz's depreciation of British Newtonian philosophy. He composed an answer to a letter from the philosopher to the Princess, starting a fivefold exchange.

Towards Clarke – whom it is impossible not to see as a disciple and supporter of Newton and difficult not to see as a friend also – Newton maintained the same overt detachment as towards Brook Taylor, de Moivre, Cotes and others.[63] Newton's relations to his British followers and defenders would make an interesting study. Fatio de Duillier is the unique man for whom (for a short space) he revealed any feeling of warmth and concern. David Gregory, Edmond Halley and later Abraham de Moivre were freely admitted to Newton's society but he has left us no sense of a strong human relationship with any one of them. Keill was the sole Newtonian (after Halley in 1686 and 1695) to correspond extensively with the Master and to leave traces of active collaboration; again, without signs of personal friendship. So far as is known Whiston and Clarke were unique early Newtonians in being (perhaps) infected by Newton's own entrenched scepticism concerning the orthodox doctrine of the Trinity; he gave support to neither man in his professional difficulties arising from these religious scruples – which of course may have been quite independent of Newton's (concealed) example. (Voltaire once jested of Samuel Clarke that he would have made an excellent archbishop of Canterbury if only he had been a Christian.) It seems that the Master of the Mint also withdrew from Fatio in his troubles, including a spell in the pillory. If we add the sceptical Halley, the Newtonian group looks strangely dissident in religion; among those lightly associated with Newton a few, such as Thomas Tenison, Archbishop of Canterbury, and Richard Bentley, Bishop of Worcester, won high preferment in the Church, but more typical was the devout, innovative and productive Stephen Hales, always a country parson.

As with several of the men classed as early Newtonians – not all mathematicians by any means – Samuel Clarke never figures directly in Newton's correspondence, nor is there more than the slightest documentary support for the belief that Newton 'not only received and studied Leibniz's papers [addressed to Clarke], but that he also collaborated fully with Clarke in his replies'.[64] However, Newton did take immense trouble in composing answers to Leibniz's criticisms, expressed in other vehicles, for example in Leibniz's letters to the Italian priest Conti, then in London.[65] Such papers by Newton may have been read by Clarke.

The nub of the exchanges was Leibniz's defence of the neo-Cartesian plenist philosophy of Nature as the only intelligible scheme of things, and Clarke's advocacy of the Newtonian world of atoms and the void filled by forces which may be expressions of the divine power. 'These documents enable us to witness the serious confrontation of two opposing philosophies – Newton's and Leibniz's – on such major issues as the nature of space, the attributes of God, miracles, the world as a machine.'

Newton's attitude to Leibniz's depiction of himself as a feeble if not absurd metaphysician was to declare it an irrelevancy to the original point in dispute, the origins of the calculus. In a neat *tu quoque* he charged Leibniz with

> calling those things miracles which create no wonder & those things occult qualities whose causes are occult though the qualities themselves be manifest . . . He prefers Hypotheses to Arguments of Induction drawn from experiments, accuses me of opinions which are not mine, & instead of proposing Questions to be examined by Experiments before they are admitted into Philosophy he proposes Hypotheses to be admitted & believed before they are examined. But all this is nothing to the *Commercium Epistolicum*.

So much is true; but it is true also that Newton had introduced novel conceptions of space and time into philosophy which were at least debatable, that he had also linked his natural philosophy with propositions about God and Nature that could no more be "examined by experiments" than Leibniz's individual metaphysics, and that in short he seemed in a number of ways to have gone counter to the course of scientific thinking as that had developed from the time of Galileo, Gassendi and Descartes.[66]

At any rate, Clarke can have been in no doubt of Newton's views about the issues raised by Leibniz, nor is it in doubt that he sought to answer Leibniz by deploying Newton's own arguments. Though now a forgotten scholar, Clarke had a considerable reputation as a dialectician in his own day and was well able to defend his Newtonian position. As for Newton, the only further contribution from him (already so advanced in years) that we need note is his endeavour, as late as 1718 in papers intended for the publicist Pierre des Maizeaux, to reconstruct both logically and chronologically the development of fluxions more than half a century before. Of a considerable mass of draft material little eventually emerged in print.[67]

Through the many years of the quarrel, which became almost a nationalist affair, the conduct of neither Newton nor Leibniz seems to satisfy the conventional canons of moral decency. To praise or blame is futile: so is it also to ask whether the participants could not have spent their time upon more profitable topics. Newton paid dearly for the effects of his own mistrustful and secretive temperament, which directed his conduct into barren paths, often against the advice of his friends. Having entered upon them, and with the misfortune of encountering so powerful a mind as that of Leibniz as his competitor, Newton could scarcely avoid protecting his ego against injury. In the debate it might be said that he had the best of it, though Leibniz had the better reason on his side. Newton's massive historical testimony, his force of argument and his powerful political sense put the heavier weapons in his hands. He spoke for common sense and Baconian empiricism against ingenious intellectual whimsicality – however much this last might appeal to the purely philosophical mind. And in consequence Newton stands higher, perhaps, in philosophy than Leibniz does in natural science. As I suggested previously, the controversy (rather than books) made Newton a dominant figure on the international literary scene, and the vindication (after 1715) of his experimental assertions gave him tremendous authority. Though the apotheosis of Newton came about long after his death he must have known that from the moment he (Master of the Mint) became King George's man, leaving Leibniz to fume impotently in Hanover, the tide was beginning to run his way. The late efforts of Bernoullis and Fontenelles to oppose it are forgotten historical curiosities. By the middle of the eighteenth century everyone understood that Newton had been the first inventor of the new infinitesimal analysis – though all Europe used Leibniz's calculus.

13

A Man of Authority and Learning, 1692–1727

Newton served the State in a variety of capacities for thirty years; he not only embodied in his own person (as Pepys had formerly done at the Admiralty and his contemporary William Lowndes did at the Treasury) a novel conception of the high-ranking, efficient, professional civil servant, but helped to established the character of the English government official in the parliamentary age: as a man of integrity, unremitting in his attention to business, thoroughly competent in matters of detail yet no mere clerk – a head of department able to realize ministerial policy in concrete terms.

As we have seen, Newton's association with government began with his election as a university Member of Parliament for Cambridge in January 1689. After the thirteen months – including a long summer recess – of the Convention Parliament no one at Cambridge seems to have wished to propose Newton for a second term. Two years after his translation to the mastership of the Mint (25 December 1699) he was again nominated and elected Member of Parliament for the university. He is known to have supported Halifax and the Whig

Junto on a vote of confidence but was in general no more obvious in the House of Commons than before. When the Parliament was dissolved after a few months (2 July 1702) Newton refused to stand again unless unopposed: "To solicit and miss for want of doing it sufficiently, would be a reflection upon me, and it's better to sit still."[1] At Cambridge as elsewhere the Tories were ascendant, even a strong canvass might fail a Whig, and there were evil rumours about the heretical religion (and worse) of Halifax's circle. A contested election might have been very disagreeable.

On the next occasion Halifax would not let Newton stand aside. During April 1705 he had to make three visits to Cambridge to pursue his canvass. One of them coincided with a visit of Queen Anne, no doubt also intended to aid the Whigs: on 16 April she knighted Newton at Trinity College, together with the Vice-Chancellor pro tem. who was Newton's friend John Ellis, and Halifax's brother. All in vain; in the following week Newton was bottom of a poll of four. 'It was shameful to see a hundred or more young students, encouraged in hollowing like schoolboys and porters, and crying No Fanatic No occasional Conformity against two worthy gentlemen that stood candidates', namely Francis Godolphin and Sir Isaac Newton. He never stood for Parliament again.[2]

Knighthoods and baronetcies had in the past been conferred upon physicians in recognition of their professional eminence and courtly discretion, but Newton was the first mathematician and philosopher to be so honoured. It would be pleasant to think that a reputation far exceeding normal academic bounds and nine years of devoted service to the Crown might have carried some weight in Newton's distinction – especially as he could not be made a bishop – but (as Westfall has said) it is more likely that the Tory strength in Cambridge and their clerical propaganda weighed more than his merits. The Queen endorsed Newton and the Whigs. The propaganda by bellowing students must have been very unpleasant to the former Lucasian professor. There is no positive evidence to show that Newton did not attend his parish church twice every Sunday and communicate frequently, but it is a natural guess that, even with so understanding a priest as Samuel Clarke, Newton's anti-Trinitarian views would have led him to undertake no more than his statutory minimum of observance of the religious forms.[3] He must have disliked and feared the non-Jurors and Jacobites, Dr Sacheverell and every other mani-festation of High Anglican Tory intolerance. This alone would have prevented any friendship with Jonathan Swift!

The Recoinage

Long before he became Master in succession to Thomas Neale, Newton had made himself the dominant figure in the Royal Mint. The Master, an expert on lotteries and a land-developer in the East End, left business to his able assistant, Thomas Hall, with whom Newton had cordial relations: Hall was one of his sureties when Newton was appointed Master. He soon found the wardenship less of a plum than Halifax had made it out to be (he and Hall received the same salary of £400 per annum, indeed a great sum in those days) nor was the Warden the Mint's chief officer; he begged the Treasury "to give such Orders for the Support of the Warden's Office as your Lordships shall think fit". Nothing came of the plea. Newton found the Mint in the throes of the great recoinage approved by Parliament: deputies were to be appointed for the various officers in five temporary provincial Mints (Halley served at Chester) and moneyers provided to do the actual work of melting the bullion, forming the blanks, and striking the coins. One of Newton's first official acts as Warden was to inform the Treasury of the progress made by June 1696 in setting up these Mints. He had already (on 2 May) taken the oath not to disclose the 'secret' machine for the 'new Invention of Rounding the Money' and milling or lettering its edges.[4]

The recoinage measure had been pushed through by the Chancellor of the Exchequer, Halifax (then still Charles Montague) in order to replace the worn-out silver money in circulation by new-minted coins of appropriate weight and face value; some of the current money had been coined under Edward VI and Elizabeth. It had lost the confidence of the public, who readily clipped it, devalued it and melted for bullion any full-weight coins put out piecemeal. Halifax had fixed the policy – which put a heavy burden on the Treasury for finding the money and on the public which lost on the face value of its coins – long before Newton went to the Mint. He had no responsibility for that policy nor for the beginning of its execution. He had indeed been consulted on the question by Halifax while still at Cambridge, as had John Locke also. Newton proposed a different and perhaps more equitable process than that which Halifax adopted, while Locke essentially proposed doing nothing at all.[5] Afterwards Halifax liked to say that he could not have carried through the recoinage without Newton, praise which must refer to his administrative ability. As for his genial powers in mathematics and science, there was no call for them at the Mint. Brewster absurdly alleged that Newton's 'chemical knowledge was of great use to the country', and perhaps if he had been able to turn the tin of Cornwall into silver it might have been so; some modern writers have alleged that Newton himself carried out delicate assays of gold and silver content. The extreme unlikeliness of the Master assuming the official functions of the King's Assayer requires no emphasis.[6]

The Mint was in the Tower of London, far from the fashionable part of town, a ramshackle collection of decrepit wooden buildings in a castle crammed with troops and horses – the Mint itself stabled thirty-three at the height of the recoinage, working its machines in relays for twenty hours a day. Then it employed 500 men. Affrays and fights, not to say robberies, within the Tower were not uncommon. No principal officer of the Mint occupied his house there – though presumably they were rented out. The officers met on Wednesdays (for bullion purchases) and on Saturdays (for general business). The Wednesday meetings clashed with the assemblies of the Royal Society, a good enough reason for Newton's rare appearance among the philosophers. When he became President of the Royal Society its day was changed to Thursday. Newton must have done much of his work at home, there preparing the Mint's letters and reports to the Treasury, his computations of currency values, in all a mass of paper drafted by him even if duly signed by the Master, Warden and Comptroller.

In any case, the new Warden was not much concerned in the great business of the recoinage, planned and carried through by Thomas Hall. True, he was involved by Halley in stormy affairs at the Chester Mint, starting from Halley's using unorthodox alloying procedures and going on to the hurling of inkpots.[7] Newton, near the end of this Mint's work, tried to interest Halley in a post at the Tower instructing Army officers, but he preferred (it seems) to stay with the Navy and began a new scientific voyage in the pink *Paramour* on 22 November 1698.

A chief duty of the Warden was to police the currency by bringing clippers and counterfeiters to justice (that is, Tyburn). A warden's clerk undertook the work, ineffectively according to Newton, until about the summer of 1696. On the retirement of George Macy from this post Newton begged that this duty be taken off the Warden, "nor enjoyned on me any longer, at least not without enabling me to go through it with safety credit and success". The currency ought to be defended by the Crown's law officers. However, in September 1696 the Treasury approved Newton's appointment of an extra clerk, Christopher Ellis, to carry on the police work into which (for all his protest) he plunged with his usual vigour. He had himself made a

Justice of the Peace in seven counties near London. He paid a man £5 to buy a suit of clothes, to qualify him for conversing with a gang of coiners. He drafted warrants for the handling of suspects turned king's evidence and for the legal protection of his own agents in pursuit of criminals. And he himself handled the affair of William Chaloner.[8]

A bold villain, Chaloner had printed a pamphlet about the recoinage in 1695 and early the next year had tried to convict the Mint of incompetence and fraud. According to Newton he was a skilled workman who had made a fortune by horse-stealing and defrauding both the Bank of England and the Exchequer; he had bought property and passed as a gentleman. In 1697 he won favour with a parliamentary committee on Mint miscarriages, despite Newton's opposition. Continuing to collect evidence that Chaloner engaged in bill-forgery and was making preparations to counterfeit the currency, Newton had him imprisoned in Newgate. Managing to escape the Mint's clutches this time, Chaloner petitioned the Commons for relief, as an honourable citizen persecuted by an evil and malicious organ of State. Another parliamentary inquiry followed (March 1698) where Newton was a principal witness against him. On the loose again Chaloner forged government bonds and counterfeited coins, or so the rather dubious witnesses appearing against him declared upon oath. He was again committed to Newgate in October 1698, convicted of illegal coining in February 1699 and executed at Tyburn on 22 March. Chaloner appealed twice to Newton ('I am murderd O God Allmighty knows I am murdered') alleging the perjury of the witnesses against him; he might well know the grisly trade, having himself 'peached' against fellow-coiners, who were condemned.[9]

Newton has been harshly judged for his keen, intelligent and successful pursuit of criminals, by which he claimed to have stopped currency crime where his writ ran. Whether, given his official responsibilities, he was more severe than many country gentlemen who protected their partridges as Justices of the Peace, others may judge. It is clear that Newton could only have saved Chaloner's neck by confessing that the witnesses against him had been suborned, while (on the contrary) Newton firmly believed Chaloner a persistent, dangerous criminal guilty of many crimes. It is not reasonable to blame Newton for the penology of his age. Manuel has written that Newton was a sadistic vampire: 'At the Mint he could hurt and kill without doing violence to his scrupulous puritan conscience. The blood of the coiners and clippers nourished him.' This is blood-tub Victorian melodrama rather than biography.[10]

Newton followed the path of duty as meticulously and relentlessly as he did all things. He was against crime and meant to end it. He was (one admits) hard on innocent button-makers and other artisans owning licensed screw-presses that could, in principle, be adapted to coining. He was hard too on the moneyers if he thought they were exorbitant in their demands for payment, and on junior officers of the Mint caught cheating. Equally, he pushed himself hard in learning the business of his office, compiling a number of *aide-mémoires* on the intricacies of Mint procedure, firm too in his mastery of its privileges and in asserting them against such rivals in the Tower as the Officers of the Ordnance.[11]

Longitude at Sea

Yet, and in spite of Newton's determination not to be distracted from the King's business, there were other things during his first years in London than the Mint and the mathematical concerns with Flamsteed and others that were his principal alternative occupations. There was a trip to meet Peter the Great of Russia, the promotion of the careers of his younger friends Halley and Gregory, Lincolnshire business of course, and – not least – the problem of finding a ship's longitude at sea. Through all the early years of the eighteenth century Newton was the Admiralty's unpaid consultant on many proposals for the solution of this problem. In this context, and in that of the supervisory function over the Royal Greenwich Observatory that Newton created for himself and the Royal Society, he made himself the initial scientific adviser to a British government.[12]

The longitude question and the Admiralty's all too frequent requests for his opinion of absurd proposals were to plague Newton almost till the year of his death. He always maintained that, like the latitude, the longitude of a place could only be found by an astronomical process, though (once found) it could be "kept" at sea by other means, such as chronometers. He by no means denied that accurate jewelled watches (Fatio's invention) could be useful at sea; they could be corrected each noon by the Sun.[13] In 1713 a minor Newtonian installed by Newton as a master at Christ's Hospital school, Humfrey Ditton, together with William Whiston, publicly proposed the maintenance of 'sound-ships' round the English coasts, by the cannon signals of which vessels might fix their precise positions under any conditions. Merchants and sea-captains petitioned Parliament for the proper consideration of this and other ideas: Newton was the leading expert consulted (in June 1714) by a parliamentary committee (others being Halley, Cotes and Whiston), the offer of a generous public reward for a proven scheme for the protection of shipping from errors in navigation being canvassed. Newton read a prepared statement of his own fixed views to the committee, declining (until badgered a little by Whiston) to pronounce on the practicality of either the Whiston–Ditton proposal or the system of rewards. Halley and Cotes favoured experiments to test the proposal. Newton was brought reluctantly to admit that the scheme would be useful near the coast, and the committee reported in favour of the rewards, adopting Newton's distinction between great and less accuracy of position.[14]

The Act of Parliament instituting the rewards also created the long-lived Board of Longitude to judge the feasibility of proposals, whose members besides the Lord High Admiral, the Speaker of the House of Commons, the Master of Trinity House and the President of the Royal Society were the appropriate university professors.

One of the well-known methods for determining longitude relied on accurate predictions of the future position of the Moon as observed from a fixed point of reference (such as Greenwich). By precisely observing the local time of the Moon's occultation of a known star (for example) at a particular place, and looking up in a table the predicted time of the event at Greenwich, the time difference of the place from Greenwich (and so the longitude) appeared at once. Newton thought that some not very great improvement of his own lunar theory might initiate the usefulness of such a method at sea. In fact, the most significant of his publications in the first years after his move to the Mint was the outline of this lunar theory which Newton allowed Gregory to print in his *Astronomiae . . .Elementa* (1702), as noted before (p. 256).[15] In this paper Newton took the solar parallax to be about ten seconds (hence the Sun's distance some 80 million miles), the length of the tropical year to be 365 days, 5 hours, 48 minutes and 57 seconds and that of the sidereal year 365 days, 6 hours, 9 minutes and $14\frac{1}{2}$ seconds. Newton's calendrical evaluations were sent by Sloane to Leibniz (who had consulted both the Royal Society and the Académie Royale des Sciences) in July 1700, when the later was concerned with the German states' adoption of the Gregorian calendar.

Master of the Mint, 1699

As a new Master of the Mint, Newton's first actions were to appoint a deputy, John Francis Fauquier, a director of the Bank of England, to cut slightly the pay of the moneyers, and to repel a challenge from his successor as Warden, Sir John Stanley, seeking to interfere in the mysteries of alloying the silver before coining. Newton overpowered him by copying thirty-four pages from Elizabethan precedents! He was also soon faced with a recurrent problem, that of the international trade in precious metals and its relation to the bullion content of coin; the gold/silver price ratio was also related to this. As Newton wrote: "this high price of Bullion has not merely put an end to the coynage of silver, but is a great occasion of the melting down and Exporting of what has already been coined."[16]

Partly for this reason, after the recoinage was finished in 1698 business was slack at the Mint. In Sir John Craig's words: 'the system of contracts and subcontracts divorced the Master from most of the staff and their activities. It is quite likely that Newton never set foot in the coining-rooms, and a visit to the melting-house during experiments on casting copper [in 1713] appears to have been quite exceptional.'[17]

Newton found work for his clerks to do by setting them to copy missing records from outside sources and to sort out old accounts. A survey and plan of the Mint buildings was made (dated 26 February 1701). He searched out old dies, destroying some and keeping others as models. He begged the gunner of the Tower to "order the Guns in such a manner that upon firing they may do least harm to the windows of the Mint." No labour was too tedious and no detail too minute for Newton. He slightly raised the fineness of the gold coined (though less than he supposed) and improved the standard of accuracy in the weight of coins, thereby reducing the price of gold bullion by £0.02183 to £3.9865 an ounce. He tried to enforce high precision in provincially stamped hallmarks and was able to postpone a move back from Britannia standard in plate (1696) to the less pure sterling silver until 1719.

At various times in his thirty years at the Mint Newton enunciated a variety of theoretical ideas about money. He realized that its value is based on confidence: "Tis mere opinion that sets a value upon money . . . and the same opinion sets a like value upon paper security." True, coins made of precious metal have a so-called intrinsic value, but even this is subject to opinion, since in the East silver is more highly valued (in comparison with gold) than in Europe; hence Europe – drawing silver from Peruvian and other mines – exports it to the East in return for gold or goods. There could be no permanent fixing of the relation between gold and silver in Europe and in Newton's view the contemporary valuation of the English gold sovereign at 21s.6d. (a 'guinea') was too high. Newton understood the problems faced by modern Chancellors of the Exchequer: too restricted a currency causes a high interest rate to prevail, which is bad for commerce "and the designs of setting the poor to work", while too large a quantity of money in circulation causes interest rates to fall, encourages luxury imports and the export of bullion. "Let it be considered therefore what rate of interest is best for the nation and let there be so much credit (and no more) as brings down money to that interest."[18] To some extent this notional concept of money conflicted with another firm principle of Newton's that coin ought to be of the intrinsic value of its metal, less the cost of manufacture. This notion he applied even to base currency and therefore consistently opposed the grant of patents for the private minting of halfpence (to the great profit of the patentees).

The practical aspect of all this theory was that Newton became the nation's authority on all matters concerning the world's money. In 1701–2 he made a thorough study of the weights, fineness and value of the coins circulating all over Europe, in Turkey and the Americas. Some of this strange coin he paid for out of his own pocket and retained. He recommended changes in the laws about the import and export of precious metals but nothing was done, nor was a coinage of copper undertaken before 1713.[19]

As the authoritative expert Newton was many times called upon to determine the values of foreign coins in circulation (as in Ireland in 1712), or to be paid to troops overseas, or to be received in payments from other countries.[20] Newton's masterpiece of this sort was a paper to which he gave immense labour, bearing strongly on the gold/silver price ratio, for he was convinced that over-valuation of the guinea drove silver from Britain:

> For as often as men are necesitated to send away money for answering debts abroad, there will be a temptation to send away silver rather than Gold because of the profit which is almost 4 per cent . . . But if only 6d. were taken off at present, it would diminish the temptation to export or melt down the silver coyn.[21]

This was done later, fixing the guinea at 21 shillings for ever, but silver was not thereby brought to the Mint. Locke had said that only one metal could serve as the basis for money, which he took to be silver; but in eighteenth-century England the basis of money was gold. Inevitably (but in vain) Newton in his papers of advice to the Treasury deplored the drain of silver to the East by the East India company, which was another factor bringing about the dominance of gold, but his mercantilist craving to ban merchants in the eastern trade from bringing costly, unnecessary luxuries into Britain could only be frustrated.

Newton constantly reiterated that no metal or little was brought to the Mint to be coined because no profit to the bringer resulted. Apart from a trickle of silver from the Welsh mines activity resulted only from such a windfall as a joint Anglo-Dutch naval force brought back from Vigo Bay in the autumn of 1702. The haul included sacred vessels, ewers and basins, crucifixes, a gilt mermaid and of course money and tableware such as plates, cups and chocolate pots. Only a small fraction of this booty came to the Mint, weighing 4504lbs. Some of the plate was set aside as "valuable for its fashion". There was a small amount of gold too, but the whole was equivalent in coin to under £15,000. The silver coins minted were stamped with the word VIGO and Newton himself produced designs for a commemorative medal of this Drake-like enterprise. Such medals were made and sold as a private speculation by the engravers, and forty types were produced under Newton's mastership. He evidently fancied himself as a designer of such things and Marlborough's victories gave him many opportunities to try his skill and ingenuity.[22]

From the autumn of 1703 until 1717 the Master was busy commercially and administratively with the disposition of the product of the Cornish tin-mines. He had assured Godolphin, the Lord High Treasurer, that the Mint could store 2,000 tons of tin without interference with a large coinage. This was rather more than the maximum weight of metal that the Crown had undertaken to buy from the miners in each year, presumably so that ministers could be sure of 'good' Parliaments. At a selling price of £76 a ton a full year's purchase was worth £121,600 for which the miners were paid £111,200. The difficulty was to sell so much tin. Newton as a salesman was resourceful and energetic, investigating the market in tin with care. He was soon informed that an increase in price would bring in competing tin from the East. He in turn continually warned the Treasury that the contract tonnage was too great, that the Mint was

stuffed with tin, and that a good deal of money was being lost in the interest charge upon the value of this unsold stock. In 1705 Newton reckoned that the tin contract would cost the Crown £270,000. But towards the end of the period trade became more brisk. A recalculation late in 1709 made the Crown's loss rather smaller.[23]

A draft of 1708 bearing on the metallurgy of tin may reflect something of Newton's own experience with melting metals, but more probably it is his digest of information given to him by commercial smelters. When the tin sales were running reasonably smoothly Newton left them to his subordinates.

A fresh problem for the Mint arose as a consequence of the Act of Union between England and Scotland in 1707: the Edinburgh Mint was to be harmonized with the London Mint and it should proceed to call in old silver money and recoin it to the English model. Gold and copper circulating were to be left alone, yet the London goldsmiths prepared new 'trial-plates' of both gold and silver for both Mints.[24] Four of the London moneyers went to Edinburgh to assist with, for a shorter time, one of Newton's own clerks to put records on a proper footing there. Newton's friend Gregory, Savilian professor of astronomy at Oxford, was put in charge of the party for the handsome fee of £250. Much equipment besides the dies had to be supplied from England before the Edinburgh Mint could work at all. Only trial and error could discover how to fire the furnaces with fossil coal rather than the charcoal used at London.[25] Gregory left, with the Mint temporarily coining £6,000 per week, late in November 1707; with £320,000 of silver coin produced the Mint in Edinburgh ceased work for ever in March 1709, leaving Newton with a tiresome legacy of complaints to tidy away.

Most of this new coin seems to have passed uselessly into the vaults of the Bank of Scotland, while in England too it was scarcer than ever. Accordingly, the Treasury decided in April 1709 to offer a small premium on silver brought to the Mint for coining, a move which brought in £79,000. When Robert Harley replaced Godolphin at the Treasury he wished to repeat the experiment, beginning on 10 May 1711, the sellers not receiving in exchange for plate a quantity of coin but certificates of subscription to a new government loan. Four days after the start the Treasury changed the rules. Tremendous confusion resulted, which Newton had to sort out; his proposed measures aroused the wrath of the Warden (Craven Peyton) – but Newton had long before noted his failure to be always on good terms with his fellow-officers. The yield of this offer was rather less than in 1709, mostly in the four days before the Treasury shuffled its feet. The

Mint of course had to be repaid for these loss-leading operations, this process too putting extra administrative burdens upon Newton just when he was getting to grips with Cotes and the second *Principia*.[26]

The Trial of the Pyx, 1710

An even worse row, in which Newton was seriously at fault (by modern assays) – or rather the Mint assayers were, and Newton took their part – broke out in 1710. In that year a 'trial of the Pyx' was due, as previously in 1701 and 1707, neither of these having caused Newton any trouble. The trial, in the government offices at Westminster, with members of the Goldsmiths' Company chosen as a jury to judge the Mint's production over a three-year period, was one of solemn splendour and also of feasting, since dinner for those concerned at the Dog tavern cost the vast sum of £1 per head (the Mint paid). Newton had of course documented the ritual and process at the beginning of his mastership with his usual thoroughness. Trouble arose in 1710 from advancing technology: the Mint itself had supplied the goldsmiths in 1707 (at the time of the Scottish recoinage) with purer gold than ever before; when the goldsmiths added the prescribed proportion of silver, the plate made was of a higher standard than the last (made in 1668) and still higher than that of 1660. Thus when Newton had written to the Treasurer and the Lord Chancellor had ordered the Goldsmiths' Company to present its jury on 21 August 1710, the jury found the Pyx coins light, substandard by a quarter of a grain (2.6 parts per thousand). Newton was furious, assuring the Treasury that the new test-plate was "finer than the last trial piece [1668] by about a quarter of a grain & that the last trial piece is something too fine by the assay". By modern methods, he argued, "gold may be refined so high as to be almost half a grain finer than 24 carats."[27] A longer paper (whether communicated or not is uncertain) added some technical detail: if goldsmiths, Newton wrote,

> when they have watered their granulated gold once or twice with Aqua fortis, . . . should dulcify it & grind it very fine as painters do their colours, & then water it once or twice more with double Aqua fortis in the same degree of heat as before & keep it longer in the water [that is, acid] then before stirring it now & then with a wooden stick to make the gold mix with fresh water: the gold would become finer than by the Assay, and by consequence finer than four & twenty carats. Chymists also tell us that Gold may be made finer by Antimony then by Aqua

fortis & by consequence then by the Assay; & gold refined by Antimony is of a better colour than Gold refined by Aqua fortis, & by reason of its fineness will go much further in gilding, as I have heard. But the Refiners of this city know not how to Refine gold by Antimony.[28]

Whether or not Newton used his science to benefit his country, he certainly used it to blind his critics!

According to modern estimations, reported by Craig, the 1660 trial-plate held by Newton to be of the true standard (916.6 parts gold in one thousand) was well below it at 912.9/1000, while the new plate of 1707, unacceptable to Newton, was indeed slightly above it at 917.1/1000, but much less than Newton supposed. Newton recommended that the currency should be measured by some physical standard constant for ever, and not by trial-pieces which might be remade from time to time, and so vary by trifling amounts between themselves. Hardly a pennyworth or two of gold was at stake in these variations, however, and no administrative alterations seem to have been made – but until 1829 the 1668 plate was always used for the trial of the Pyx, thus meeting Newton's requirement. His gold coin was a little below this fineness.

I find no hint anywhere in these discussions that Newton himself made experiments on the refining of gold by any of the methods stated, or that he was drawing upon his own chemical experience. His notebooks contain nothing on the refinement of gold above 24 carats, nor is it easy to see why any such ambition should have guided him during his academic days.

Copper Coinage

The recent history of the copper currency – I ignore an ill-fated attempt to introduce tin – was related by Newton in a Treasury letter of December 1710, after (typically) having "Enquired into all the Coynages of that sort since the yeare 1672". From then to 1694 bronze coins at 20 pence to the pound weight, of Swedish metal, had been issued by a commission; then for seven years patentees had loosed 700 tons of money that was "light, of bad [English] copper, & ill coyned" upon the public. Newton condemned this patent and affirmed that there was no shortage of halfpence and farthings in the country (there were no copper pennies as such, though there was still a formal issue of silver pence). The public may have thought otherwise, and certainly there were plenty of businessmen with propositions for supplying its needs in ways immensely profitable to themselves and the Crown. The tale never omitted, that Newton was offered a bribe of £6,000 to promote one of these consortia, is probably false and certainly he opposed them all. But when in 1712 Lord Oxford and Mortimer himself set this ball a-rolling, Newton could not oppose him. He merely opined that copper coins should be made as near as possible to the intrinsic value of the metal, that they should be issued by the State, and that 600–700 tons would be quite enough. When would-be patentees and officers of the Mint were called to Whitehall by Oxford on 8 May 1713 Newton repeated his advice: good copper, he added, "will bear hammering when it is red hot" and should cost less than twelve pence a pound. The Mint, he insisted again and again, should do the work. The principles of procedure were drafted by Newton.[29]

In the summer of 1713, presumably, the Treasurer ordered the Mint to go ahead on the lines indicated by Newton, who (for once) was only too willing to drag his feet. Little seems to have happened by January 1714 when the Mint submitted a paper (perhaps written by Newton) on the quality and malleability of copper. Even by March the preparation of Mint buildings for the coinage, the setting up of a furnace and so on were still to do. Experiments showed that the Mint's horse-powered rolling-mills were not adequate to cope with thick bars of copper; in the end Newton had to buy strip copper (from which the blanks would be cut) from various copper-mills. Some halfpence and farthings were coined – perhaps a few hundred – but none were ever issued to the public. A few years later the project was resumed: slowly in the autumn of 1717 Newton examined specimens of copper submitted for coinage at the Mint and when again summoned to the Treasury on 8 April 1718 could report that six tons of copper halfpence had been issued by it. Newton had also, as ordered, made preparations to coin quarter-guineas of gold (these minute coins were not a success). Intermittently, coinage of copper continued until 1725, by which date just over £30,000 of small coin had been issued.[30]

A number of manufacturers of copper from whom strip was not bought voiced grievances that caused Newton trouble, among whom was William Wood of Wolverhampton. But this row was nothing compared to that over the Irish base currency. In August 1722, despite Newton's protests, this Wood was allowed to establish a Mint at Bristol – of which Newton was appointed Comptroller (acting by deputy) – in order to coin 360 tons of copper farthings and halfpence for Ireland. Wood's coins were issued from April 1723 to March 1724, when the outcry from every quarter of Ireland forced the Ministry to halt the issue (it was resumed later, to a smaller total). None was fiercer against 'Wood's ha'pence' than one 'M. B. Drapier', whose real name was Jonathan Swift.[31] 'Wretch', 'public enemy', 'little impudent hardwareman' were among Swift's kinder descriptions of Wood. Newton's name figures in the Drapier's *Second Letter*, for he and two others had, under government orders, tested the quality of Wood's production which they found to be of full weight and of good copper well coined. The money was superior to what had passed in Ireland before. To Swift this was irrelevant: if Newton's report could be read as vindicating Wood's performance of his contract, Swift retorted that the Irish people had never been a party to it. Rightly, they had detested, abhorred and rejected Wood's coins 'as corrupt, fraudulent, mingled with dirt and trash'. Half-crazed with disease, rage, and resentment of his rejection by the Whigs, he urged the people of Ireland to revolt:

> I will shoot Mr. Wood and his deputies through the head, like highwaymen or housebreakers, if they dare to force one farthing of their coin upon me in the payment of an hundred pounds. It is no loss of honour to submit to the lion but who, with the figure of a man, can think with patience of being devoured alive by a rat.[32]

Swift's object was to raise among the Irish a demand for sovereign control over their own affairs and so to damage the hated Whigs. This reckless aim he pursued with all the power of his malicious pen.

Did he then recall Leicester Fields? That the *Drapier's Letters* were condemned by proclamation is hardly astonishing. One wonders what Mrs Conduitt thought of them.

John Conduitt

By this time Newton was very old. In his eighty-third year he was still writing letters about the right of Samuel Rollos to follow the late Samuel Bull as chief engraver to the Mint, though by now his nephew-in-law John Conduitt had been brought in unofficially to assist the Master, and later to succeed him. (Conduitt, who had married Catherine Barton on 26 August 1717, is virtually unknown before his emergence in Newton's family. A Hampshire gentleman, with an estate at Cranbury Park near Winchester (whither Catherine withdrew with him) he had graduated from Trinity College, Cambridge, and perhaps served in the British Army in Portugal, 1711–12; he was some nine years junior to his wife.) Although much of Newton's private financial business had previously been conducted by Francis Fauquier, his deputy as Master, in the last years of Newton's life it seems that Conduitt assumed the functions of his business manager. Especially he understood the intertwining of Newton's private finance with the Mint accounts, and was much involved in the settlement of his estate.[33]

The twenty-two known letters written by Newton to the Treasury during the last decade of his life are almost all routine. In December 1721 the Mint received from one Jacob Rowe, a patentee for 'fishing of wrecks', 618lbs. of silver bullion and Spanish coins, taken from a galleon wrecked off Ireland, being the Crown's one-tenth share of the salvage; in June 1722 slow inflation forced an increase in the salaries of the clerks of the Mint; in September 1723 Newton was threatened with prosecution. (It is not clear whether the case ever came to court – it arose from the discovery of a small silver-mine in Scotland in 1716 – both then and now the affair cost Newton a good deal of labour.) The final Mint letter was from the weigher and teller, Hopton Haynes, a Unitarian in belief with whom Newton had had cordial relations over many years.[34]

The Royal Society revived

Newton was not only one of the longest-serving presidents of the Royal Society but also (it is commonly held) one of the most diligent, dictatorial and intolerant of opposition. It may be that Newton's character as a president mirrored his character as a man; but it must also be remembered that the tradition of strong, attentive presidents, such as William, Viscount Brouncker, had been during the first thirty years (almost) of the Society's existence, had long been broken, that the Society was dreary and unenterprising during the 1690s and that energetic and novel measures were required to deal with the problems facing it during the early part of Newton's presidency. His critics seem to have been mostly mediocre men – though some, like John Harris, were active enough in publication – who wished the Society to remain a peasant club, while the new President and his friends wished to raise the professional level of its meetings. As I have pointed out already, Newton did not seek to devote the lion's share of the Society's attention to the physical sciences, nor much change the discursive form of its assemblies, but he did enliven them by reintroducing both physical experiments and anatomical demonstrations.

For practical scientific work Newton was now largely if not wholly dependent upon others' hands. Denis Papin, who returned to furnish matter for the Society's meetings from 1708 to 1712 (when he died) is never mentioned by Newton, who none the less was certainly stimulated by Papin's contributions. The latter was eager to win material support for a number of inventions, including an improved steam-pump, of which Newton's native caution may well have rendered him sceptical. Of the two great Curators of Experiments of this time, successively Francis Hauksbee and John Theophilus Desaguliers, a good deal is heard in Newton's books and correspondence, as well as in the *Philosophical Transactions*.[35] These two men, who both earned their bread by presenting natural philosophy to the public, were by no means merely Newton's technical assistants, but their Royal Society work was subservient to Newton's direction. He certainly gave money to the former and possibly to Desaguliers also. A demonstration of air-pump experiments by Hauksbee at Newton's home in 1705 (his first appearance on the scene) was worth the generous sum of two guineas. (Yet one might suppose Newton's getting together of "some philosophical persons . . . who will otherwise be difficultly got together" was for Hauksbee's advantage.)[36] Among the persons to be present on this occasion were the Earls of Halifax and Pembroke. The latter, briefly President of the Society before Newton, was something more than his acquaintance since he introduced Newton to Locke (it is alleged) and was certainly a pall-bearer at Newton's funeral in Westminster Abbey. Newton called him "a lover of stone dolls" because he collected statuary.

Desaguliers, an Oxford graduate and an independent investigator of a less mechanical type than Hauksbee, was brought to the Royal Society by Newton in 1714. His importance greatly increased with his successful vindications of Newton's optical experiments in 1715 and later; these had a great effect in bringing over the French and other sceptics. His cometary observations of July 1719 and experiment on air resistance to falling bodies in the same year both appeared in the third *Principia*. Unfortunately, no correspondence between Newton and Desaguliers survives, apart from a single letter in which the latter appears with a dignified subservience to the former. Desaguliers had to ask leave to miss a meeting of the Society when he went to observe an expected solar eclipse at Bath, and to look after the bricklayer and smith when they came to fix a stove to warm the repository at Crane Court, but he could also stand in for an absent secretary.[37]

Desaguliers, whose major book still lay far in the future, was too young to be a rival to Newton's old friend Halley, to substitute for the lost Fatio de Duillier, or to replace those friends who died early in the time of Newton's greatness, like Locke (d.1704) and Gregory (d.1708). Richard Bentley, twenty years younger than Newton, was to outlive him without real intimacy ever growing between them, despite Bentley's importance in Newton's history. His brilliant protégé, Roger Cotes, forty years younger than Newton, resembled Desaguliers in being unable to strike warm relations with the septuagenarian whom he revered so greatly; Newton consistently treated him with distant politeness. Bentley, constantly in the background of Newton's life since the *Principia*, appears in full light again in June 1708 with the news that he was to begin printing the second edition of the book at Cambridge, the paper, type and layout being chosen according to his own elevated ideas. Meanwhile, with Newton's concurrence (but against Flamsteed's wish) he had secured the new Plumian professorship of astronomy at Cambridge for Cotes.[38]

The Ancient History of the World

I shall return to Cotes and the revised editions of the *Principia* in the next chapter, devoted to Newton's later scientific publications following *Opticks*. One might have supposed that his official duties, his distant superintendence of his Lincolnshire estate, his participation in the quarrels between British and continental mathematicians, his presidency of the Royal Society and his improvement of his own scientific writings would have been occupation enough for an elderly man. Besides all these activities he was engaged in a great deal of committee work; he was a member of the 'Society of the City of London of and for the Mines, the Mineral, and the Battery Works'; of the commission for building fifty new churches in London and of the commission for finishing St. Paul's Cathedral, besides the parochial responsibilities mentioned already.[39] But all these tasks were not enough for Newton. He had still other work to do. In his London years he perfected, with enormous labour, his own system of the world's chronology. Like every Christian of his age Newton believed that the universe had been created by God in six days – periods commonly given a wide allegorical meaning – and he did not meddle with Archbishop Ussher's dictum that the creation had occurred on 23 October in the year 4004 BC nor did he seek greatly to vary from standard interpretations of Hebrew chronology as related in the Old Testament, despite the variant traditions of which he was aware. Outside that holy chronicle Newton presumably believed that nothing could be known of the first millennia of the world's history; the problem began with the emergence of a non-Jewish history (though we might call it mythology) late in the second millennium before Christ. When had these events of Egyptian, Greek, Roman and Phoenician history occurred, and how could they be related in time to the divinely chronicled episodes of Hebrew history? To cope with this problem the prehistorian of Newton's age had to make what he could of classical stories, the lists of kings and genealogies traditional to the Greeks, Egyptians and others, and the accounts given by the earliest writers among the Greeks and Romans.[40]

It is more than likely that Newton's interest in the scholarly problems of ancient history, and above all its essential chronological skeleton, was at least as old as his undergraduate days in Cambridge. As soon as he knew anything of astronomy he must have realized that in principle astronomical events could present absolute benchmarks within historical time; nor was he the first to do so though he was the first author to employ the slow rotation of the equinoxes for this purpose. Manuel cites one manuscript dated 1680 but most explicit or deducible dates seem to be much later. The working-out of the astronomical chronology cited (according to the same writer) first the star positions printed by Hevelius in his *Prodromus* of 1690, and later those communicated by Flamsteed to Newton. The datable letters and drafts used by Newton as rough paper for his chronology belong to the first quarter of the eighteenth century. Thus, though ancient history was far from being an interest of only the second half of Newton's life, the preparation of the *Chronology of Ancient Kingdoms Amended* does seem to fall in that period. The book was sketched, drafted and modified many times, for as always Newton spread himself over endless sheets of paper. And, as almost invariably, he meant to keep his work to himself.[41]

There were two chief purposes in Newton's investigations. One was to demonstrate that the antiquity of well-known events in early classical history (most conspicuously, the voyage of the Argonauts in search of the Golden Fleece) had been grossly exaggerated by previous chronographers. So had the antiquity of the lists of kings and geneaologies; in this regard Newton made the purely rational point that it must be wrong to attribute to the rulers in such lists reigns of thirty years or more when so few died in their beds. He proposed an average regnal length of twenty years, thus at a stroke reducing prehistory by a factor of three to two! His second purpose, by no means an unusual one in Newton's age, was to prove the superior antiquity of the Jews, as apparent in the biblical chronology, with consequential confirmation of their unique historical importance as the source from which other peoples drew their knowledge of arts and letters, science and learning. I find it strange that Newton seemingly did not possess a copy of Theophilus Gale's vast book, *The Court of the Gentiles* (1669–77) which so strongly asserted the case for the Jews as the pioneers of Mediterranean civilization. Both Gale and Newton invoked the authority of patristic writers for their historical thesis, and possibly Newton like Gale believed in the force of an historical link or analogy between Judaism and Puritanism, paganism and papalism.[42]

At any rate, Newton's chronology clearly demonstrated that the parallel events in Jewish history to the Argonaut expedition and other major incidents in Greek history were three centuries later than the Greeks and modern scholars had supposed: "*Solomon* Reigned in the times between the raptures [rapes] of *Europa* and *Helena*, and *Europa* and her brother *Cadmus* flourished in the days of *David*." In terms of dates Newton put Solomon's Temple at 1015 BC; the Argonauts and the release of Prometheus by Hercules at 937; and the capture of Troy by the Greeks at 904. The Roman republic was founded in 508 after the expulsion of the kings. How Newton argued by dead reckoning may be seen from these examples: Hercules and Aesculapius were Argonauts, and Hippocrates of Cos was descended from both at a remove of eighteen or nineteen generations. Here allowing the usual thirty years to a generation, we have about 507 (!) years between the Argonauts and Hippocrates, who "began to flourish" about the beginning of the Peloponnesian War (431 BC according to Newton).[43]

But Newton did not rest his case only upon such dead reckoning, for he also invoked astronomy. His statements about its earliest evolution are interesting, if enigmatic. Ammon, king of Egypt, was the first to make sea-going long-ships with sails, a generation or more before Daedalus introduced carpentry into Europe. So, before 1000 BC, the Egyptians "in order to cross the seas without seeing the shore" began "to observe the Stars: and from this beginning Astronomy and Sailing had their rise". Knowledge of astronomical navigation then spread to the Greeks, among whom the wise Chiron formed "the Constellations for the use of the *Argonauts*" (the Egyptians having presumably managed without), placing the solstitial and equinoctial points "in the fifteenth degrees [of each quadrant] or middles of the Constellations of *Cancer, Chelae, Capricorn* and *Aries*". Since Meton found the summer solstice in the eighth degree, the precession from Chiron's to Meton's time amounted to seven degrees, requiring a lapse of 504 years. QED according to Newton: the Argonauts had sailed about 936 BC.[44]

There is much more in the *Chronology*, including a laborious digest of Old Testament books to shape Newton's chronology of the Assyrian empire. Some of its weaknesses, like internal inconsistency, are obvious. Genesis hardly suggests that the enslaved Jews were more civilized and powerful than the Egyptians in the time of Moses, as Newton's thesis seems to require. And again, how could it be established, from such uncertain materials as Newton possessed, just where the equinoxes were really placed on the globe in the time of Chiron and Jason?

Newton's replies to such possible criticisms were largely rhetorical. The Jews had the longest unbroken record of history and their greatest apologist, Josephus, also had access to the lost records of Tyre, since the time of Hiram who was Solomon's ally. Josephus's history, translated into English by William Whiston, was indeed a great fount of authority and information to Newton and like-minded historians. Another favourite author, equally convinced of Jewish primacy, was Clement of Alexandria, who had asserted that this primacy was admitted by the Greeks themselves.[45] The same Clement was a source of the story that Chiron first drew the outlines of the constellations, naming them after heroes and heroic events of Greek 'history', down to ARGO but no later, as Newton sought to prove at some length. Now the earliest celestial globe recorded among the Greeks was that of Eudoxus, a well-known astronomer, according to the report of Hipparchus, who stated that its equinoctial points were in the middle of the constellations of Cancer, Chelae and so on. Newton simply adopted the assumption that Eudoxus's sphere was the same as that of Chiron, as there was no reason for the men of that time to suppose any change in the heavens. For it was Hipparchus himself who first discovered the precession of the equinoxes. Worse arguments have been proposed, if only we could believe that Chiron was a practical astronomer, as Newton declared.[46] "After the *Argo-nautic* Expedition," he went on, "we hear no more of astronomy till the days of *Thales*. He revived Astronomy, and wrote a book of the Tropics and Equinoxes, and predicted Eclipses." From Pliny (a little later!) Newton took Thales's observation of the heliacal rising of the

Pleiades, whence he computed a precession equivalent to 320 years between Chiron's sphere and Thales, thus again making the Argonauts about forty-four years after the death of Solomon.

Criticism of Newton

Such reasoning, such computations, such an abbreviation of Antiquity, could not pass unchallenged when known to conventional scholars. This came about through not untypical lack of principle among those to whom Newton imparted his ideas. In 1716, at the time of Samuel Clarke's correspondence with Leibniz, Caroline, Princess of Wales, had heard (perhaps from the former) that Newton had in hand a great work on chronology. The learned Princess was curious to see it. Newton however, declining to part with a jejune and incomplete treatise, prepared a précis of it for her. The Abbé Conti, under promise of keeping the tract secret, begged the favour which the Princess obtained from Newton of making a personal copy. The unworthy and slippery Italian allowed friends to read and transcribe this copy so that by April 1725 it had fallen into the hands of a French bookseller, Guillaume Cavalier. Recognizing a prize, Cavalier impudently wrote to Newton seeking a copy corrected by the author, with a view to its publication. Newton, furious, made no reply even when the bookseller made it plain that he would translate and print his imperfect copy of the "Chronological Index" (as Newton called it: in French it became the *Abrégé de Chronologie*). Conti, fairly enough, was at the focus of Newton's wrath, Newton taking the trouble to compose a denunciatory article for the *Philosophical Transactions*. At last, when Newton finally wrote refusing Cavalier permission to publish, the printing was already complete. A copy came to Newton's hands in November 1725.[47]

The translator from English of the *Abrégé* was Nicolas Fréret, Secretary of the Académie des Inscriptions et Belles-Lettres of Paris, a scholar who had himself been imprisoned in the Bastille for expressing heterodox opinions on the origin of the French people. In introducing the *Abrégé* Fréret reaffirmed the antiquity of the Egyptian and other ancient civilizations, but it was a Jesuit, Fr Etienne Souciet, who vigorously and at length led the attack upon Newton's revisionism. Since the aged man made no reply – as he had declared he would not – we may leave Souciet with his polemics.

However, Newton had now accepted the opinion of his friends that the full *Chronology* must be published in justification of its improperly published abridgement. The last 'portrait' of Newton at work, a few days before his death in Kensington, due to Zachary Pearce then Rector of St Martin-in-the-Fields, reveals him toiling upon his manuscript, in half-light yet without spectacles. He told Pearce that the *Chronology* was the work of thirty years in study and more in composing and rewriting. After Newton's death the manuscript of it was valued at £250 in his estate, and was printed by Conduitt with a dedication to Queen Caroline. His assertion that the book was to Newton 'diversion only, and amusement!' is hardly to be taken seriously.[48]

Newton's counter-traditional and critical review of ancient history, given its assumptions and purposes, is not merely absurd. Neither archaeology, nor classical epigraphy, nor the elucidation of 'lost' ancient languages and the inscriptions written in them had as yet freed scholarship from the toils of mythology. Newton was indeed credulous as to the veracity of certain ancient traditions preserved by his authors and unreasonably sceptical of others, but his methods were rational ones founded upon immense learning. 'The habits of the Master of the Mint and the physicist were not absent in the Bible commentator and the chronologist.'[49] As always Newton aimed for complete knowledge, judiciously analysed. A mind as remarkable for its clarity as its fullness is at work on every page of the *Chronology*. Only hindsight makes the book read like an immensely clever fairy story, as if it were the work of a seventeenth-century Tolkien.

The Classical Scholia

The *Chronology of Ancient Kingdoms Amended* was not Newton's only important exercise in history.[50] His study of the pre-Socratic philosophers has already been mentioned more than once. The date of its beginning is impossible to establish, since his interest only became apparent after the publication of the *Principia*, but the fact that he drew on Ralph Cudworth's *True Intellectual System of the World* (1678) – a book one might well suppose to have been in Newton's possession, though it is not so listed – suggests that his collection of materials began in the 1670s, when he was certainly reading both classical and religious history. For reasons that are less than clear, in the 1690s when Newton was sketching a revised *Principia* he resolved to attach the fundamental natural-philosophical ideas of the book such as atomism, the concept of force, the intrinsic mathematical harmony of Nature and the idea of universal gravitation itself to the most ancient tradition of Greek philosophy (as he supposed it), that is, the tradition of Pythagoras and Plato. Newton reveals no overt interest in the neo-Platonism descending through Plotinus, nor in the 'practical' mathematical sciences of the Ancients (as exemplified by 'Aristotle's' mechanics, the writings of Hero, Vitruvius, Ptolemy (in part) or Pappus). Indeed, in the preface to the first *Principia* he explicitly distinguished his own enterprise in mechanics from theirs; but he believed – or some part of his mind believed – that the ancient sages had possessed and applied an essentially correct idea of Nature to which he himself (first of the moderns) had returned. This historical interest was therefore different from that of the *Chronology* in relating immediately to his own researches in mathematical philosophy. Not that Newton would ever have doubted that such scholarly studies as he pursued bore upon his investigations carried out by the mathematical-experimental method, since he surely 'shared the belief, common in the seventeenth century, that natural and divine knowledge could be harmonized and shown to support each other'.[51]

That Newton had in mind the remote classical antecedents of the Renaissance revolution in astronomy, and of his own dynamical investigations which had subsequently justified it, when he was in the full flow of writing the *Principia*, is evident from the grandiose opening to the original Book III of that book, suppressed in 1686 but printed immediately after Newton's death as *The System of the World*:

> The most ancient opinion of the Philosophers was that the fixed stars stood motionless in the highest parts of the world, and that the planets revolved about the Sun beneath these stars; that the Earth likewise is moved in an annual course, as well as with a daily motion about its own axis, and that the Sun or hearth of the Universe rests quietly at the centre of all things. For this was the belief of Philolaus, of Aristarchus of Samos, of Plato in his riper years, of the sect of the Pythagoreans, and (more ancient than these) of Anaximander and of that most sage king of the Romans, Numa Pompilius.

It is not to be expected that Newton or any man should probe the astronomical theories of this nebulous monarch, but Newton states his reason for his assertion: "The latter erected a temple to Vesta, round in form, and ordained perpetual fire to be maintained at its centre, to symbolize the round shape of the Orb with the solar fire at its centre." Since the Egyptians, "the oldest observers of the stars", also embraced the heliocentric hypothesis (according to Newton) he was able to voice here a common view about the transmission of knowledge: "It seems that the Greeks, a race more given to philology than to philosophy, obtained this philosophy which was the oldest and soundest of all from the Egyptians and neighbouring peoples."[52]

A certain similarity between Newton's use of his reading among ancient authors – probably through most of his life – in relation to chronology on the one hand and natural philosphy on the other is apparent from notes made by David Gregory after his visit to Cambridge in 1694: 'He [Newton] will spread himself in exhibiting the general agreement of his philosophy with that of the ancients, and principally that of Thales . . . It is clear from the names of the planets given by Thoth (the Egyptian Mercury) . . . that he was a believer in the Copernican system.' Newton believed, as all in the tradition of esoteric knowledge believe, that within the gnomic legacy of myth and legend from the past profound truths lay disguised, to be reached by a correct process of decipherment. So, no doubt, it was with the alchemists too. In history, just as Newton confidently reconstructed the sphere of the Argonauts, so he felt that he understood the physics of Thales and the astronomy of Thoth. More specifically, Fatio de Duillier reported in February 1692 to Christiaan Huygens:

> Mr Newton believes that he has discovered pretty clearly that the Ancients like Pythagoras and Plato &c. possessed all the demonstrations that he gives of the true system of the world, and which are based upon gravity diminishing inversely as the squares of the increasing distances . . . certain fragments [indicate] that effectively they had the same ideas that are spread throughout the *Principia*.[53]

This opinion, strange to us, seems to have rested upon the supposition that the Pythagoreans used acoustics as a code for mechanics; in Gregory's words: '*Pythagoras* . . . applied the [harmonic] Proportion he had thus found by Experiments, to the Heavens, and from thence learned the Harmony of the Spheres.'[54] Thus the oft-discussed music of the spheres was for Newton a covert way of expressing celestial mechanics! Analogously, he 'revised' the notions of the ancient atomists, Lucretius especially, so that they anticipated his own more sophisticated theory of matter.

Why did Newton think of propounding these curious anachronisms (as we see them), apparently diminishing his own originality, on the basis of such tortuous textual evidence? If he had ever looked into Copernicus's *De revolutionibus orbium coelestium* (1543) – which is unlikely – he would have seen that his great predecessor had made a similar claim about the Pythagoreans in relation to his own discovery. As Copernicus sought to excuse or justify his paradoxical innovation by reference to classical precedent, so did Newton. It was very much a commonplace still in the seventeenth century, inherited from the Renaissance, that Europe was undergoing a revival of learning, mathematics and science, not a creation of new learning.

> However antithetical to the idea of progress, indeed preposterous, such a way of thinking inherently seems to us – to us and to so many who from the end of the seventeenth century already were convinced of the superiority of the Moderns and of the unambiguous progress of knowledge – it was not repugnant to Newton, who thought of the book of Nature as a palimpsest, long ago deciphered by the Ancients, but whose meaning had been obliterated by time and so had to be recovered by the experimental and mathematical method.[55]

Many scholars still half-believed (at least) that the present race of men had degenerated, along with Nature itself, from the state of the Earth long before, when it was still new-made. Newton may have been one of these.

Though the form and application of Newton's ideas in these proposed 'historical' additions to the *Principia* were new, the underlying principles were not. The notion that ancient myth and legend contained profound truths disguised was expounded by several authors before Newton; some passages from Natale Conti's *Mythologiae, sive explicationis fabularum libri X* (1612) were merely paraphrased by Newton for his own purposes.[56] As with his biblical and chronological studies, his alchemical investigations and his interpretation of holy prophecy, what the passage of centuries has rendered strange, not to say bizarre, in Newton's work would not have surprised his contemporaries Boyle, Locke and Leibniz. The commonplaces of learning in the late seventeenth century were very different from those current three hundred years later. Because he accepted them, we do not need to call Newton a mystic or a magus, a Platonist or a Pythagorean, as though such labels in the twentieth-century significations had any meaning for Newton's age.

Since David Gregory published Newton's ideas on this subject as his own, without a blush, in the preface to his *Elementa . . . Astronomiae* (1702), creating no uproar thereby, there is hardly room to doubt that Newton could have done the same. Yet he chose in the end to take a narrower line in the second *Principia* and in the Queries developed in the later *Opticks* also, for which cognate passages were prepared. Perhaps innate caution or timidity prevailed with him. Perhaps Newton thought this parade of scholarship would overbalance the introductory sections of a mathematical treatise. Perhaps he did not wish so openly to remove his support from the moderns. We do not know.[57]

14

Later Books, 1706–1726

The Queries in Opticks

The last quarter of Newton's life was distinguished from the first three-quarters by the profusion of his written work that was published by himself or by others for him. As always with Newton, none of this publication was fundamentally new: his intellectual effort was devoted to making more perfect what had been published before (as with the *Principia* and *Opticks*), to bringing to light from his files writings and correspondence of long ago (as with *Commercium Epistolicum*), or to polishing for the benefit of posterity the results of studies that he had pursued for decades (as with *The Chronology of Ancient Kingdoms Amended* and *The Prophecies of Daniel and John*). In making this point I do not mean quite to assert that, had Newton died in 1705, it would have been possible to reconstruct from his papers all that was subsequently published under his name. Only of the posthumous *System of the World* would this be exactly true. On most of the late books Newton was always at work, modifying, adding new material from experiments and reading, extending and revising.

Of the Queries appended to the successive editions of *Opticks* this is particularly true. The main text of this book was not greatly changed when it was republished in Latin in 1706, nor indeed in the subsequent English editions.[1] The Queries, however, developed into the most important series of statements that Newton released to the public about the deepest questions of natural philosophy: the concepts of matter and force, the possible role of an aether and the relation of God to the universe.

Opticks in its first edition (1704) was wholly Newton's own production, as the *Principia* never was, and once published he seems – so far as in him lay – to have regarded it as a record of past researches that needed no reworking. The Latin text (1706), translated by Samuel Clarke, was we may be sure carefully reviewed by its author and some joint adjustments were made to the main text. Abraham de Moivre, the first great student of probability (FRS 1697), a Huguenot refugee mathematician who supported himself by teaching private pupils, saw this text through the press. But we have, seemingly, no evidence that any significant change was introduced by either man. Newton identified as his own, for example, the new passage near the end of Book I, Part I, explaining why the light of stars occulted by the Moon vanishes and reappears instantaneously; another, closing Part I, on atmospheric disturbance of vision; and another still at the end of the first paragraph of Proposition II, Theorem II, later simplified in the English.[2] But the most important additions were made to the Queries. The first English edition contained sixteen of these, opening with rather staccato questions springing from the preceding experimental discourses on diffraction: "Qu. 2. Do not the Rays which differ in Refrangibility differ also in Flexibility; and are they not by their different Inflexions separated from one another, so as after separation to make the Colours in the three Fringes above described?" The first group of Queries assumed unhesitatingly that light is particulate, that there is dynamic interaction between matter and light which may set either into vibration, and that force is the intermediary between them. Only in physiological explanations (Queries 12 to 16) relating to light did vibration play any role as a mechanism of *transmission* (in nerves) "the most refrangible rays [exciting] the shortest Vibrations for making a Sensation of deep Violet". Vibrations in matter, on the other hand, are causes of the *emission* of light. Even these initial sixteen Queries were revised later: to Query 8 in the Latin edition (1706), on the emission of light from the "agitated" particles of bodies, were added numerous examples, while in the next English edition (1717) more still was adjoined on the triboelectric light discovered by Hauksbee. Queries 10 and 11 were also much increased in the Latin, while Queries 12 to 16 were never altered in any significant way.[3]

Where the first group in English stopped, the Latin edition continued with seven more Queries, which (to avoid confusion) I shall here number from 25 to 31, as in the later English editions. The first two Queries in this 1706 group dealt with the phenomenon of double refraction, explained previously by Huygens (in his *Traité de la Lumière*, 1690) and now by Newton using his own very different theory. It is hard to understand why this extraordinary and then unique phenomenon of calcite had not been discussed in the first edition of *Opticks*, for we know that Newton's theory of it was far from new to him. Query 27 is very terse, claiming that the phenomena of light arise only from the permanent properties of its rays, and not from any modifications of them (caused by their interactions with matter). Query 28 similarly scouts pulse and wave theories of light, as incorporating such modifications, and Newton backtracks to castigate again Huygens's explanation of double refraction in terms of his wave theory. Waves, if infinitely swift, would require an infinite force; if of finite speed, they must bend into the shadow; in any case waves could not heat matter, as light does, nor could Huygens's waves explain the "sides" that light-rays have, or the fits of easy transmission and reflection, unless *two* aethers were involved. Newton regarded such a duality as inconceivable; indeed, he went on – in 1706 – no compelling reason enforces our belief in a single aether. Rather, the celestial motions suggest that the celestial spaces are empty of all sensible, resisting matter. Newton continued with a short essay on the properties of resisting fluids and the diminishing density of the atmosphere with height, concluding that the interstellar space must be void and that the supposed aether [of Descartes and Leibniz] filling it "is in no way useful for the explanation of natural phenomena, when the motions of the planets and comets by means of gravity, are better accounted for without it; and gravity has not yet been explained by its means."[4] So much for Fatio's favourite hypothesis of a decade earlier! Having suppressed such allusions before, Newton here introduced as sound models "the most famous and ancient philosophers of Greece and Phoenicia, who made a void, atoms, and the gravity of atoms the principles of their philosophy, tacitly attributing gravity to some other cause distinct from matter". This Query ends with an important passage emphasized by italic type in which Newton made his first attempt (in print) to set God into Nature: how could its perfection arise save from the creative wisdom of a "Being incorporeal, living, intelligent, omnipresent, who in infinite space, as it were in his Sensorium, sees the things intimately, and perceives them thoroughly, and comprehends them wholly by their immediate presence to himself"?

Having disposed of light as a mode of motion, Newton turns next in Query 29 to the positive affirmation that it is a substance: "Are not the Rays of Light small corpuscles emitted from shining Substances and refracted by certain attractions, by which Light and Bodies act upon each other?" With the hypothesis of light which he had espoused for forty years at last unequivocally set before his readers, Newton in the rest of this Query summarized in its context the chief features of his life-long investigation:

> Nothing more is requisite for producing all the variety of Colours and degrees of Refrangibility, than that the Rays of Light be Bodies of different Sizes, the least of which may make violet the weakest and darkest of the Colours, and be more easily diverted by refracting Surfaces from the right Course; and the rest as they are bigger and bigger, may make the strongest and more lucid Colours, blue green, yellow and red, and be more and more difficultly diverted.

Such corpuscles, by their attractive powers, could "stir up Vibrations" which are the source of the fits, and by their quasi-magnetic polarity account for double refraction. Newton was of course aware that these were no more than in-principle hypotheses, and that there were grave difficulties in framing a dynamical theory of light agreeable to them, but he was after all only framing questions for examination.[6]

Material Transmutations

The last two questions are both concerned with the transmutations of substances, and are the main evidence in Newton's printed works of his interest in this major topic, and in the chemical art by which it might be investigated, that had been present in his mind since the late 1660s. Transmutations are universal in both the organic and the inorganic worlds, he observes: all living things are formed from watery substances and salts, returning to the same after death; water becomes earth, metals become vapours or soluble salts. Among "such various and strange Transformations, why may not Nature change Bodies into Light, and Light into Bodies?" This thought again implies the materiality of light – if only Newton had known of photochemistry! The very long Query 31 (and last) investigates in great detail the combination of three concepts: those of transmutation, differential attractive and repulsive forces, and atomic structure. With many examples drawn from his familiarity with chemical reactions, Newton explained the theory that had been in his mind for many years, that the gross properties of substances arise from their molecular structure (to use our language) and that the structures of molecules are made stable by a pattern of forces acting between the corpuscles composing the molecule. When outside forces of the same kind intervene, changing the internal pattern, rearrangements of the component particles may occur within the molecule, creating a different substance. A limited notion of this hypothesis had been conveyed in *De natura acidorum* (1710). "Have not the small Particles of Bodies certain Powers, Virtues, or Forces, by which they act at a distance, not only upon the Rays of Light for reflecting, refracting, and inflecting them, but also upon one another for producing a great Part of the Phaenomena of Nature?"[7] In a well-known passage, Newton laid down the concept of elective affinity, in chemistry perhaps the most significant concept relating to reactivity that had yet been introduced:

> When . . . a Solution of Copper [in nitric acid] dissolves Iron immersed in it and lets go the Copper, or a Solution of Silver dissolves Copper and lets go the Silver . . . does not this signify that the acid Particles of the *Aqua Fortis* [nitric acid] are attracted more strongly by Iron than by Copper, and more strongly by Copper than by Silver . . . ?

Rejecting Cartesian and Leibnizian notions Newton inferred from the cohesion of hard bodies that "their Particles attract one another by some Force, which in immediate Contact is exceeding strong, at small distances performs the chymical Operations above-mention'd, and at slightly greater distances from the particles (which never the less are perceptible to the sense) does not operate at all." From this dynamic account of the transmutations of matter, Newton concluded, first, that his oft-repeated metaphysical maxim "Nature is ever simple and conformable to herself" was again verified, and that the analogy between the micro-chemical force operating at the molecular level and the macro-force of gravity prevailing at the scale of the Earth and the solar system was complete and perfect. Secondly, he deduced that because forces are the operative agents in Nature, they cannot remain constant (or be conserved *grosso modo* as Descartes had supposed motion to be) but must be maintained against secular decay by "active Principles":

> such as are the cause of Gravity, by which Planets and Comets keep their Motions in their Orbs, and Bodies acquire great Motion in falling; and the cause of Fermentation, by which the Heart and Blood of Animals are kept in perpetual Motion and Heat; . . . For we meet with very little Motion in the World, besides what is owing to these active Principles.[8]

In the final analysis, then, the "active Principles", by their injection of fresh force or vigour into the universe, are the factors preventing its rapid degeneration and disintegration and (by analogy and by what here follows) are close to the divine principle.

For, in Newton's thought, these last intermediate causes in Nature are not occult, even though they be unknown, just as the person of God is unknown but is not in the philosophical sense occult. Such causes are purposeful – to maintain the universe – and comprehensible only in terms of a planned creation. As before, Newton invoked natural theology: a planned universe "can be the effect of nothing else than the Wisdom and Skill of a powerful ever-living Agent, who being in all Places, is more able by his Will to move the Bodies within his boundless uniform Sensorium, and thereby to form and reform the Parts of the Universe, than we are able by our Soul, which is in us the image of God, to move the Parts of our own Bodies". And so Newton at last closed his book, with a little homily to the effect that if natural philosophy could be so far developed by his method as to become a perfect science, the bounds of moral philosophy and even of religious observance would themselves be much enlarged also.[9]

The composition of all this new material in the Queries, of which earlier antecedents can readily be found, thus disproving George Cheyne's claim (to David Gregory) that Newton had stolen it from him, probably occurred in 1705. The essence of the last of the 1706 Queries was certainly conveyed by Newton to Gregory on 21 December 1705, though the latter may have misunderstood Newton's meaning. He suggested to Gregory that the omnipresence of God, within matter as in space, was the immediate cause of gravity and that the Ancients had shared the same opinion.[10] In the printed text Newton said nothing of the sort, but spoke of "active Principles" as the causes of forces.

Revisions after 1706

In this crucial passage of Query 31 as in many others the changes made to the Queries between 1706 and 1717 very much altered the character of Newton's thought as there presented (see chapter 8). To take this latter place first, Newton deleted the comparison between the human soul and God, while adding new sentences denying that God is to be considered the soul of the world, or the world God's body. The universality of God means that he has no need of organs of sensation. This modification was provoked by Leibniz's scornful criticism of Newtonian metaphysics during the decade after 1706, which irritated Newton at least as much as the calculus priority claim had done before. However, it was the insertion of eight completely novel Queries (numbered 17 to 24 in the second and subsequent editions of *Opticks*) that deeply modified the character of the whole set and made Newton's message uncertain. He seemed henceforth to speak with two voices, both for and against aetherial physics.

As mentioned before, Newton's change of mind was brought about, with almost indecent haste, by Hauksbee's demonstrations of the visible, almost tangible "electric fluid". He added a long passage about it (as a strange source of light) to Query 8. In the new Query 17 he insidiously abandoned optical dynamics to postulate a "refracting or reflecting Medium" in which rays of light can propagate vibrations. In Query 18 he cited that deceptive experiment of Hauksbee's, which purported to show that a 'void' transmits heat as fast as air does (as well as light) and therefore can be no true void. It must contain "a much subtiler Medium than Air, which after the Air was drawn out remained in the *Vacuum*". Is it not, Newton now asks, this aether that puts light rays into fits of easy transmission and easy reflection, pervading all bodies and filling all space? Being less dense in matter, denser in space – Newton's favourite old idea – this aether (not optical force any longer) is the cause of refraction and reflection, and of diffraction too. Query 21 – Fatio vindicated at last! – makes the variations in density of this material the cause of gravity. Newton is simply serving up his speculations of 1675. Displacing the fertile framer of hypotheses, the calculating Newton reappears for a moment to argue that the extreme velocity of the vibrations in this aether (faster than the light rays since they overtake light, and light travels about 10 million miles in one minute) proves that it is highly elastic, in fact 4.9×10^{11} times more elastic than air, but also 6×10^{8} times less resistant than water. Planets and comets can swim through such a fluid without losing a noticeable part of their motion in ten thousand years. For good measure, in our bodies the same aether performs vision and hearing (Query 23) and connects the brain to the muscles (Query 24).

There can be no doubt that when Newton felt the electricity from Hauksbee's whirling globe tingling in his finger-tips, something snapped in his brain and he underwent a sudden Gestalt switch. The duck which Newton had laboriously transformed into a rabbit flipped back into being a duck again. Indeed, the world was full of ducks, not a rabbit to be seen.[11] Newton's thoughts went back, as they often did in other areas of his mind, to familiar grooves that he had explored from his boyhood onwards, back into the safe and familiar world of neo-Cartesian mechanist hypotheses. The dynamical adventure, the universe of forces explored with his intensest mathematical concentration between 1680 and 1687, had failed to resolved the problem of causation.[12] Newton now felt himself awakened from the metaphysical nightmare of action at a distance in which he had dwelt for a quarter of a century.

Of course, numerous consequential changes had to be made in the Queries, and even to a word or two in the text, to weaken Newton's former declared preference for dynamics, for atom, force and void. He now wished to be read as rejecting the coarse, gross aether of Descartes and Leibniz while advocating a quite different, incredibly refined and tenuous aether of his own. Changes were made silently and without explanation. For example, Newton sacrificed the wonderfully ingenious idea expressed in *Optice* Query 22 [= Query 30 in later English editions]. Always and forever Newton thought of a shining light ray as a particle, therefore a very high-speed projectile. Now if we know the speed of a projectile and the curvature into which it is bent by a force, we can establish the magnitude of the force even while ignorant of the mass of the projectile. As we have seen, Newton took the speed of light to be ten million miles a minute, and he guessed the incredibly small radius of curvature of a reflected light ray as 10^{-5} inches. From these estimates – but his calculation was badly flawed – he declared that the optical force in matter is 10^{15} times greater than the force of gravity at the Earth's surface. Such an enormous force acting upon an incredibly minute particle is also compatible with the vast speed of the radiated light ray.[13]

All such traces of optical dynamics Newton eliminated from the 1717 edition, in order to avoid contradicting himself within the book, though contradiction between the new edition and *Optice* was unavoidable. Subsequent readers were thereby deprived of very important aspects of Newton's thinking, and indeed deeply misled as to its nature in its dynamic phase. Newton was not to be the only progenitor of the great aether delusion which lasted well into the present century, though it is now conveniently forgotten, but it would be hard to deny Newton's pre-eminence in authorizing the delusion, on the basis of two or three badly analysed qualitative experiments. For their sake he undermined, and turned into logical nonsense, the great mathematical and conceptual architecture which had been the master-work of his life. He had already made an incomprehensible enigma of his *Principia* by the final paragraph that he had added to it (hastily) in 1713. To the extent that the aged Newton so transformed his own thoughts as to hint that forces are mere arbitrary conveniences invented by mathematicians so that they can work out their sums, rather than physical realities, he surrendered the game to Leibniz and others like him who at least maintained the identity of mathematical and physical ontology.

Late Publications

I have mentioned before that *Optice* had a considerable continental influence; it was reprinted by Newton in 1719, again (in Switzerland, with a dedication to Johann Bernoulli) in 1740, and thereafter. *Arithmetica universalis* too, of which Whiston had been the midwife (1707), became freely available through further London and continental printings before 1750 (see p. 173). Its English translation (by Joseph Raphson) was also reissued several times. It was the book that first convinced the world of Newton's general mathematical excellence. Not that Newton was delighted by Whiston's publication of his Lucasian lectures of a generation before, almost without his consent, or thought the task well done. Inevitably, when he came to a reissue under his own direction (1722) he devoted labour to effecting (during 1720 to 1721) changes and additions, mostly trivial. More interesting, because closer to Newton's mathematical originality, was the publication at last in 1711, together with letters to Collins and Leibniz from Newton and the two mathematical essays printed by their author with *Opticks* in 1704, of the tracts so long suppressed: *De Analysi* and *Methodus differentialis*, this latter 'in the greatest part a word-for-word reissue of the scheme of interpolation through central finite differences which [Newton] had set down in his Waste Book around the autumn of 1676'.[14] William Jones, who compiled this volume, was a rather dim Newtonian, elected FRS in 1712 at the age of thirty-seven, a protégé of John Harris of the *Lexicon Technicum* (wherein he was assisted by Jones), a teacher of mathematics and navigation. He had the great good fortune to buy the manuscripts and books of the late John Collins, which passed from him to the future Earl of Maccles-field, whose tutor he had been.[15] Jones was during some time a correspondent of Roger Cotes, but nothing survives of his personal relationship with Newton. The last fluxions tract, *De methodis serierum et fluxionum* (1671) was never printed in any form in New-ton's lifetime, making its first appearance in two English translations (by John Colson in 1736 and a rival in 1737). The tract was also translated into French by the great Buffon (1740), by which time it had long been a work of historical interest only.

For the purposes of this biography, enough has been said already in chapter 10 of the 'Royal Society's' – that is, Isaac Newton's – *Commercium Epistolicum* (1712) and of the anonymous 'Account' of that miserably produced little book that appeared from the same hand in the *Philosophical Transactions* three years later. Perhaps, with Leibniz dead, Newton should have stayed his hand, but he did not. In a variety of ways he himself pursued the continuing conflict with Johann Bernoulli, in which his supporters were even more active. When in his seventies, for all his success in overcoming Bernoulli's problem of 1697, Newton began to write that he had "left off the study of Mathematicks" with his departure from Cambridge in 1696, yet it is evident that he maintained, even into the last decade of his life, a lively capacity to tell the story of his own mathematical odyssey again and again, to the discomfiture of his opponents. In 1716 he responded in print, it must be said feebly and without success, to a second challenge problem launched by Johann Bernoulli. In 1717 or 1718 he added fresh justificatory material to unsold copies of Raph-son's *History of Fluxions* (1715), which were then reissued; in 1718 and subsequently he meddled a good deal with the collection of learned correspondence (in which he was involved) that the Huguenot refugee Pierre des Maizeaux wished to print;[16] at the same time he prepared material to be added to the third *Principia* on the role of his analytical method of quadratures in that book (it was not used), reconsidered the fluxions Scholium in Book II, and made ready to add to it yet another edition of *De quadratura curvarum* (the plan was given up).[17] In 1721 – and this was to be his last stroke of the pen in the priority dispute – Newton drafted an abortive new and lengthy preface to the new edition of *Commercium Epistolicum* that appeared in the following year.

Thus, if not in mathematics, then in mathematical history (or autobiography!) Newton remained active till near the end; as we shall see in the next section, his revision of the *Principia* for its third publication went on through the years of illness before his death. Meanwhile, the last edition of *Opticks* to appear in his lifetime also occupied a good deal of his attention. This was the second text in French (*Traité d'Optique*, Paris, 1722), revised and edited by Pierre Varignon from the translation by Pierre Coste (Amsterdam, 1720). It is a handsome book, crowning the recognition of the real importance of Newton's investigations of light and colour in France; an engraved vignette of the "crucial experiment" at each chapter-head was made from a sketch provided by Newton in the winter of 1721–2. The printing was finished in July of the latter year, Varignon expressing particular gratitude to de Moivre for his improvement of the mathematical parts of the translation. One may infer from Newton's letter of thanks to Varignon for his labours that de Moivre had acted generally as Newton's assistant in this business.[18] And as a token of esteem, Newton presented Varignon with an oil portrait of himself by Kneller (see Appendix B).[19] Johann Bernoulli was expressly denied such a privilege.

The Second Principia, *1713*

If the story of the Paris *Traité d'Optique* ends with a portrait of Newton, that of the second *Principia* more than a decade earlier began with one. As a fruit of his revival of close relations with Newton in 1708–9, Richard Bentley had persuaded Newton to sit for a likeness by Sir James Thornhill, to hang in the college of which Bentley was Master.[20] Long before the picture reached Cambridge, Roger Cotes had taken over from the Master of Trinity as the effective editor of the work in progress, though still submitting to Bentley's direction. For all Bentley's initial enthusiasm and masterful conduct, progress was not swift. Nor, despite some fussy dispute about the choice of words, was Bentley a man equipped to accelerate it or assist Newton in his task.

Presumably Newton had approved Bentley's choice of an 'assistant' to take on the drudgery of proof-correction and verification of the text. Probably he had met the young mathematician in the course of his visits to Cambridge in 1705. At any rate, in February 1709 Cotes paid Newton a visit in London, when the latter ordered a high-class clock to be made for the new observatory to be built at Trinity College, of which Cotes was to take charge. Cotes possibly repeated this visit in July. By the summer of 1709 some fifty pages of Bentley's edition had been printed. When Newton advised of a mistake to be corrected on page 3 (11 October 1709) Cotes, it seems, had already made the correction. In the same letter Newton announced his sending (by Whiston's hand) another large batch of copy running into Book II.[21]

Cotes was flushed with gratitude to Newton and longing to 'take all the Care I possibly can that [the new edition] shall be correct'. At the same time he pointed out (from his own re-calculation) two minor errors in Newton's *De Quadratura*. Thus from the first a certain pernickety tactlessness appeared; knowledge of these oversights (irrelevant to the matter in hand) could hardly gratify Newton. Perhaps Bentley, writing a couple of months later: 'we will take care that no little slip in a Calculation shall pass this fine Edition' fretted the rub with salt! At the same time, with magisterial loftiness, he told Newton 'not to be so shy of giving Mr Cotes too much trouble: he has more esteem for you, & obligations to you, than to think that trouble too grievous.' Were these younger men taking the new edition out of Newton's hands?[22]

During the following winter roughly the first half of the new *Principia* was safely set in type without problems, though Cotes with some trepidation explained (on 15 April 1710) that

> I have ventured to make some little alterations my self whilst I was correcting the Press, such as I thought either Elegancy or Perspicuity or Truth sometimes required. I hope I shall have Your pardon if I be found to have trusted perhaps too much to my own judgment, it not being possible for me without great inconvenience to the Work & uneasiness to Yourself to have Your approbation in every particular.[23]

In other words, proof-sheets were not sent to Newton: they were printed off after Cotes had passed them so that the type could be re-used. At about halfway, in Proposition 10 of Book II, Cotes – quite rightly – found difficulties in Newton's copy.[24] So began a cycle of events, causing considerable work for Newton and delays in the printing, in which Cotes pointed out illogicalities, mistakes or inconsistencies in Newton's text, and (quite often) proposed alterations to overcome them. Where, as was quite often the case, Cotes's suggestions were both sufficient and correct, Newton accepted them and the work went on; but the change even so sometimes made Newton dissatisfied with his text – the subject of so many pains over more than twenty years – so that reconstruction became necessary. Where Cotes felt that he could make no clear suggestion for an amendment or his suggestion did not please Newton, there was obviously even more work for the latter to do. After that, Cotes was not always satisfied with Newton's amendments, and reiterated his own view of the point at issue, to which Newton sometimes capitulated, as when he wrote to Cotes (13 May 1710): "I have reconsidered the 15th Proposition [of Book II] with its Corollaries & they may stand as you have put them in your Letters." Thus in this second version the latter part of the *Principia* was unlike the former in that it was produced by a dialogue between Newton and his editor. The dialogue, in which Cotes was difficult to shift on essentials, not only corrected the text but enriched it at many points. In the words of Bernard Cohen: 'It is clear, I believe, from a reading of the Newton–Cotes correspondence that Newton had originally intended a far less drastic revision of Books II and III than he eventually produced. The credit is Cotes's.'[25]

By the end of June 1710, through many debates, Cotes had slowly printed all Newton's copy, to page 296 of the new edition; he then took a summer holiday. By early September he was ready to begin again, by the middle of the month he had Newton's next batch of copy but at once found (in Propositions 36 and 37 of Book II) a new obstacle: he could not follow Newton's argument about the flow of water through a small hole, under a head of pressure, and made an experiment to confirm his difficulty. Newton, preoccupied with moving house from Chelsea to St Martin's Street, Mint matters, buying Crane Court for the Royal Society, and putting Flamsteed in his place, abandoned his book for months. Not before half a year had passed did Cotes receive fresh copy for these propositions. He still found further difficulties in their new version, but was pleased by a public compliment paid him by Newton (who was to expunge it from the printed text), paralleling the appreciative private remarks in Newton's letters. Only late in June 1711 was this hydraulic issue at last settled.[26] Both men now hoped for smooth and swift progress. Alas! That was not to be achieved, for Cotes soon found further severe difficulties in Propositions 47 and 48, which he persuaded Newton to transpose in the second edition. These are part of Newton's mathematical treatment of sound vibrations. Newton's first reply did not satisfy Cotes, and then through the summer and autumn Newton left him without further word. Not until 2 February 1712 did he write to Cotes, ruling that the copy as it was should stand and proposing one brief change only – and this Cotes rejected, printing a form of words of his own. A small difficulty at the beginning of Book III was cleared away but graver obstacles lay ahead. It may be that at this time Newton began to think Cotes unnecessarily critical and to tire of the mental labour that his questions imposed. He began to concede a number of points to Cotes, but this may not have suited his proud spirit. Discussions continued through Book III: the shape of the Earth and the theory of the Moon, both complex topics, inevitably raised doubts in Cotes's mind both about the logic of Newton's arguments and the particular quantities he chose to introduce into them. In Book III, indeed, Cotes saved Newton from countless errors of computation and all kinds of minor inconsistencies, mistakes and omissions: for here many quantities were to be established from theory and then justified – often only too precisely! – by experiment or observation. Cotes was just as ready as Newton himself to 'fiddle' with numbers to obtain exact agreement between theory and fact.[27] In March and April 1712 major propositions were thoroughly reconstructed, Newton confessing (on 18 March), after all these years, "I have not yet been able fully to settle the Theory of the 19th, 20th, 36th, 37th and 39th Propositions" of Book III. In fact, revision of the theory of the Moon continued till September, and fresh sheets came but slowly from the press. While Newton did not take up Cotes's proposal that he might print a companion volume of Newton's published mathematical writings, he did gratify Cotes with his portrait – probably John Smith's 1712 mezzotint of Kneller's 1702 oil portrait. Cotes said he preferred Thornhill's first.[28]

In mid-October 1712, when the work had at last reached the discussion of comets closing Newton's main text, he sent Cotes an important addition that he had planned some time before, and added bad news: Proposition 10 of Book II must be reset and a cancel inserted into the piles of printed sheets.[29] As related in chapter 12 (p. 314) Johann Bernoulli's discovery of Newton's mistake in the first edition – a geometrical mistake – brought to London by his nephew Nikolaus had been put right by Newton after immense labour. Several days of early October were thus consumed, until (on the 13th) he could write an urgent letter on Woolsthorpe affairs.[30] The new material was not, however, ready for Cotes until January 1713.[31]

The General Scholium and Cotes's Preface

Newton had now come to the point of sending Cotes his new General Scholium, the crowning touch to the *Principia*, and of ordering Cotes to write an additional preface of his own to the book – or rather he allowed Bentley so to command Cotes.[32] This preface has become important as an early apologia for Newtonianism and an open counter-attack upon the natural philosophy of Descartes and Leibniz which had resisted its acceptance. Neither Newton nor Bentley had a hand in it, and though Samuel Clarke reviewed it for Cotes in draft, it was all the latter's production. It has not, in its readiness to see attractive force as a natural property of matter, won much favour from metaphysicians then or now, nor was this an opinion favoured by Newton. Cotes had tried to wriggle out of the necessity to write a defence of Newtonianism against its continental critics without help – this being necessary, as he judged – and so said a great deal more than he need have done. For Cotes this was by no means a happy time: burdened besides the preface with a table of contents and an index, in poor health, dragging tediously through the volume's final throes. It is curious that two of Newton's best-known utterances appear for the first time as afterthoughts, in a letter to Cotes of 28 March 1713:

> it surely does belong to natural philosophy to discourse of God from the phenomena [of Nature] . . .*and*
> whatever is not deduced from phenomena is to be called an hypothesis, and hypotheses of this kind whether metaphysical or physical, whether of occult qualities or mechanics have no place in experimental philosophy.

But at last the whole was done by mid-June 1713, almost exactly twenty-six years after the completion of the first edition in Halley's hands.[33]

Cotes may have sensed, through the autumn of 1712, that Newton was no longer so well pleased with his work. Compliments and thanks from Newton ceased; his letters became more formal. In March 1711 Cotes had thanked Newton for the 'undeserved Honour' of a compliment in a proposed addition to the text, but nothing of the sort appeared anywhere in the printed book. Newton, perhaps in the summer of 1712, ended a long draft preface with a compliment to Cotes. As printed, however, the preface became brief in the extreme, without thanks to Bentley, Cotes or anyone else. Before 1713 was well advanced Newton was incensed at Bernoulli for his attacks upon himself, and had resolved not to acknowledge publicly Bernoulli's correction to Proposition 10, Book II; in consequence, it may be, Cotes too received no word of acknowledgement in public or private for his countless improvements to the new edition. Instead, nearly six months after its publication, Newton sent to the University Press a long list of corrections and additions. This of course came to Cotes, who deservedly wrote Newton a very stiff letter. All but twenty of these corrections, he wrote, were in the table of corrigenda printed in the book by himself; some of the twenty he had thought too trivial to be included. The one noted by Newton for page 191, he declared sternly 'wants no correction. I cannot understand by what reasoning You make one: You will be pleased to reconsider it.' In this 'very correct' printing, he pointed out, he had made hundreds of small improvements never reported to Newton.[34]

Cotes was cross and injured. In my view, Newton's treatment of Cotes is one of the worst blemishes upon his moral character. It is impossible to excuse Newton's breaches of the established and proper conventions of scholarly life, still unrepaired in the third edition (1726) – where Pemberton was lavishly thanked! Newton's anecdotal praise of Cotes after his death made no amends for this great piece of injustice.

As in the first edition, so in the second Newton omitted much material of a general nature that he had drafted for the book. The draft preface just mentioned, for example, not only contains sentences about the differences between the first and second editions, but outlined ideas about the universality of magnetic and electric forces, in addition to the gravitational force. Here too Newton alludes to a mathematical piece he proposed to add to the book in order to make plain its analytical foundations, which he entitled "Analysis by fluent quantities and their [fluxional] moments". Drafts show that he married to propositions from *De Quadratura curvarum* a well-worked-out account of his method of expansion into power series, here borrowing from his *First Letter* to Leibniz of 1676. The tract contained nothing not already published and its main purpose was specious: that of inviting his readers "to infer that all the propositions in his *Principia* were discovered by the fluxional analysis". Indeed, the tract opens with the words: "The analysis by which I sought out propositions in the books of the *Principia* I have decided now to subjoin so that readers instructed in it may be the more easily able to assess the propositions delivered in those books." We have noted already that in his late years Newton often tried deliberately to inculcate the idea of an antecedent, analytical *Principia*.[35]

Very brief mention may also be made here of a more interesting but no less stultified mathematical contribution to the book, wherein Newton was on the verge of letting the public see for the first time his method of handling the radius of curvature of curves – first written half a century before into his 1666 tract on fluxions. He thought of adding this to page 240 of the second edition, the date of the drafts being perhaps 1715–16, Whiteside suggests. He remarks: 'Had Newton lived to put this *addendum* into print, his successors might have known something.'[36]

In his letter to Cotes enclosing the revised text of Proposition 10, Book II, Newton wrote of a possible appendix to the book, "concerning the attraction of the small particles of bodies", about which he was not yet resolved. He had rejected this topic previously from the first edition and he was to reject it also from the second. Its place would have been in the *Scholium Generale* completing the book. Appropriate drafts do exist. Most of the material in them was transferred to the Queries in the new *Opticks* edition of 1717. Another draft, entitled by Newton himself "De vi electrica" ('On the electric force') is printed in the *Correspondence* and was the obvious basis for the last few lines of the *Scholium Generale* about the electric and elastic spirit.[37]

The Third Edition, 1726

The last of Newton's young disciples, Henry Pemberton, may be quickly dismissed. He was probably introduced into Newton's circle by Richard Mead in 1722, when he was twenty-eight and Newton eighty years of age. As after the first edition, so after the second Newton had begun at once to compile corrections and new material for a third. De Moivre wrote to Johann Bernoulli (28 June 1714) that there would soon be another printing, more accurate and with better typography than the second *Principia* and on better paper (so much for Bentley!). It was to be supervised by Newton himself and Bernoulli would surely receive a gift copy. Newton alleged Bentley's meanness as a reason why Bernoulli had received no free copy of the second – nor (if we may continue to believe de Moivre) had either Halley or himself. Bentley certainly did make a substantial profit from the book but he had specifically reminded Newton to send a presentation copy to Halley.[38] If there was such a scheme it came to nothing, as did another scheme (1719) for a fresh edition combined with *De Quadratura*. When Newton at last addressed himself seriously to his formidable task for the third time (1723) he presumably chose the industrious but not very bright Pemberton, a professional physician and amateur mathematician, because he was docile and unlikely to cause trouble. No letter from Newton to Pemberton about their joint work survives, though a goodly number from the latter indicate that his proposed changes to the copy he was given were of a trivial kind, and a few of them downright foolish. Newton approved some, rejected many. In this edition Newton as well as his editor read the proof pages, on which he indicated his final decisions. Pemberton is of some interest as the author of a lost English translation of the *Principia* and a useful popular exposition of Newtonian science, *A View of Sir Isaac Newton's Philosophy* (1728); as a figure in Newton's last years he is negligible. Of his book, read by Newton before publication, he is said to have remarked that Pemberton "evidently had more in him than he imagined".[39]

Last Friends

Close personal disciples evidently did little to mitigate the isolation of Newton's old age. With her marriage Mrs Conduitt left St Martin's Street, where Newton remained for a further eight years with his servants, his crimson hangings and his ever-active pen. Occasionally, as through all the years before, he was called upon by distant poor relations seeking his bounty and, it would seem, commonly receiving it. Of Newton's associates in recent years Keill died in August 1721, thereby effectively ending the dispute with Johann Bernoulli. Brook Taylor had withdrawn to his country seat in Kent but was still marginally in touch with mathematics and Newton's circle.[40] On the other hand, Abraham de Moivre outlived Newton to write some interesting recollections of him for Conduitt, and gave Newton much moral and intellectual support during these lonely years. According to a late story (1754) de Moivre used to join Newton at a coffe-house – perhaps the Grecian off the Strand where Fellows of the Royal Society gathered after their assemblies, or Slaughter's in St Martin's Street – whence the two men would repair to Newton's house to dine and 'y passer la soirée dans des tête-à-tête philosophique [*sic*]'. Another late addition to Newton's circle was John Machin (FRS 1710) whom Newton recommended (1713) as Gresham professor of astronomy, having previously engaged him for the work of preparing for the press 'Halley's' *Historia Coelestis*. Newton praised his mastery of the *Principia*, into whose third edition he incorporated two propositions of Machin's on the motion of the Moon's nodes. Of Newton's older friends Halley had removed to Oxford as Savilian professor; Samuel Clarke was still in London and we find these three men dining together in company on 14 December 1721. Halley, Clarke and de Moivre, with the Conduitts, were Newton's only real friends in his last years, to whom we may add William Stukeley on a rare visit.[41]

In the matter of discipleship Leibniz, with his followers and successors forming a talented and active band spreading right across the Continent, was more fortunate than Newton. Colin Maclaurin, reckoned the finest British successor of Newton in mathematics, had but the barest place in his life: they never met.[42] The same may be said of Robert Smith, Cotes's much younger cousin and Newton's chief successor in optics, who wrote at least twice to Newton (December 1718 and August 1720) about his posthumous publication of Cotes's writings. W. J. 'sGravesande, Stephen Hales and above all Voltaire, all authors who figured prominently in the early formation of Newtonianism, had insignificant or no contact with the man himself. Newton lived to see and enjoy an immense increase of his reputation among foreigners during the last dozen years of his life, but those who promoted this reputation (and in the case of Pierre Varignon exchanged many letters with Newton, so establishing a cordial relationship) cannot be reckoned as members of Newton's circle and most did not accept the foundation of his physics, the concept of attractive force.[43] In the end, it was Newton's niece, Catherine Conduitt, who (with her husband) gave greatest support to Newton in his years of decay and took immediate steps to preserve his memory.

15

Kensington, 1725–1727

When his work ceased, the story of Newton's life ended. William Stukeley and John Conduitt – the latter in a memoir sent to Fontenelle, then writing his *éloge* of Newton as a *membre étranger* of the *Académie Royale des Sciences* – agree that his decay began with 'a relaxation of the sphincter of the bladder; so that he was oblig'd to make water frequently'. Newton curtailed his social life and dined simply: 'chiefly upon broth, vegetables, and fruit, of which he ate very heartily'. This was about 1722. In August 1724 he passed a bladder-stone, without much pain, and in the following winter suffered bronchitis (or worse). Temporarily, he was much improved in general health by moving out to the village of Kensington, in country air, though he suffered an attack of gout. Through two good years he worked on the third *Principia* with Pemberton. It was on 7 March 1725 that he had a 'curious conversation' with Conduitt about his notion of a secular revolution in the cosmos, whereby matter from the Sun was condensed into bodies, becoming planets, then comets which ultimately fall into the Sun, and so falling may destroy the Earth by fuelling the Sun to excessive heat. Man's duration on Earth Newton took to be brief: for all arts and sciences had been discovered 'within the memory of history, which could not have happened, if the world had been eternal; and . . . there were visible marks of ruin upon it which could not be effected by a flood alone.' It was at Kensington too, and also in talking to Conduitt, that Newton voiced one of his best-known sayings: "I do not know what I may appear to the world; but to myself I seem to have been only like a boy, playing on the sea-shore, and diverting myself in now and then finding a smoother pebble or prettier shell than ordinary, whilst the great ocean of truth lay all undiscovered before me." A strange parable from a man who almost certainly never sat upon a sea-shore in his life!

Conduitt blamed a visit to preside over a session of the Royal Society (28 February 1727) as the proximate cause of Newton's last illness, a prolonged agony of the stone from which he could not recover. Stukeley opined that the disease was 'an inflammation of the neck of his bladder, with the most excruciating pain that can be ima- gin'd . . . I have no scruple in imagining it to be gout.' Both men agree that Newton bore the intense pain, which at times caused the bed to shake in his torment, 'with a most exemplary and remarkable patience, truly philosophical, truly Christian'. He was attended in vain by Drs Mead and Cheselden, remaining conscious until the evening of Saturday 18 March, dying two days later at about two in the morning. Until his mortal illness he had 'all his senses and faculties strong, vigorous, and lively, and he continued writing and studying many hours everyday.'[1]

Nevertheless, the last portraits show him as decidedly frail. New- ton had not been regarded as handsome or distinguished in appear- ance. The antiquary Thomas Hearne said he was 'of no very promising aspect', Bishop Atterbury thought he 'did not raise any expectations in those who did not know him'. John Conduitt, writing after his death of Newton's appearance in old age, recorded: 'He was of middle stature, and plump in his latter years: he had a very lively and piercing eye, a comely and gracious aspect, and a fine head of hair, as white as silver, without any baldness . . . [he] never wore spectacles, nor lost more than one tooth to the day of his death.'[2] In the last years of his life Newton abandoned the massive wig so prominent in his middle-age portraits, and wore his own silvery hair; as in the portrait (1725) by Enoch Seeman, an attractive likeness. Conduitt, Humphrey Newton and Stukeley all remark on the quality of Newton's hair. Stukeley adds that he had full and protuberant eyes, was near-sighted in youth but was corrected by presbyopia in middle age. He goes on to write that 'his natural disposition was of a chearful turn, when not actually engag'd in thought. He could be very agreeable in company, and even sometimes talkative.' For his size, his chest was large and 'his voice was of deep tone, but pleasant enough'. At the end of the 'curious conversation' with his 'nephew',

Newton made a last remark 'with a laugh'. Perhaps too much has been made of the view, supported by Henry More (see p. 000), that Newton had the melancholy air of a thinker. True, it was a trope of the time that thought and melancholy go together, and it seems true of More himself from his only portrait. On the other hand, the majority of Newton's portraits seem to me to show strength, a notable absence of the facial fat so common at the time, determination and courage rather than marked melancholy.

No deathbed wisdom, philosophical or religious, is known to have come from the lips of the dying Newton, who well understood his condition. It is said that he did not receive the sacraments. On his last conscious day he conversed with Richard Mead; no doubt he was also comforted by visits from his clerical friends, such as Samuel Clarke, a neighbour in Kensington.

In fact, though Newton wrote so many thousands of pages on the early history of Christianity, prophecy, the nature of Jesus, and other religious topics, we know little of his personal religious beliefs and practices. Gregory tells us that he used a grace before meat; he was at least once formally certified as a communicant. Whether he offered personal prayers, whether he took Communion often or seldom, these and many other points are uncertain. To Humphrey Newton during the preoccupied days of the *Principia* he was careless in worship, scarcely knowing the house of prayer.[3] But such opinions are relative. What is certain is that Newton's version of Christianity and specially of its triune God was singular. He believed, from a most thorough study of the Gospels and all surviving materials bearing on the early history of the church, that there is but one God, God the Father, Creator, Pantokrator. As he wrote in the General Scholium added to the second *Principia* (1713): "He rules all things, not as the Soul of the World but as its Lord. For God is a relative word and it relates to servants. Almighty God is a Being eternal, infinite, absolutely perfect; but a being without ruling power, however perfect, is not the Lord God."[4] To Him (in Newton's theology) Jesus Christ and the Holy Ghost were alike subservient and lesser divine beings. Jesus was certainly God's special creation, not coeval with Him, neither a holy man nor prophet but the Lamb of God risen from the dead, ascended into Heaven and sitting at the right hand of the Father in judgement on the quick and the dead. Orthodox Nicene Christianity was to Newton merely the weak work of men's minds, of men within the fourth-century Church from whom the Papists descended. Frank E. Manuel found in the British Library a Book of Common Prayer belonging to Samuel Clarke in which every reference to the Trinity had been scored out; if less rabid in his position, Newton would surely have sympathized with such a view of the Anglican liturgy. He took 'Thou shalt have no other God but Me' in the strictest sense.[5]

If God be the Ruler, the Lord of Hosts borne on chariots of wrath, Jesus is to most Christians the God who loves, supports and saves. But to the Newtonian anti-Trinitarians God's love counted for less than his power. In Manuel's words: 'Of course there are passages on divine mercy in Samuel Clarke's sermons and in Newton's manuscripts; but they are minimal if compared with glorifications of God's omniscience and omnipotence.' And of course the divine love has little to do with natural theology.

A manuscript presented by Brewster examines the relationship of Father and Son more minutely and radically. It contains twelve articles, of which the first reads: "There is one God the Father, ever living, omnipresent, omniscient, almighty, the maker of heaven and earth, and one Mediator between God and man, the man Jesus Christ." Then, in Article 3: "The Father hath life in himself, and hath given the Son to have life in himself." To Jesus Christ, wrote Newton, future knowledge is given by God such as no other being is worthy to receive, and therefore the testimony of Jesus is the spirit of prophecy. Christ came not to diminish the worship of the Father. When we pray to God or thank God for his goodness we should do so in the name of the son. Finally, Article 12 reads:

> To us there is but one God, the Father, of whom are all things, and one Lord Jesus Christ, by whom are all things, and we by him. – That is, we are to worship the Father alone as God Almighty, and Jesus alone as the Lord, the Messiah, the Great King, the Lamb of God who was slain, and hath redeemed us with his blood, and made us kings and priests.[6]

Newton, then, was no deist. He believed that Christianity, the unique true religion, is consistent with natural theology and, like so many Englishmen of his time (More, Boyle, Ray, Bentley, to name only a few) he went further to argue that natural theology was valuable because it gave rational support to Christian theology; but in itself it was neither essential nor sufficient for proper Christian belief. To Newton's mind the Bible was essential; he had studied it solemnly from his undergraduate days at least, and thanks to his prodigious memory knew it as few men before or since. The true religion could not be known without revelation, through the history of the Jews, through the Gospel story, and especially through the continuing historical testimony of the prophets.

Deciphering Prophecies

The decipherment of the mysterious visions and parables of both Testaments, which so strongly inspired Mede, More, Locke and Newton, no longer has much appeal. Their value is judged to be poetic, rather than that of a coded message directly inspired by God, in which each word carried deep esoteric significance. Few today share Newton's early opinion that "no book in all the scriptures [is] so much recommended & guarded by providence as this", the Book of the Revelation of St John the Divine, which Newton supposed to have been written in the time of Nero. Who cares whether Newton was correct in maintaining that the prophecy of the seventy weeks in the Book of Daniel referred to the interval of 490 years after Ezra's leading the Jews from Babylon back to Jerusalem (457 BC) to the Crucifixion in AD 33/34? Because the majority even of Christians have ceased to view the Bible as not only the most important repository of history accessible to men, but also as the unique key to their future, these matters of decipherment are no longer of concern. Newton was fully immersed in those now outmoded views; he was fully in the Protestant and millenarian tradition – but in his own singular version. Newton did not believe that the end of the world was nigh (though he did expect the downfall of the papacy to occur soon), rather it was still some time in the future (perhaps about the year 1867); and though he agreed with so many earlier Reformers that the Church of Rome was the Beast of the Revelation, he regarded the great apostasy as being not the rise of papalism but the ascendancy of Nicene Trinitarianism. From this grand error Protestantism had made but a partial recovery.[7]

The book, much rewritten in old age, containing the results of these studies, *Observations upon the Prophecies of Daniel and the Apocalypse of St. John,* was published by Newton's scapegrace nephew, Benjamin Smith, in 1733. It has proved one of Newton's best sellers, being soon translated into Latin and German, and reprinted many times since in English, down to 1922. The more elaborate and biting earlier drafts for it extend over half a century. In revision, by bowdlerization he made it innocuous, to the extent that Richard S. Westfall affirms that 'it is quite impossible to find any point in the rambling chronologies that compose' the book. For once, as Westfall also discovered, a precise early date can be attached to this passionate interest of Newton's, for the draft of a letter to Oldenburg written in January 1675 carries on the reverse 'material very similar to his early full-scale treatise' on prophecy, whose manuscript is now in the Jewish

National and University Library, along with much else of the same kind.[8] We have no reason to believe that Newton's engagement in these exegetical exercises was then novel; rather the reverse, because we may well conclude – from earlier extracts of books that Newton read – that his entrance into them took place in the 1660s. Such studies continued to the last days of Newton's life, though his boldness in recording his sincere but unorthodox convictions gradually diminished. The core of everything he wrote was his belief that Christianity had taken a wrong direction since the fourth century AD.[9]

Newton was confident that a precise technique must be mastered and strictly followed in interpreting prophecies: he studied it as he also studied the establishment of an accurate text of the *Revelation.*[10] Interpretation should be consistent, simple and natural. "It is the perfection of God's works that they are all done with the greatest simplicity," he declared.[11] The same symbols – seals, trumpets, frogs, eaten books – should always be given the same sense. This should be established from the common value of the symbol in ancient literature, for the ancient peoples commonly held to the same ideas of things; Newton believed that the Jewish prophecies followed a pattern common to all peoples. Regard should also be paid to the metaphorical use of natural events as well as physical objects: so when a prophet speaks of fire he means war. The result of proper procedures of interpretation was, Newton claimed, rigorously accurate:

> For as of an Engin made by an excellent Artificer a man readily believes that the parts are right set together when he sees them joyn truly with one another . . . so a man ought with equal reason to acquiesce in that construction of these Prophesies when he sees their parts set in order according to their suitableness and the characters imprinted in them for that purpose.

Further, he denied that there could be any ambiguity in so fitting the clear to the cipher version of prophecy, "becaus God who knew how to frame it [*sc.* the Apocalypse] without ambiguity intended it for an article of faith."[12]

Newton had his own firm views about the processes by which the dissolution of this present world would occur. In a paper entitled "The Synchronisms of the Three Parts of the Prophetick Interpretation", amid massive quotation from the Scriptures, the nub of his account seems to be that in a time when the world is engulfed in war, Christ will come to rule his kingdom with a rod of iron. He will judge the raised dead and those yet alive "and . . . as many as are found written in the Book of life are adjudged to life and saved by being either caught up into the air to be with the Lord or left below on earth in the [continuing] kingdom of the mortals" which will last forever under Christ's rule. The rest are condemned to eternal death and cast into the lake of fire.[13] Thus, in Newton's vision, after the Resurrection, "the sons of the Resurrection live among [the remaining mortal men] like other men and reign over them in the beloved city," which is the new Jerusalem, whose dimensions Newton defines. This city, like none other, stretches up into heaven itself, for "I say that as fishes in water ascend and descend, move whither they will and rest where they will, so may angels and Christ and the Children of the resurrection do in the air and heavens. 'Tis not the place but the state which makes heaven and happiness." After this bizarre contemplation of a city, or a world, inhabited by angels, the occasionally visible Saved, and mortals under Christ's stern rule, it is perhaps not strange that Newton was also willing to populate "these immense spaces of the heavens above the clouds" with immaterial living beings: "For in God's house (which is the universe) are many mansions, and he governs them by agents which can pass through the heavens from one mansion to another."[14] Perhaps it was inevitable that the imagination which in youth was fertile and original in comprehending the natural world by means of experiment and mathematics should seem so banal, in old age particularly, when facing the imponderables of mortality and immortality.

The Anglican Church

However, in his papers to do with religion Newton did not always occupy the sublime ground of prophecy, theology and sacred history. There are more practical manuscripts which nevertheless illuminate his own positions. One, in Newton's hand, appears to be a draft for a proclamation or an Act of Parliament, written before the accession of George I, since it asserts that the Church of England is said by some to be endangered by that event. It provided for each member of the population to declare formally his or her detestation of the Church of Rome, as a "false, uncharitable, and idolatrous church, with whom it is not lawful to communicate" while affirming the Lutheran and Calvinist Churches to be true and lawful Churches. There is of course no proof that Newton devised such a plan or drafted the proposed affirmation. Another paper, entitled "Irenicum: or ecclesiastical polity tending to peace" seems faithfully to represent Newtonian Erastianism. In the spirit of Henry VIII it declares (Thesis 10) that "the King is supreme head and governor of the Church in all things indifferent, and can nominate new bishops and presbyters to succeed in vacant places, and deprive and depone them whenever they may deserve it." Ordinary members of the Church are all those baptized into it. Such membership of the whole Church a man cannot lose, though he cease (for some sin) to be a member of some local congregation. But no cause of sin should be trivial or formal: "For we were to be saved not by the works of the law, but by faith in Jesus Christ" and "articles of communion" cannot be prescribed by men as necessary to salvation. "All such impositions are teaching another Gospel."[15] Here surely we may see Newton's opinion of the Anglican Church of his day and its treatment of those who dissented from its narrowly formulated articles of faith. As always, Newton was a simplicist. And in the same spirit he seems to say that the civil power acts justly in enforcing the basic elements of Christianity, unjustly in punishing men who reject arbitrary niceties of doctrine.

The absolute sincerity of Newton's unorthodox religious belief, constant throughout his life, cannot be questioned, nor can it be doubted that the history and meaning of Christianity were topics studied long and deeply. Though he may not have taken to his step-father, Barnabas Smith, upbringing in a clerical family did not render theology a tedious topic in Newton's eyes. Yet his interest was long concealed from all about him, including young Humphrey, and it left few traces in his early natural philosophy. The major exception is the unfinished sketch *De gravitatione et aequipondio fluidorum*, which invokes God frequently as the creator of the natural world. But the tract reveals little of the author's religion. The *Principia* and *Opticks* reveal even less in their original editions. Newton did not at first rest his natural philosophy upon any declared metaphysical or religious foundations. He was satisfied to start from physical axioms, exploring Nature by experiment and mathematical analysis. Only when in advanced age, starting with *Optice* in 1706, did Newton begin to inject into his scientific writings his system of natural theology, and his notions about the constraints in the universe that arise not because things must be so (as in logic) but because the divine will made them so. Thus, for example, the passage in Query 31 about the divinely appointed order manifest in the bilateral symmetry of animals was lifted from a theological manuscript. Such ideas in Query 31 were reworked for the General Scholium of the 1713 *Principia*, Newton's most familiar statement of his view that natural philosophy must be written with an eye to divinity and that the scientific study of causation must bring one at last to a final cause.

This change in character in Newton's books after 1706 may be attributed to various causes: to his increasing age and concern with religion; to the need to satisfy critics who had objected that his physical principles were absurd and lacked all philosophical foundation; to a sense that it was now time to reveal his honest view of human knowledge on the grand scale and (like others before him) to demonstrate the ineffable harmony of religion and science. Whatever our opinion about Newton's thoughts on these matters as revealed during the early eighteenth century, it does not follow that the same thoughts were already present in his mind during the 1680s and 1690s. Surely, Newton would always have agreed that at some level the writing of *Principia* and *Opticks* and his years of mathematical study were for the greater glory of God and understanding of his creation. But I am less sure that Frank E. Manuel was quite right to claim that Newton was 'making a last great attempt at one and the same time to keep science sacred and to reveal scientific rationality in what was once the purely sacral. The coupling of the two realms – the religious and the scientific – is the syncretist fantasy of a scientific genius and a God-seeker.' True, Newton was both of these; but he did not dwell in each realm with equal intensity at all times. It is my conviction that Newton was not thinking intently about religion when he was immersed in his optical experiments or his mathematical theory of motion, when he was carried along by a passionate absorption in unravelling (by one method or the other) a series of interlocked and evolving problems to their ultimate resolution in new comprehension, and when he was driven by an intense desire to disclose the inner workings of Nature through eliciting the pattern and law-like simplicity of its construction. In day-to-day, week-to-week terms the pursuit of genius and the search for God could never proceed simultaneously along identical lines, however devout the natural philosopher. And however clearly Newton's notebooks may prefigure a godly outlook upon knowledge, the mature formulation of his natural theology in print can only be judged a *post facto* rationalization of his career of scientific investigation.

Last Days

Newton's last academic action was the epistolary support he gave to Colin Maclaurin at Edinburgh in November 1725. Few details remain of the activities of his last months, during which he resigned attendance at the Mint to his deputy, Conduitt. The last extant letter from Pemberton about the third edition of *Principia* was written on 9 February 1726: Newton must have decided against several changes there proposed by his editor. In May 1726 he contributed £3 to the reflooring of the church at Colsterworth; to the incumbent of this living, Thomas Mason, he addressed his last known letter, about the unprofitable assay of a sample of 'ore' dug at Woolsthorpe.[16] Thus to the end Newton maintained his connection with the soil of his birth, though he never returned to Lincolnshire after 1696. Estate management had been a continual task for Newton since (as in everything he did) he neglected no detail: the good repair of his tenants' buildings (by them!), their not cutting timber without leave, the value of a dung-heap, were all dealt with in the day's work.[17] Scanty evidence also survives of intermittent correspondence with his Lincolnshire half-nieces, Hannah Tompson and Mary Pilkington, who occasionally sent presents of game.

Immediately upon Newton's death, intestate, Conduitt wrote to Thomas Mason asking him to inform the natural heir, John Newton (1707–37), of the event. 'God knows,' Mason replied, '[he is] a poor Representative of so great a man, but this is a case that often happens.' John hastened to London, bearing Mason's assurance that Sir Isaac Newton had no closer male relative. He inherited the family lands at Woolsthorpe and Sewstern reckoned to be worth £80 per annum. Also independent of Newton's personal estate at death were the properties bought for the unfortunate family of the late Robert Barton and, in Kensington, for the infant Catherine Conduitt ("Kitty" to Newton). Other cousins soon hurried to London to join John Newton, for everything was to be divided among Isaac Newton's eight living half-nieces and half-nephews. Of these six signed the papers connected with the valuation of his personal estate.[18]

Attempts have been made to compute Newton's wealth in his last years. Never a poor man, always sparing in his personal life, besides his private resources he derived from 1696 onwards a substantial income and a free house from the Crown – though he had soon ceased to live in it. Villamil, adding in his profits from the coinage after he became Master of the Mint, reckoned his annual income in his last years as £4,000 per annum, perhaps exaggerating. This was noble if not princely. Conduitt put his personal estate at £32,000 while Stukeley guessed that his eight heirs received £3,500 each – 'but all soon found a period, as if to shew the fleeting vanity of riches, family and secular acquirements . . .'[19]

In detail, Newton's household goods and plate were valued at the considerable sum of £424, his medals, books and manuscripts together at £627. All this implies a high standard of comfort and display: even the female servants' garret was embellished by rugs, a picture and six prints. Mrs Conduitt's elegantly furnished room was valued at £39. Newton's investments in the Bank of England and the South Sea Company, however, amounted to £28,330 only, to which should be added £2,151 due to him from the Mint by Conduitt's account, and £201 of personal debts, that is (with cash in hand) a total of £30,760. Thus Conduitt's figure for his total estate is confirmed.

However, Newton's debts have to be set against these assets. Conduitt explained in the inventory that Newton (or rather his heirs) had 'to account with the Crown for' £34,325 (Conduitt made a mistake!) impressed to him over recent years for the expenses of running the Mint, part of which sum had been expended for that purpose while the residue lay upon his hands. This residue, according to the final Mint account with Newton, amounted to £22,278, a debt which Conduitt assumed as the next Master of the Mint.[20]

In the practice of that age, the holder of funds imprest to him by the Crown held them personally, and could invest them to his personal profit until they were expended upon their proper purpose; conversely, if the Crown owed him money, the loss of interest was his till the Crown chose to pay up. Just before his death Newton might have drawn £1,000 a year in interest on the balances he held. And so might Conduitt. But it is highly unlikely that the latter undertook to burden himself with a massive debt in order to benefit Newton's eight heirs, and it seems obvious that when Conduitt became Newton's Deputy Master of the Mint, Newton would have made over to him a sum at least equal to his own indebtedness to the Crown. This would be the fund from which Conduitt would ultimately discharge that debt, and so avoid a charge falling upon Newton's personal estate. As with much else in Newton's financial history, there is no record of such a transaction, and therefore, though it might be possible (if unprofitable) to establish Newton's income in a particular year, it is impossible to work out his capital accumulation during his long life. The antithetical stories that he either gained, or alternatively lost, a great sum when the South Sea Bubble burst are both likely to be false. It may well be that had Newton sold his investment at the highest point of the madness he would have realized a great gain; but many Stock Exchange speculators could claim to have lost such a profit as that! All that can certainly be said is that Newton became a very wealthy man and was generous to others with his wealth. Miss Catherine Conduitt thereby became a very marriageable heiress, who evidently also had the good sense to maintain the vast intellectual heritage that she also brought to the family of Wallop (see Appendix E).

In Newton's lifetime a grateful nation awarded to the victor of Blenheim, Ramillies and Oudenarde a princely estate and a perpetual revenue; in later ages the victors of Trafalgar and Waterloo similarly received, besides honour, great tangible rewards from their people. Not so with Newton. Only of recent years has a national organization taken charge of his modest birthplace and made it a place of pilgrimage. But honour was heaped upon Newton. His victories over human ignorance and Nature's subtlety did not go unrecognized in his own time. He was buried in Westminster Abbey with such pomp as befitted a great commander. Though it was Queen Anne who had touched his shoulder with a sword, it was under her successor that Newton rose to become a national hero. Probably few waited at Crane Court to see the great man pass, and his walking along the streets must often have gone unremarked, but the ruling ranks of Britain, like the writers, the clergy and the artists fully recognized his merit and his authority. Pope composed the celebrated epitaph that does as much credit to Newton as to himself. According to Conduitt and the *London Gazette*:

On the 28th [March] past [1727], the corpse of Sir Isaac Newton lay in state in the Jerusalem Chamber, and was buried from thence in Westminster Abbey, near the entry into the choir. The pall was supported by the Lord High Chancellor, the Dukes of Montrose and Roxborough, and the Earls of Pembroke, Sussex, and Macclesfield, being Fellows of the Ryoal Society. The Hon. Sir Michael Newton, Knight of the Bath, was chief mourner and was followed by some other relations, and several eminent persons, intimately acquainted with the deceased. The office was performed by the Bishop of Rochester, attended by the Prebend and choir.[21]

Voltaire, who was present in the Abbey, called it the funeral of a king who had done well by his subjects.

The memorial to Newton in Westminster Abbey, whose design symbolizing Newton's discoveries was due to John Conduitt, though its execution in detail was by William Kent and Michael Rysbrack, was completed in 1731.[22]

Really, the oddest thing in Newton's life is the fact that the story of the apple, known to everyone, is true. It is as though the archaeologists at Winchester had found Alfred's payment for the burnt cakes.

A nineteenth-century Japanese print of Newton in his orchard, watching the apple fall. Courtesy of Stillman Drake, Toronto.

APPENDIX A

Newton's Alchemical Studies and his Idea of the Atomic Structure of Matter

In response to my request, some years ago, Professor Karin Figala prepared an elementary account of the results of her investigation of Newton's alchemical papers and related documents. This was communicated to me privately in March 1984. In the following paragraphs Professor Figala's summary is printed, with her permission. For her technical paper on this material, see 'Die exakte Alchemie von Isaac Newton' in *Verhandlungen der Naturforschenden Gesellschaft Basel*, **94**, 1984, pp. 157–228.

The search for a combination of the exact sciences with magical thought was a central motive of the seventeenth-century attitude of mind: Newton was no exception to this rule. His personal attitude towards the problem is best illustrated by his (al)chemical research, for he seems to have seen in traditional alchemy – which he considered to be of divine origin – a possible synthesis of seemingly divergent lines of thought. He was convinced that God had arranged everything in Heaven and on Earth according to number and measure, thus sharing the firm belief of the ancient Pythagoreans.

Harmonic proportion seems to be a key concept in Newton's thought, as mirrored in his theory of the material world. In fact, his theory of the composition of matter – laid down in principle in *Opticks* – centres on the material world; accordingly, contemporary alchemy immediately presented itself to his mind. It is in Newton's 'rational alchemy' that we can get hold of at least a small part of his attempt to reconcile magic and science.

In Newton's theory of the composition of matter, seemingly impenetrable particles are built up from cubic elementary cells, some of which hold matter, some of which do not. An example of the type of composition considered by Newton is given in the illustration: different stages of composition are clearly visible. A material body of stage (or order) n contains a certain number of empty elementary cells, another certain number of full elementary cells. The ratio between full and empty is given by $(2^n - 1) : 1$, as becomes clear from reading the appropriate paragraph in *Opticks* (reproduced below the illustration).

In this paragraph Newton seems to accept the common notions of 'atom' and 'void' (or vacuum); in addition to these classical concepts, forces (*passive* or *inherent* = the force of inertia; and *active*, such as the forces of gravitation, fermentation and cohesion) as well as a hierarchical arrangement of the particles enter into his consideration.

All this would be mere speculation, were it not for certain practical consequences to be deduced from Newton's scheme. At each stage of composition, if we look for a generating principle of the hierarchy of matter, we can reach the next stage by injecting additional void into a particle of order n; as the illustrations show, the number of elementary cells in a particle of order n is 8^n, when $n \geq 0$. $4n$ of these cells are full, the remainder are empty. Thus we arrive at the ratio mentioned above: void/matter = total matter/matter =

$$\frac{8^n - 4^n}{4^n} = 2^n - 1$$

A critical quantity is the factor of two that occurs in every transition to a higher stage; that is, Newton has decided in favour of a model doubling the length of each edge of the material cube. Alternative models employing (integral) factors of 3, 4, . . . and so on have been indicated by David Gregory. But let us stay with Newton's choice: the series of 'rarefactions', $a_n = 2^n - 1$, can be interpreted as a recurrent series; the recursion formula is given by

$$3a_n = 2a_{n-1} + 1a_{n+1}$$

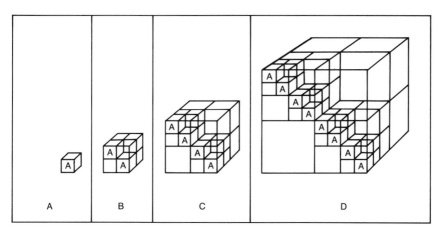

A = one perfectly hard solid atom

B = one corpuscle, Order 1, comprising 8 atoms, matter/void = 1:1

C = one corpuscle, Order 2, comprising 8^2 atoms, matter/void = 1:3

D = one corpuscle, Order 3, comprising 8^3 atoms, matter/void = 1:7

Therefore a corpuscle of Order N, comprising 8^n atoms, has a matter/void ratio of $1:2^N-1$. Atoms are shown cubical for convenience only.

APPENDIX A.1 Newton's idea of the atomic structure of matter

"if we conceive these Particles of Bodies to be so disposed amongst themselves, that the Intervals or empty Spaces between them may be equal in magnitude to them all; and that these Particles may be composed of other Particles much smaller, which have as much empty Space between them as equals all the Magnitudes of these smaller Particles: And that in like manner these smaller Particles are again composed of others much smaller, all which together are equal to all the Pores or empty Spaces between them; and so on perpetually till you come to solid Particles, such as have no Pores or empty Spaces within them: And if in any gross Body there be, for instance, three such degrees of Particles, the least of which are solid; this Body will have seven times more Pores than solid Parts. But if there be four such degrees of Particles, the least of which are solid, the Body will have fifteen times more Pores than solid Parts. If there be five degrees, the Body will have one and thirty times more Pores than solid Parts. If six degrees, the Body will have sixty and three times more Pores than solid Parts. And so on perpetually. And there are other ways of conceiving how Bodies may be exceeding porous. But what is really their inward Frame is not yet known to us."

(*Opticks*, 1952, Book II, Part III, Proposition VIII, pp. 168–9)

We may detect a trinitarian interpretation: three times the middle term $(3a_n)$ unites in a certain way its parent ancestor $(2a_{n-1})$ and its son successor $(1a_{n+1})$; a more 'substantial' interpretation would be that a_n mediates between 'solid' (Earth) and 'thin' (Heaven), a_{n-1} being more solid than a_n and a_{n+1} being rarefied with respect to a_n. Thus a_n also mediates between 'down' and 'above' and so on. Seen in an alchemical way, the 'soul' a_n mediates between 'matter' (the full) a_{n-1} and 'void' (the empty, without matter = spirit) a_{n+1}. The mediator is part of what it joins together; its nature is hermaphrodite. In traditional alchemy Mercury plays this role and on yet another level Mercurius is Hermes, messenger of the gods, who mediates between the gods and mankind. Thus Newton's scheme has the additional advantage of conforming to magic-alchemical religious ideas.

Newton's theory of composition not only conforms to ancient magical thought, but allows also numerous applications. Most prominent is the application to specific weights, which results immediately from the theory and has also already been laid down in the *Opticks*. First, we imagine an isosceles triangle in the shape of a Lambda (see the second figure thus recovering, by the way, the Great Pythagorean Tetraktys); to the left we write the relative proportion of matter $(1:2^n)$, to the right the corresponding proportion of void

$$\left(\frac{2^n-1}{2^n}\right)$$

in a body of order n. The line joining the corresponding points defines a pair

$$\left(\frac{1}{2^{n'}}\cdot\frac{2^{n-1}}{2^n}\right)$$

which in turn defines the nth stage of composition. Increasing n means increasing rarefaction. For water Newton has found a rarefaction near n=6; specifically, he has water with 1 part of matter against 65 parts of void. Conversely, this defines the pair (1/66;65/66) on our scale; (n=6) defines the pair (1/64;63/64)). Once *one* point of reference has been determined, the specific weight of any other material can be derived under the assumption that its ratio of rarefaction is known; vice versa, the specific weight being known, this ratio is determined. For example, Newton found gold to have a specific weight of 19 (relative to water); the proportion of matter in gold is thus 19 times that of water, i.e. $19 \times 1/66$, or nearly 1/3.5. The proportion of void

APPENDIX A

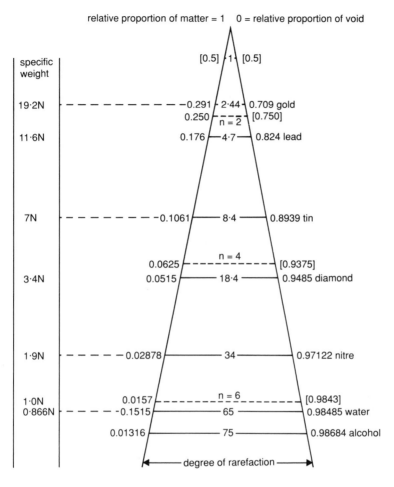

relative proportion of matter = 1   0 = relative proportion of void

APPENDIX A. 2 Solid matter, void and density related (Dr Figala's diagram simplified)

has to be $(66-19)/66 = 47/66$; the degree of rarefaction being (void:matter) = $(47/66):(19/66)$, or nearly 2.5. Now $1/3.5 = 2/7$, and thus gold has 2 particles of matter in a total of 7 particles, leaving 5 particles of void. It is exactly this distribution (void:matter) = $5:2 = 2.5:1$ that Newton gives for the composition of gold. A further deduction is the proof of existence of a maximal specific weight, which is also easy to derive from Newton's theory of composition. An a-priori consideration has led Newton to the following consideration:

— 385 —

every natural body on earth has at least order 1, because order 0 (as well as infinity) is reserved to the ultimate. Order 1 is defined by the distribution matter:void = 1:1, or, in terms of relative proportion to the total volume, $\frac{1}{2}$ matter, $\frac{1}{2}$ void. Now let m be the specific weight (relative to water) of any material body; according to the scheme developed with gold, the proportion of matter in this body, measured against the total volume, is m × $\frac{1}{66}$; since this product may not surpass $\frac{1}{2}$, m cannot be greater than 33, which is an upper bound on specific weight. Starting the same argument not from the composition of water (1:65 = matter:void), but from gold (2:5; relative proportion of matter 2:7), we obtain y × $\frac{2}{7}$, which may not surpass $\frac{1}{2}$; here y is the specific weight measured against gold. The maximal value of y is therefore 1.75, and this multiplied by the specific weight of gold, 19, gives us 33.25 as the upper bound of the specific weight, measured against water. (The slightly different values are due to rounding.) Obviously, this upper bound depends on the 'basic factor' 2 chosen by Newton. But in fact, no metal nor mineral – these being the main objects of Newton's research – surpass a specific weight of about 33, so that his theoretical deductions are in accordance with practical observations. There is still the implicit relation between specific weight, rarefaction and order, e.g. water with order 6.044 ~ 6, gold with 1.796 ~ 1.8; the relation is reciprocal.

How do metals and minerals, i.e. their inherent particles, stick together? Here Newton decides in favour of a sulphuric principle, occurring in three different forms, which can be identified with certain orders: (a) raw, unclean (order 3); (b) 'metallic' (order 2); (c) clean, not coagulating, fiery (order 1, that is, the ultimate natural order). The 'sulphur' of type (c) is, according to Newton, responsible for the transmutation of metals.

APPENDIX B

Newton's Portraits

Newton was painted at least seventeen times by seven or more artists: Sir Godfrey Kneller depicted him four times, Sir James Thornhill and John Vanderbank twice. In addition one may examine a death mask, ivory medallions made by David Le Marchand at various dates and an excellent medal in profile (1726) by Newton's Mint artist, John Croker. All these images of Newton, except the 1689 Kneller (of which at least two versions exist) were made after Newton's move to London, hence after the age of sixty (Kneller, 1702). Many of the portraits were engraved and widely circulated.

Three of especial interest are the 1689 Kneller painting of Newton with a youthful, alert, intellectual appearance, of which the original has always been in the family; the 1720 Kneller painted for Pierre Varignon; and the 1725 Vanderbank, commissioned by the Royal Society and still hanging in its rooms.

Of the second, showing Newton at the age of seventy-eight, Varignon remarked that he appeared no more than fifty years old, yet Brook Taylor had assured him of its excellence as a likeness. Stukeley, present at the 'sittings' for this portrait, wrote that 'it was no little entertainment to hear the discourse that passed between these two first men in their way. tho it was Sir Isaac's temper to say little, yet it was one of Sir Godfrey's arts to keep up a perpetual discourse, to preserve the lines and spirit of a face' (*Memoirs*, 1936, pp. 12–13). This portrait was later acquired in Paris by Lord Egremont and now hangs at Petworth.

Newton's portraiture is markedly inconsistent. The 1689 Kneller shows a thin, long, lively face, markedly aquiline. In subsequent portraits by the same artist, in which Newton appears wigged, the face is recognizably the same but heavier, fuller, more stolid and even stern in aspect. The autocrat of the Royal Society appears, complete (in time) with sword, hat and gloves. But the division of the upper lip is always slightly askew, the hands delicate and fine, the nose long. Thomas Murray's portrait of Newton aged seventy-five (in 1718) shows similar characteristics and gives him (but for Varignon one might suppose) an excessively youthful freshness (Westfall, *Never at Rest*, 1980, fig. 13.6). The exaggeration of heavy jowls in the lithograph of a portrait by William Gandy (1706: ibid., fig. 12.3) produces a face quite unrecognizable as Newton's. The squareness of Newton's lower jaw and the strong bony chin, slightly cleft, emphasized by Le Marchand (ibid., figs 14.6, 14.7) and seen in Kneller's heads, are clear also in Vanderbank's 1726 portrait, which gives Newton, despite his age, a powerful and haughty look. My own portrait sees an older Newton much as Kneller (1720) and Vanderbank did, and in the same garb.

All these are wigged portraits. Another group of pictures showing Newton in old age and his own hair, notably that by Enoch Seeman (Newton with the third *Principia*) especially as engraved by James Macardel, 1760, show a much decayed, weaker individual with a round rather than a long face. Thornhill's portrait at Trinity College, Cambridge, has the long nose and square jaw mentioned above, while the Portsmouth family head by the same painter shows its curiously long, rounded back clearly brought out in Le Marchand's bust. The obviously damaged late Kneller (Westfall, *Never at Rest*, fig. 15.10) is quite on its own: only the large nose and hair look like Newton.

I thank Mr H. A. Chancellor for enabling me to read his dissertation on 'The Influence and Iconography of Isaac Newton in the Eighteenth and early Nineteenth Centuries'.

APPENDIX C

Newton's London Homes

Newton first took up residence in a house in Jermyn Street (now number 88) towards the socially superior western end, in which he lived four years, then from 1700 to 1709 in the next-door house (87). Sir Robert Gayer was a near neighbour (letter 710). Leases for these properties had been granted by the trustees of the Earl of St Albans in 1665, and the buildings were put up about ten years later. Number 87 is at present undergoing total reconstruction and number 88 is a shop ('James Bodenham') with the original brick structure preserved.

After his brief sojourn in Chelsea (where he lived at the east end of Paradise Row) Newton moved to 35 St Martin's Street, a few yards south of Leicester Fields (now Square), an area just newly developed by Lord Leicester. There he lived in a house built c.1695 from 1711 to 1725. It had three storeys with basement and was built of brick with a tiled roof; there was a bracketed hood over the front door. On the main floors were rooms front and back with a projecting 'closet' wing at the rear. The ground-floor front room was panelled. According to a footnote in the *Survey of London* Newton had a small observatory built at the top of the house.

In Kensington he lived (in lodgings?) in Orbell's Buildings, a little north of Kensington High Street and west of Church Street, a site now occupied by Bullingham Mansions. He already owned property south of the High Street. In 1720 he paid £1712 for three acres now covered by Cheniston Gardens, and he also acquired a further four acres in the Marloes Road/Wright's Lane area. This was the land that ultimately went to Kitty. The Conduitts briefly lived in this neighbourhood in 1719.

Sources:

Survey of London, vol. II, 1909, p. 17; vol. XX, 1940, pp. 107–8 (with plan of St Martin's Street house and plate 96 showing two of its chimney-pieces); vol. XXIX, 1960, p. 279; vol. XLII, 1986, pp. 2, 100.

APPENDIX D

Jonathan Swift and Catherine Barton

Swift was in England from November 1707 to May 1709 and from September 1710 to the middle of 1713, on Irish Church business. His first known reference to Catherine Barton is in his terse accounts, where he notes a loss of two shillings at ombre, at 'Barton's' on 4 December 1708. Twelve days later he was there again, suffering very bad 'fitts' of his habitual giddiness; he was forced to take a chair home, instead of walking as usual, and next day rewarded Mrs Barton's servants with the large present of 7s.6d in gratitude for their attentions. In April, July and August 1709 he recorded letters from Catherine, and on 26 May 1709 one written to her.[1]

During his second and longer visit to England Swift noted in his journal for 'Stella' (Esther Johnson) that he dined with Mrs Barton three times in thirty-three months (28.ix.1710; 30.xi.1710; 7.iii.1711); he records visits to her on nine further occasions (19.xii.1710; 23.i.1711; 2.iv.1711; 10.iv.1711; 6.vii.1711; 18.vii.1711; 14.x.1711; 25.x.1711; 20.xii.1711). It is notable that by his own record Swift saw nothing of Mrs Barton during the whole of 1712, nor during the six months of 1713 that he spent in London.

During his first visit to London Swift associated with the Whigs and became friendly with Halifax, among others. He never in any way connected Halifax with Mrs Barton, though he continued to meet both during his second visit to London, when he supported the Tory interest (and therefore disputed with Mrs Barton, who like her uncle was a Whig). I find it unimaginable that Swift would have been such a close friend of hers if she had been a notorious woman, or indeed other than fully respectable. It is of course possible that the same consideration explains Swift's (recorded) lack of social exchanges with her in 1712 and 1713. On the other hand it is certain that he continued always to regard her as a friend.

Most of Swift's journal notes are barren of interest, but on 18 July 1711 he noted that Mrs Barton was going into the country; nothing proves that Newton remained in town. On 9 October he told Stella that he was moving to lodgings in Leicester-Fields, and a fortnight later called on Mrs Barton, his near neighbour. On 3 March 1711 he rehearsed a common quip: 'I love Mrs Barton better than anybody here, and see her seldomer', possibly as a result of a visit that he failed to record since he recounted to Stella Mrs Barton's tale of the virgins. Swift was quick to rebuke Stella when (in a lost letter) she made a joke about Swift's friendship with this young unmarried woman (her own status too, of course). 'I'll break your head in good earnest, young woman [he told her], for your nasty jest about Mrs Barton. Unlucky sluttikins what a word is there? Faith, I was thinking yesterday, when I was with her, whether she could break them or no, and it quite spoiled my imagination' (30 November 1710). But he makes Catherine lacking in feeling: she mourned her brother Robert only for form's sake; he was a 'sad dog' (14 October 1711).

Halifax had apartments in Old Palace Yard, Whitehall. I have not found the address of his town house, but De Morgan could not find him residing in the Leicester-Fields/Soho area.[2] If not, then *he* could not have been a neighbour of Swift's.

Personally, I see no reason why Catherine should not have had her own rooms in her uncle's house, no doubt attended by the same servants (apart from her own maid). Swift never mentioned Newton in the *Journal to Stella* but a late letter indicates that he had conversed with him at least once, about the longitude![3] Swift's latest biographer believes that Catherine was living at Newton's house in the days when Swift used to meet her.[4] Brewster (*Memoirs*, vol. II, pp. 492–5), reviewing the same evidence, came to the same conclusion. But the fact remains that Catherine Barton disappeared from sight for four years – from the end of 1711 until after Halifax's death. By 1716, at any rate, she was again acting as her uncle's hostess.

Only more than a quarter of a century later did Catherine Barton reappear in Swift's correspondence, during which time Halifax had died, Catherine had married John Conduitt and Newton had died too. On 19 April 1730 Swift wrote to Lady Worsley: 'How is our old friend Mrs Barton? (I forget her new name). I saw her three years ago, at court, almost dwindled to an echo, and hardly knew her.' He did indeed make a last visit to England in 1726 and 1727. About three years later he wrote a letter of introduction to Mrs Conduitt (now lost) brought to London by his friend Mary Barber, also asking the Conduitts to support her new book of poems. To this Mrs Conduitt replied (from George Street) on 29 November 1733, assuring Swift that her husband had already subscribed for the book of verse. She continued, banteringly:

> I should have guessed your holiness would rather have laid than called up the ghost of my departed friendship, which since you are brave enough to face, you will find divested of every terror, but the remorse that you were abandoned to be an alien to your friends, your country and yourself. Not to renew an acquaintance with one who can twenty years after remember a bare intention to serve him will be to throw away a prize I am not now able to repurchase; therefore when you return to *England* I shall try to excel in what I am very sorry you want, a nurse; in the mean time I am exercising that gift to preserve one who is your devoted admirer.[6]

The letter ends with some literary news. The original is lost; perhaps the enigmatic sense may in part arise from some errors of transcription. It is clear that Catherine meant to reproach Swift gently for his long retreat to Dublin from London society; and that she felt herself guiltless with regard to the long lapse in their friendship.

Sources

1 Paul V. Thompson and Dorothy J. Thompson, *The Account Books of Jonathan Swift*, 1984, pp. li, lxvii, 60, 61, 63, 72, 83, 84.
2 Augustus De Morgan, *Isaac Newton: his Friend and his Niece*, 1885, p. 115.
3 Jonathan Swift, *The Correspondence of Jonathan Swift*, ed. Harold Williams, vol. IV, 1965, p. 240, September 1727.

4 Irvin Ehrenpreis, *Swift: the Man, his Works, and the Age*, vol. II, 1983, p. 307.
5 Swift, *Correspondence*, vol. III, 1963, pp. 391–2.
6 Ibid., vol. IV, 1965, pp. 213–14.

APPENDIX E

The Fate of Newton's Papers and his Library

As will be evident from this biography, Newton carefully preserved his notebooks and papers, including his correspondence, throughout his life, accumulating a mass of folded bundles whose contents were not in all cases readily identifiable. After his death, as part of the settlement of his intestacy, John and Catherine Conduitt received undivided rights to this great mass. Only two manuscripts (*The Chronology of Ancient Kingdoms Amended* and *The System of the World*) were found suitable for immediate sale to booksellers for publication; otherwise little from this mass reached print during the eighteenth century, even in Samuel Horsley's collected edition of Newton's writings (1779–85). While Horsley had been able to make only a cursory examination of the papers and remained unperceptive of their significance, Newton's first real biographer, Sir David Brewster, amended and amplified a book already in proof from biographical (and some other) materials selected for him by Henry Arthur Fellowes in 1837.

Fellowes was a nephew of the then Earl of Portsmouth, owner by inheritance of the Newton papers, which were preserved at his seat, Hurstbourne Park in Hampshire. His family name was Wallop; the younger Catherine Conduitt (Newton's 'Kitty') had married John Wallop in 1740, becoming titular Viscountess Lymington by the creation of the earldom of Portsmouth for her father-in-law in 1743.

Meanwhile, before Brewster's time many letters and documents sent by Newton to others (including Collins, Boyle and Locke) had reached print, as had extracts or paraphrases of the memoirs of Newton compiled by John Conduitt before and after Newton's death. Some of these Newtonian scraps have since been rejected as false. Brewster and other authors – Francis Baily (1835), Stephen Rigaud (1838), Augustus De Morgan (1840 onwards) and Joseph Edleston (1850) in particular – aroused interest in Newton's processes of discovery, the development of his genius and his character. In circumstances whose detail is unclear the Duke of Devonshire, founder of the Cavendish Laboratory at Cambridge and Chancellor of that University, together with the Vice-Chancellor, agreed with Lord Portsmouth in 1872 that two Cambridge mathematicians, John Couch Adams and Sir George Stokes, should visit Hurstbourne Park to look over Newton's manuscripts. Quickly gauging that little useful could be accomplished in a brief stay there as Lord Portsmouth's guests, Adams and Stokes with his consent bore the whole mass back to Cambridge for thorough examination. For this they brought in H. R. Luard and G. D. Liveing, the former dealing with the personal papers, the latter with those of chemical interest. The four men constituted a syndicate of the university; besides transcribing many papers and letters they published a summary catalogue of them all (*A Catalogue of the Portsmouth Collection of Books and Papers written by or belonging to Sir Isaac Newton*). The papers were physically reordered into subject categories such as 'Mathematics', 'History', 'Letters' and so on.

The catalogue was published in 1888. In the same year all material in categories of a non-scientific character (including letters) was returned to Lord Portsmouth, as previously agreed, while the scientific categories were retained by the university on permanent loan and made available to scholars. For half a century or more few looked into them, of whom the most effective was W. W. Rouse Ball.

In 1936, after the emigration of the then Lord Portsmouth to Kenya, the whole of that part of the Newton papers that had been returned to his ancestor was put up at Sotheby's to be sold 'by Order of the Viscount Lymington'. The two-day sale brought in less than £10,000. Although a few lots were ultimately reserved by the family, some three million words of Newton's autograph were in consequence scattered from Jerusalem to Los Angeles. Lord Keynes was the principal buyer, for the benefit (within a few years) of King's College, Cambridge. Some other considerable collections were formed, now at Trinity College, Cambridge, the Hebrew National and University Library, the Babson Institute in Massachusetts and various other university libraries in the USA. However, the location of a few lots is still not known, and some in private possession are inaccessible.

Unlike his papers, Newton's books did not remain in the family; perhaps Conduitt did not choose to buy them. After Newton's death, according to the inventory of his estate, his library consisted of 362 folios and 1534 other volumes, not counting pamphlets; the whole (including his manuscript papers) appraised at £270. In a private sale, the heirs gained £300 by disposing of the printed material to William Huggins, 'cruel and grasping' Warden of the Fleet prison. He in turn passed the library to his son Charles, for whom he also acquired the patronage of the living of Chinnor in Oxfordshire. Charles Huggins took the books to the rectory at Chinnor when he installed himself there as rector in 1728. The list prepared for this sale is the first account of Newton's library.

A few printed volumes, heavily annotated by Newton, though listed, were withdrawn by Conduitt and so in the end passed into the Cambridge University Library with the scientific portion of the Portsmouth Collection, or (in at least one instance) were sold in 1936.

The great bulk of Newton's library remained at the rectory, Chinnor, gradually assimilating other books. In 1750, on Charles Huggins's death, the whole was bought by his successor rector, James Musgrave, for £400. Musgrave had the Newton books recatalogued, with the rest of the Chinnor library, in 1767; this catalogue (with appropriate abstraction of later acquisitions) agrees well with the Huggins list. In 1778 when Musgrave died his son, another James, transferred all the Chinnor books, of whose Newtonian affiliation his father had been justly proud, to Barnsley Park in Gloucestershire, a seat he had inherited from his mother's side of the family. There Newton's books were shelved in a library designed by John Nash, and remained forgotten for 140 years.

It is not surprising that a few of Newton's books disappeared between the Huggins (1727) and Musgrave (1767) lists. Far more were put on the market in 1920, without any record of their provenance and association, when a great sale took place at Thame Park, Oxfordshire. This estate had been in common ownership with Barnsley Park so, presumably to unburden the shelves there, nearly 3000 volumes were sent to Thame Park to be disposed of – about one-third of them formerly Newton's. The disposal was carelessly handled: the books were auctioned in large, uncatalogued lots, and in some cases one volume of a set was sold while its fellows stayed on the shelf. For a mere £170 more than half of Newton's library entered the antiquarian book trade, whence many books have passed (with due recognition) into academic libraries.

In 1928 Richard de Villamil drew public attention to his discovery of 'Newton's library' at Barnsley Park – a house easily enough traced from the bookplates in volumes sold off in 1920. The owners then entrusted this remaining Newtonian collection to Henry Sotheran Ltd. (who had acquired many scientific books from the Thame Park sale) for disposal *en bloc* as a library full of unique Newtonian materials. Fifteen years later in 1943 the Pilgrim Trust paid the £5,500 asked for the 856 volumes which eventually proved (with fair certainty) to have been Newton's, and presented them to Trinity College, Cambridge.

Sources

J. Harrison, *The Library of Isaac Newton*, 1978, ch. 2.
R. de Villamil, *Newton the Man* [1931?], pp. 2–7
D. T. Whiteside in *Mathematical Papers of Isaac Newton*, vol. I, 1964, pp. xvii–xxxv

NOTES

Chapter 1

1 K. A. Baird, 'Some influences upon the young Isaac Newton', *Notes and Records of the Royal Society*, 41, 1987, pp. 169–79.
2 *Mathematical Papers of Isaac Newton*, vol. I, 1964, pp. 8, 112 and plate I. The computation is correct only to 28 decimals.
3 Samuel Butler, *The Way of All Flesh* (1903), ch. 20 at end; cf. Somerset Maugham, *Of Human Bondage* (1915) painting a similar childhood with different detail.
4 The list of 'sins' was decoded and annotated by Richard S. Westfall in *Notes and Records of the Royal Society*, 18, 1963, pp. 10–16. Cf. Frank Manuel, *A Portrait of Isaac Newton*, 1968, 2nd edn, 1980, pp. 61–5.
5 *Notes and Records of the Royal Society*, 5, 1947, pp. 34–6.
6 William Stukeley, *Memoirs of Sir Isaac Newton's Life*, ed. A. Hastings White, 1936, p. 34.
7 Among Newton's surviving books are copies of Stephanus's *Pindar* (1560) and Ovid's *Metamorphoses* (1593) owned by him in 1659: see John Harrison, *The Library of Isaac Newton*, 1978, nos 1224, 1317.
8 Whiteside in *Mathematical Papers*, vol. I, pp. 3–4.
9 Stukeley, *Memoirs*, p. 51. The same story may be found elsewhere in equivalent words.
10 Ibid., p. 46. Stukeley wrote 'Miss Storey' but this is universally taken to have been a lapse of memory.

11 Louis Trenchard More, *Isaac Newton*, 1924, pp. 491–3; Sir David Brewster, *Memoirs of Sir Isaac Newton*, 1855, vol. II, p. 211. The letter is not included in the modern printed edition of Newton's *Correspondence*.

12 Stukeley, *Memoirs*, p. 39.

13 Ibid., p. 43.

14 D. E. Smith, 'Two unpublished documents of Sir Isaac Newton' in W. J. Greenstreet (ed.), *Isaac Newton, 1642–1727*, 1927, p. 28. The same notebook in the Pierpont Morgan Library, New York, contains a page or more on "streight lined triangles" and others on the "Systema mundanum secundum Copernicum". Nothing proves that this material was not entered *after* 1659.

15 Richard S. Westfall, *Never at Rest*, 1980, p. 61, note; Frank Manuel, *A Portrait of Isaac Newton*, p. 39.

16 Manuel, ibid., pp. 57, 58; in his introduction Manuel gives an excellent account of Newton's four earliest notebooks.

17 Westfall, *Never at Rest*, p. 74; the story is told by Wickins's son.

18 Ibid., p. 72. Too much can easily be made of all this. Was a sizar so very different from a school 'fag', the fate of every new boy? And even in my day few grammar-school boys, who came up with exhibitions or scholarships, mixed socially with those whose names appeared in the *Peerage* or the *Landed Gentry*.

19 Loans are recorded in both the Trinity College notebook (for 1661–5) – fifteen men – and the Fitzwilliam Museum notebook (1665–9) – six men. There is no indication whether Newton charged interest or not. The undergraduate notebook records no card-playing or tavern-drinking.

20 Harrison, *Library of Isaac Newton*, p. 5. Proof that Newton had access to this chained library is lacking.

21 Ibid., nos 1224, 1317, 199, 1264, 181.

22 Ibid., nos 1640, 335.

23 Ibid., nos 793, 1559, 629.

24 Ibid., nos 1472, 1442.

25 Ibid., no. 1521. Another early purchase of a historical book noted by Westfall (*Never at Rest*, p. 88 n. 62) but not listed by Harrison is Edward Hall, *The Union of Lancaster and Yorke* (1548).

26 I do not know why Newton first went up to Cambridge in early June, matriculating on 8 July 1661, then presumably riding home for the long vacation. At this time scholarship elections took place triennially. The election of April 1664 was the only one at which Newton could have been chosen.

27 Westfall, *Never at Rest*, p. 102.

28 Stukeley, *Memoirs*, p. 53.

29 Edmond Turnor, *Collections for the History of the Town and Soke of Grantham, containing authentic Memoirs of Sir Isaac Newton*, 1806, p. 159. The story was not used by Fontenelle.

30 Mordechai Feingold, *The Mathematicians' Apprenticeship, 1560–1640*, 1984, pp. 26–7.

31 A. R. Hall, 'Sir Isaac Newton's Notebook, 1661–65', *Cambridge Historical Journal*, 9, 1949, pp. 239–50; Westfall, *Never at Rest*, p. 84. Newton's annotations of books rarely if ever persist to the end of the book.

32 C. E. Raven, *John Ray Naturalist*, 1950, p. 28. If Raven's date is slightly out it matters little. He paints a detailed picture of Cambridge just before Newton's arrival. Ray left Trinity College in 1662 and probably Newton met him, if at all, only in later life.

33 Ibid. p. 24 n. 3, quoting from Ray's *Wisdom of God* (1691) pp. 125–6. I do not mean to decry the excellence of the education provided in some schools.

34 Mark H. Curtis, *Oxford and Cambridge in Transition, 1558–1642*, 1959, esp. ch. IX. There had been more mathematicians at Cambridge earlier in the century; see Feingold, *Mathematicians' Apprenticeship*, pp. 60–110. As a professor Newton could not take undergraduate pupils. The fact that William Whiston was conscientious as Newton's deputy (1696–1701) does not excuse the system.

35 J. C. Scriba, 'The Autobiography of John Wallis', *Notes and Records of the Royal Society*, 25, 1970, p. 27.

36 The Gresham College professors from Cambridge were Samuel Foster, Laurence Rooke, Walter Pope, Henry Briggs and Isaac Barrow.

37 Charles Webster, *The Great Instauration*, 1975, p. 134.

38 Marjorie Nicolson, 'The early stages of Cartesianism in England', *Studies in Philology*, 26, 1929, pp. 356–74.

39 Feingold, *Mathematicians' Apprenticeship*, p. 97.

40 Hall, 'Newton's Notebook', p. 240 n. 6; *Mathematical Papers*, vol. I, pp. 7–8. A senior sophister was a fourth-year undergraduate.

41 Cambridge University Library MS Add. 3968, no. 41, Hall, 'Newton's Notebook', p. 241; I. Bernard Cohen, *Introduction to Newton's Principia*, 1971, p. 291. This document is part of a rejected draft letter, about 1720.

42 *Mathematical Papers*, vol. I, pp. 25–121.

43 Ibid., p. 22.

44 Ibid., pp. 5–6. The story is inconsistent with Newton's having advanced far in geometry while at school.

45 Ibid., pp. 29, 63 and note. Full bibliographical details are given on pp. 20 and 22, nn. (5), (6) and (13). In the first place, Newton studied the second (1659–61) edition of Van Schooten's *Geometria*, but the copy he bought in December 1664 may have been that of the first edition in one volume still in his library. The copy of the *Exercitationum . . .* bought at the same time is there too.

46 Although one cannot be certain, it seems likely that Newton never read anything directly of the work of many eminent mathematicians: Apollonios (conics), surprisingly, and nearer his own time Briggs (logarithms), Fermat (infinitesimals) or Kepler (optics, astronomy).

47 In later life Newton more than once admitted a debt to Barrow in "considering figures by motion" which Whiteside (after Hofmann) plausibly traces to the third of Barrow's 1664 mathematical lectures (*Mathema-*

tical Papers, vol. I, 1964, p. 344 n. (4) also *ante* p. 11 note and p. 150). Whiteside takes the date of Barrow's Prefatory Lecture, '14 March 1664' to be New Style rather than the Old Style year which we would render as 1665.

48 Whiteside, ibid., p. 12 n. (28).
49 Raven, *Ray*, p. 27.
50 C. Webster, 'Henry More and Descartes: some new sources', *British Journal for the History of Science*, 4, 1969, pp. 359–77, on pp. 361–2. The honour of introducing Descartes to Newton has been assigned to both Barrow and More.
51 Wallis's autobiography records the special interest taken in him by Benjamin Whichcote, Master of his college, Emmanuel.
52 Westfall, *Never at Rest*, pp. 102–3. Further, in 1671 Barrow brought Newton's telescope to the Royal Society. But I am conscious of reviving an hypothesis currently unfashionable.
53 J. E. McGuire and M. Tamny, *Certain Philosophical Questions: Newton's Trinity Notebook*, 1983, pp. 127–8.
54 Westfall, *Never at Rest*, p. 90.
55 John Harrison, *Library of Isaac Newton*, no. 509, mistitled.
56 Cambridge University Library MS Add. 3996, fol. 88–135; see Hall, 'Newton's Notebook' and R. S. Westfall, 'The foundations of Newton's philosophy of Nature', *British Journal for the History of Science*, 1, 1962, pp. 171–82.
57 Hall, 'Newton's Notebook', p. 245.
58 Probably the tag came from p. 3 of Walter Charleton's *Physiologia* (see n. 60 below); its history has been traced by Henry Guerlac, 'Amicus Plato and other friends', *Journal of the History of Ideas*, 39, 1978, pp. 627–33.
59 Hall, 'Newton's Notebook', pp. 244, 245, 247.
60 In more detail, the books evident are (besides Descartes): Robert Boyle, *New Experiments Physico-Mechanicall touching the Spring and Weight of the Air* (1660); *idem, Experiments and Considerations touching Colours* (1664); *idem, New Experiments and Observations touching Cold* (1665); Joseph Glanvill, *The Vanity of Dogmatizing* (1661); Kenelm Digby, *Two Treatises . . . The Nature of Bodies . . . The Nature of Mans Soule* (1644); Galileo Galilei, *Dialogue on the Two Chief Systems of the World* (in one of the Latin editions or in the English version of Thomas Salusbury, *Mathematical Collections*, I, 1661); Henry More, *The Immortality of the Soul* (1659); Thomas Hobbes, *De corpore* (1655); Walter Charleton, *Physiologia Epicuro-Gassendo-Charletoniana* (1654).
61 Aristotle had never dominated philosophy in Antiquity, as is proved by the Roman Lucretius's composition of an Epicurean poem in the time of Julius Caesar.
62 This point and other philosophical matters considered by Newton are elaborately analysed by McGuire and Tamny in their edition of the *Quaestiones* (*Certain Philosophical Questions*)
63 Hall, 'Newton's Notebook', p. 243; McGuire and Tamny, *Certain Philosophical Questions*, p. 340.

64 Westfall, 'Foundations of Newton's Philosophy', p. 173 note.
65 McGuire and Tamny, *Certain Philosophical Questions*, p. 123.
66 Westfall, 'Foundations of Newton's Philosophy' p. 182.
67 Ibid., p. 173.
68 M. Feingold, *Before Newton*, 1990, p. 36; Raven, *Ray*, pp. 46–7 makes Nidd the leader of the group, comparative anatomy its chief concern. Others regard Nidd as primarily a chemist or alchemist, perhaps with less reason.
69 Raven, *Ray*, p. 148; Oldenburg, *Correspondence*, vol. VII, 1971, p. 149, vol. IX, 1973, p. 661, vols X–XIII see indices.

Chapter 2

1 Newton, *Correspondence*, vol. I, 1959, p. 2
2 Richard S. Westfall, *Never at Rest*, 1980, p. 142
3 Newton to Oldenburg, 6 February 1672, *Correspondence*, vol. I, 1959, pp. 95–6, first printed in *Philosophical Transactions*, no. 80, 19 February 1672, p. 52.
4 *Mathematical Papers*, vol. I, 1967, pp. 91, 96–9.
5 Ibid., p. 103.
6 Ibid.
7 Ibid., p. 99
8 Ibid., p. 110.
9 *Correspondence*, vol. II, 1960, p. 21.
10 Ibid., p. 111. Newton's Latin is highly compressed. I have expanded the English to make the sense clearer.
11 Whiteside in *Mathematical Papers*, vol. IV, 1971, p. 673, n. 57. While late in life Newton would copy out whole treatises, in youth his notebook entries were laconic. A paragraph on optics might represent weeks of work (in sunny hours only). Similarly, a few lines of mathematics might represent many hours of work.
12 *Mathematical Papers*, vol. I, p. 145.
13 Ibid., p. 213.
14 Ibid., p. 248.
15 Ibid., p. 272.
16 Ibid., p. 280
17 Ibid., p. 344. We can at once rewrite the example in other notation as $3ax^2dx + a^2ydx - y^3dx + a^2xdy - 3xy^2dy + 4y^3dy = 0$.
18 Ibid., p. 385 and notes. The treatment in the October 1666 tract (p. 414) is virtually identical.
19 Ibid., vol. I, pp. 400–48. Obviously, we do not know that Newton was *not* studying mathematics during those periods of the years 1664–6 when he was not (seemingly) developing calculus. In fact, during these years he was also concerned with other, different branches of mathematics, as shown by the miscellaneous, undated notes on trigonometry, algebraic

equations, the division of the musical octave and "Nauticall Questions" printed at the end of Whiteside's first volume.

20 Newton, *Correspondence*, vol. II, pp. 214–15, 220–1 (my translation). Leibniz's letter was of course intended for transmission to Newton.
21 Whiteside in *Mathematical Papers*, vol. I, pp. 326–7 and n. 25; vol. II, p. 206.
22 Ibid., vol. I, p. 403.
23 Ibid., vol. I, p. 410 n. 30; p. 414 n. 44.
24 Though Collins laid open to Leibniz's inspection during his second visit to London in 1676 his large collections of British mathematics, he disclosed neither Newton's earliest notes nor the 1666 tract.
25 *Correspondence*, vol. II, p. 163 (to Oldenburg, 26 October 1676). And Newton wrote to Collins a few days later "I could wish I could retract what has been done [i.e. printed], but by that, I have learnt what's to my convenience which is to let what I write ly by till I am out of the way" (ibid., p. 179). By this time Newton had been soured by the protracted, irritating disputes following the first publication of his optical researches in 1672. There was a period, as we shall see, when Newton steadily contemplated publishing his mathematics.
26 Quoted by Westfall, *Never at Rest*, p. 157.
27 A. R. Hall, 'Sir Isaac Newton's Notebook, 1661–65', *Cambridge Historical Journal*, 9, 1949, p. 246; John Hendry, 'Newton's Theory of Colour', *Centaurus*, 23, 1980, pp. 230–51.
28 A. R. Hall, 'Further Optical Experiments of Isaac Newton', *Annals of Science*, 11, 1955, pp. 27–43.
29 Hall, 'Newton's Notebook', p. 246. Perhaps it may be needless to add that many other notes in this manuscript testify to his reading of Descartes.
30 Ibid., p. 247; Hendry, 'Newton's Theory of Colour', pp. 235–6. Note that the prismatic spectroscope opened vast worlds of knowledge little more than a hundred years after Newton's death.
31 Blank in the manuscript.
32 Hall, 'Newton's Notebook', p. 247.
33 Ibid., pp. 247–8.
34 *Opticks*, 1952, pp. 20–5.
35 Hall, 'Newton's Notebook', p. 248; the calculation is explained in Hendry, 'Newton's Theory of Colour', pp. 240–2.
36 Hall, 'Further Optical Experiments', pp. 28–9.
37 *Correspondence*, vol. I, pp. 92, 95. Whiteside (*Mathematical Papers*, vol. I, p. 559) dates the 'glassworks' to 'Winter 1665–66'.
38 *Unpublished Scientific Papers*, pp. 400–13.
39 Hooke – not without reason – called Boyle's Law ($pv = k$) 'Mr Towneley's hypothesis' (*Micrographia*, 1665, p. 225) and so does Newton in his note.
40 Johannes Castillioneus, *Is. Newtonus Opuscula*, 1744, vol. II, p. 55.
41 *Unpublished Scientific Papers*, p. 400.
42 Ibid., pp. 402–3.
43 Ibid., p. 404; the square brackets here are Newton's.
44 Ibid., p. 403.

45 Ibid.
46 *Mathematical Papers*, vol. I, p. 551.
47 *Ibid.*, pp. 575–6. Geometrical optics generally is dealt with on pp. 549–85.
48 W. Stukeley, *Memoirs*, 1936, pp. 19–20.
49 Papers by Christiaan Huygens, Christopher Wren and John Wallis settling the matter were published in the *Philosophical Transactions* in September 1669.
50 J. W. Herivel, 'Sur les premières recherches de Newton en dynamique', *Revue d'Histoire des Sciences*, 15, 1962, pp. 105–40, p. 106.
51 J. W. Herivel, *The Background to Newton's Principia*, 1965 deals with all the material before 1687; cf. also Westfall, *Never at Rest*, pp. 144–55. The date 20 January 1665 may be taken (in the calendrical confusion of the time) as Old Style (1666) or New Style (1665), but the difference is hardly crucial.
52 Herivel, *Background*, p. 153; Westfall, *Never at Rest*, p. 145.
53 Herivel, *Background*, p. 141; Westfall, *Never at Rest*, p. 146.
54 This was first elucidated by Herivel in *Isis*, 51, 1960, pp. 546–53.
55 Ibid., p. 549.
56 Again, first elucidated by Herivel, in *Isis*, 52, 1961, pp. 410–16.
57 After an unsatisfactory attempt to calculate g by timing a conical pendulum 81 ins. long, Newton timed the swings of vertical pendulums. He then – using a (correct) rule of his own – found the length of the 45° circular pendulum revolving in the same period. Then it is obvious that (as he stated) gravity is equal to the centrifugal force. The latter he computed from the length and radius of the imagined circular pendulum. The $\pi\sqrt{l/g}$ rule was first published by Huygens in 1673; it was stated by Newton in *Principia*, Book I, Prop. 52.
58 A. R. Hall, 'Newton on the calculation of central forces', *Annals of Science*, 13, 1957, pp. 62–71; *Correspondence*, vol. I, pp. 297–303.
59 Hooke also borrowed the term *conatus*, but was no more successful than Descartes in rendering it a mathematical concept.
60 In both this document (CUL MS Add. 3958.5, fol. 87) and in the antecedent vellum sheet (3958.2) Newton stated the central force as that which, in the period of one revolution of a revolving body, would propel it through a distance $2\pi^2 r$, r being the radius of revolution. This is equivalent to the usual v^2/r.
61 Hall, 'Newton on central forces', p. 68; *Correspondence*, vol. I, p. 300.
62 Presumably by Newton's instruction, the sentences on mechanics in this letter were not in the end transmitted to Huygens: *Correspondence*, vol. I, p. 290, 23 June 1673.
63 Ibid., vol. II, pp. 436, 446, 20 June and 27 July 1686.
64 Ibid., vol. I, p. 301, n. 1.
65 *Mathematical Papers*, vol. VI, p. 5.
66 Needless to say that such a formulation produces an egg-shaped orbit, already rejected by Kepler, not the symmetry of the ellipse.
67 In validation of Newton's calculation, it may be worth confirming that employing more accurate values for the Moon's mean distance (60.39

Earth radii), the Earth's mean diameter (3,959 miles) and terrestrial gravity (32.2 ft/sec^2) the 'corrected' Newtonian ratio of $g/3600$ is repeated.

68 In the planetary context the precise $1/r^2$ relationship appeared immediately to Newton from the combination of Kepler's Third Law with his own law of rotational acceleration, the acceleration of a planet from the Sun being as v^2/r where, by the Third Law, $v^2 = k/r$, k being a constant for the solar system. Hence the acceleration is proportional to $1/r^2$.

69 The major exception is Newton's dynamical proof – very different from Huygens's kinematical reasoning – that motion in a vertical cycloid under constant gravity is isochronous.

Chapter 3

1 See D. T. Whiteside's general introduction to *Mathematical Papers*, vol. II, 1968, and Richard S. Westfall, *Never at Rest*, 1980, pp. 164–208.

2 W. Stukeley, *Memoirs*, 1936, pp. 58–9.

3 *Correspondence*, ed. J. Edleston, 1850, pp. xlvi–xlvii, quoting a source of 1780. Perhaps the tale is too tall to be believed?

4 The first piece printed in Newton's correspondence is not (I believe) a genuine letter. The second is the note from his mother already mentioned. The third letter exists only in Collins's copy but no one has ever doubted its authenticity. *Correspondence*, vol. I, pp. 3–5. The fourth is that to Aston.

5 *Correspondence*, ed. J. Edleston, p. lxxxv gives most of the known dates of Newton's departure from and return to Cambridge while a Fellow, but I here follow Westfall, p. 182, from the Fitzwilliam Notebook.

6 *Correspondence*, vol. I, 1959, p. 3. However, Galileo discovered all these features with a three-foot tube! In Newton's instrument most of the magnifying is done by the powerful glass eye-lens.

7 Maurizio Mamiani, *Isaac Newton Filosofo della Natura*, 1976, pp. 83–5; *Il Prisma di Newton*, 1986, pp. 17ff.

8 Marci's work appeared in *Thaumantias: Liber de arcu coelestis*, Prague, 1648. His recognition of the inequality of the refractions of the various coloured rays was rendered of less effect by his geographical and intellectual isolation.

9 *Correspondence*, vol. I, pp. 92–102.

10 A. R. Hall, 'Further Optical Experiments of Isaac Newton, *Annals of Science*, 11, 1955, p. 35.

11 The square brackets here are Newton's.

12 Hall, 'Further Optical Experiments', p. 33.

13 On the same page are notes on a book of Boyle's first printed in 1673.

14 Richard S. Westfall, 'Isaac Newton's Coloured Circles twixt two Contiguous Glasses', *Archive for History of Exact Sciences*, 2, 1965, pp. 181–96. Collins was allowed to read and copy *De Analysi* and even MS notes including the 1666 fluxions tract.

15 No doubt at this stage Newton supposed the "optical density" (refractive index) of a transparent substance to be proportional to its physical density.

16 Newton seems here to hint at his method of reuniting the rays dispersed by passing through a prism at the focus of a lens.

17 Westfall, 'Isaac Newton's Coloured Circles', pp. 195–6, n. 14.

18 Untypically, the essay is the sole content of this notebook, which still contains many blank leaves. Newton did not commonly waste paper! See *Unpublished Scientific Papers of Isaac Newton*, 1962, pp. 89–156.

19 The book is J. Harrison, *Library of Isaac Newton*, 1978, no. 509, containing several philosophical and scientific works. For Newton's shift to Latin see *Mathematical Papers*, vol. II, p. 11, n. 4. Claude Clerselier first printed Descartes's letters between 1657 and 1667.

20 *Unpublished Scientific Papers*, p. 144, n. 18

21 *Opticks*, 1952, p. 403.

22 To be more exact, it is that the common centre of gravity of the solar system lies within the Sun; about this all the planets revolve. Descartes had been scrupulous to deny that he proposed a formal motion of the Earth in the terms of his own definition of motion.

23 *Unpublished Scientific Papers*, pp. 148–9, n. 18.

24 Ibid., p. 149. Not surprisingly, Newton always had difficulty with language in such discussions as these. The one word "powers" seems to embrace very different things, such as force and pressure. Similarly, in *Opticks*, we find "powers, virtues or forces" equated.

25 Ibid., pp. 149–50.

26 Ibid., pp. 151–2.

27 Westfall, *Never at Rest*, pp. 303–4.

28 J. E. McGuire, 'Neoplatonism and Active Principles: Newton and the *Corpus Hermeticum*' in Robert S. Westman and J. E. McGuire, *Hermeticism and the Scientific Revolution*, 1977. I do not here debate the possible influence of Walter Charleton's *Physiologia Epicuro-Gassendo-Charltoniana* (1654) on Newton, still less the results of Newton's putative reading of Gassendi's *Syntagma*.

29 See further A. Rupert Hall, *Henry More: Magic, Religion and Experiment*, 1990, in the same series as the present volume.

30 See Whiteside in *Mathematical Papers*, vol. I, p. 344, n. 4; also p. 11, n. 26.

31 *The Mathematical Works of Isaac Barrow*, ed. William Whewell, 1860, vol. I, pp. 84–5.

32 *Correspondence*, vol. I, pp. 13–14. I should record that D. T. Whiteside takes the opposite view (private communication).

33 The considerable mass of Collins's papers is now in private possession. A selection from it was edited by S. P. Rigaud in *Correspondence of Scientific Men of the XVIIth Century*, 1841. Should the whole of these papers ever be transcribed and published, it would be of great benefit to learning. Meanwhile, other documents have been printed by Turnbull in the *Correspondence* and by Whiteside in the *Mathematical Papers*.

34 *The Correspondence of Henry Oldenburg*, vol. VI, 1969, p. 227. Collins's
words (unlike Barrow's quoted earlier) suggest that Newton had taken
Mercator's method and generalized it, which of course was not at all the
case.

35 *Correspondence*, vol. I, pp. 53–4.

36 This may go back to a piece in the *Journal des Sçavans*, September 1669, sent
to Oldenburg (Newton, *Unpublished Scientific Papers*, p. 449, n. 2). From
the related material printed in the *Philosophical Transactions* (no. 57, 25
March 1670) by Oldenburg Newton could have learned, if he had not
before, of the 'empty focus' equant treatment of planetary motion in an
ellipse.

37 *Correspondence*, vol. I, p. 16.

38 *Mathematical Papers*, vol. II, 1968, pp. 163–71, 206–47.

39 Ibid., p. 163.

40 See a draft letter of Collins to Wallis, ? 1677–8, in *Correspondence*, vol. II,
1960, pp. 241–4. Some material in the Kinckhuysen additions came from
Newton's study of equations in the 1660s. In turn, Newton was to draw
upon them in writing his own Lucasian lectures on algebra, which were
printed as *Arithmetica universalis*.

41 In Leibnizian equivalent, where $y = ax^{m/n}$, then

$$\int y.dx = \frac{1}{m/n + 1} x.^{m/n + 1}$$

42 *Mathematical Papers*, vol. II, p. 213. Newton was to stress the use of this
symbol in making his case against Leibniz; see ibid., p. 273.

43 Ibid., p. 227. The peculiar dot notation for partial derivatives has already
been noted in ch. 2.

44 Ibid., p. 241. The editor's translation is slightly modified.

45 J. E. Hofmann, *Leibniz in Paris: his growth to mathematical maturity*, 1974,
pp. 278–9.

46 *Mathematical Papers*, vol. II, p. xiii. *Sotheby Catalogue*, 1936, p. 53. The book
is extant in Philadelphia; J. Harrison, *The Library of Isaac Newton*, 1978, no.
93.

47 Humphrey was brought to Trinity from Grantham School by Newton, but
he claimed no relationship beyond that of a common surname.

48 L. T. More, *Issac Newton*, 1934, pp. 247–8; *Sotheby Catalogue*, 1936, p. 59.

49 Lord Keynes, 'Newton the Man' in *Royal Society Newton Tercentenary
Celebrations*, 1947, pp. 27–34; plate between pp. 30 and 31. Another of
Loggan's prints in *Cantabrigia Illustrata* (1690), of Great St. Mary's church,
is dedicated to Newton who presumably paid for this privilege.

50 C. E. Raven, *John Ray Naturalist*, 1950, pp. 38–9. Though the Loggan print
of Trinity College shows 'Newton's' garden as bounded by the chapel,
there may have been a short section of wall beyond it.

51 More, *Isaac Newton*, p. 249; *Sotheby Catalogue*, p. 59.

52 Raven, *John Ray*, pp. 38–9, n. 50. Cf. above, ch. 1, p. 28.

52 Raven, *John Ray*, pp. 38–9, n. 50. cf. above, ch. 1, p. 00.
53 B. J. T. Dobbs, *The Foundations of Newton's Alchemy*, 1975, pp. 97, 98.
54 He may have been Herbert Thorndike (till 1667) and after him Thomas Gale.
55 Westfall, *Never at Rest*, pp. 253–4. Wickins was soon to spend much of his time away from Cambridge, though retaining his fellowship until 1683. Could the 'cellar' here mentioned be the 'shed', as I have described it, leaning against the chapel wall, which has gone down to history as the 'laboratory'?
56 The date 1668 for the move, proposed by C. D. Broad (Dobbs, *Newton's Alchemy*, pp. 98–9, n. 53) seems highly unlikely.

Chapter 4

1 The first version of the lectures was reproduced in facsimile by D. T. Whiteside (Cambridge University Library, 1973); both versions are printed in *Optical Papers*, vol. I. *Arithmetica Universalis* is in *Mathematical Papers*, vol. VI. *De motu corporum* is reproduced in facsimile by D. T. Whiteside in *Preliminary Manuscripts*. It is printed in *Unpublished Papers* (pp. 239–92).
2 *Mathematical Papers*, vol. III, 1969, p. xviii, the statutes themselves, pp. xx–xxvii.
3 H. Newton to John Conduitt, 17 January 1728 in D. Brewster, *Memoirs*, 1855, vol. II, p. 92; *Mathematical Papers*, vol. III, p. xix, n. 39.
4 Flamsteed preserved an algebra 'hand-out' given by Newton at this lecture, which may suggest that an audience was expected by the professor. Flamsteed stayed in Cambridge six weeks in order to receive a degree by royal mandate. He had previously met Newton about the end of 1670.
5 A. Rupert Hall, 'Further Newton Correspondence', *Notes and Records of the Royal Society*, 37, 1982, pp. 11–12. Newton presented a copy of the 1687 *Principia* to De Volder; Zimmerman may be mentioned in the third edition.
6 R. Westfall, *Never at Rest*, 1980, p. 406, quoting Whiston's *Memoirs* (1749) p. 36.
7 Ibid., p. 210, n. 94.
8 *Unpublished Papers*, p. 370. This MS was first printed by W. W. Rouse Ball in the *Cambridge Review*, 29 October 1909, pp. 29–30.
9 *Correspondence*, vol. I, 1959, p. 252.
10 M. Feingold, *Before Newton*, 1990, p. 336; *Mathematical Papers*, vol. III, p. xiv. Feingold argues that Newton bought many books from the sale of Barrow's library in 1677; he was concerned in making the sale catalogue.
11 *Mathematical Papers*, vol. III, p. xiii.
12 A. E. Shapiro, 'The Optical Lectures' in Feingold, *Before Newton*, p. 113.
13 *Mathematical Papers*, vol.III, pp. xiv–xv.

14 Ibid., p. xxi and note. Newton was to argue that his personal academic stipend should be free from tax!

15 *Correspondence*, ed. J. Edleston, 1850, pp. xliv–xlv, n. 16.

16 There is some confusion about Newton's election (or transfer) to a major fellowship. Westfall (*Never at Rest*, p. 180) dates it 7 July when Newton was admitted MA. Edleston, usually reliable, makes it 16 March 1668. Either way the fellowship would have run out in 1675, requiring Newton to take orders before it could be extended. See *Correspondence*, vol. III, 1961, pp. 146–7 (evidently the editing was at fault here); vol. VI, 1976, p. 387. Also *Mathematical Papers*, vol. III, 1969, p. xxiii, n. (10), needlessly hostile to Barrow; Westfall, *Never at Rest*, pp. 330–4; *Correspondence*, ed. Edleston, pp. xlviii–l. I find it inconceivable that Newton would have sought the injunction in opposition to Barrow.

17 *Correspondence*, vol. I, 1959, p. 67. One must presume that Collins's copies from Newton's MSS had been made earlier.

18 Newton's trip into Bedfordshire began after 11 June 1672 and ended before the 17th (ibid., pp. 193, 194); for the rest, see ibid., pp. 210, 215.

19 Newton's introduction to his lectures refers to Barrow's optical course as "not so long ago"; the course may therefore have been concluded in the previous Lent term of 1669, or in the Michaelmas term opening the following academic year, still in 1669. Cf. Feingold, *Before Newton*, p. 70.

20 *Isaac Barrow's Optical Lectures 1667* [*sic*], trans. and ed. H. C. Fay et al., 1987, p. 7.

21 Shapiro, 'The Optical Lectures', pp. 112–13, n. 12; Feingold, *Before Newton*, pp. 68–70.

22 Lecture XIV; Barrow, *Optical Lectures*, (n. 20), p. 180. Newton himself never republished this construction in its most complete form – see Shapiro, 'The Optical Lectures', p. 151.

23 Lecture XIII; Barrow, *Optical Lectures*, p. 166; Shapiro, 'The Optical Lectures', pp. 147–8. Newton did use this construction in his own optical lectures and prove it (*Optical Papers*, vol. I, 1984, pp. 410–15; *Mathematical Papers*, vol. III, 1969, pp. 488–94); Part III of this volume gave the first modern study of Newton's optical lectures from a mathematical point of view.

24 Lecture XII, p. 14; Barrow, *Optical Lectures*, p. 150.

25 My translation; compare *Mathematical Papers*, vol. III, pp. 438–9, 444 and *Optical Papers*, pp. 46–7.

26 Newton prevented the printing of the lectures in his lifetime but several differing copies were made from the Latin text (the later version) in Cambridge. One was printed in 1729 and there were several later issues.

27 *Optical Papers*, p. 281.

28 *Correspondence*, vol. I, 1959, p. 92.

29 *Optical Papers*, p. 283.

30 Ibid., pp. 283–5. Note that at first Newton had ignored the shade purple, calling the most refracted colour blue.

31 A. R. Hall, 'Further optical experiments of Isaac Newton', *Annals of Science*, 11, 1955, pp. 27–36.

32 *Optical Papers*, pp. 17–19. Some topics promised for later discussion do not figure in the shorter set of lectures. Very little of any rough draft of them remains; ibid., p. 597, n. 16.

33 Barrow, *Optical Lectures*, pp. 152–4, n. 20.

34 It might be held that there is a modification of white light when some of its rays are absorbed by passing a beam through coloured glass, or when it is reflected from a 'coloured' surface. But both Hooke and Newton meant by 'modification' not simple addition or subtraction but an alteration in the physical structure of the light.

35 *Opticks*, 1952, pp. 34–45.

36 A rather simpler treatment is given in the concluding portion of the first set of lectures.

37 Chromatic aberration – a fuzziness in the image associated with coloured fringes – is caused by the inability of ordinary lenses to focus all the coloured constituents of white light to the same precise point; it arises from the physical nature of visible light. Spherical aberration is a fuzziness in the image caused by the inability of a spherical lens or mirror to focus all incident monochromatic light to the same precise point. It is caused by the geometry of the lens surface, as Descartes discovered.

38 However, Newton also anticipated that all samples of water, crystal, olive oil, glass etc. would possess the same typical refractive index of each medium. He did not realize that clear glasses could be made of very different chemical compositions, so possessing diverse refractive indices.

39 *Optical Papers*, vol. I, p. 429. The idea is developed in a draft of 1672, *Correspondence*, vol. I, pp. 191–2 and *Mathematical Papers*, vol. III, pp. 512–13.

40 *Papers and Letters*, 1978, pp. 116–17 from *Philosophical Transactions*, 1672, pp. 5084–5; *Correspondence*, vol. I, p. 172; *Mathematical Papers*, vol. III, pp. 442–3. In the contextual discussion of lenses, I myself think that Newton is responding to Hooke's discussion of his own problem, and is therefore writing of spherical, *not* chromatic aberration.

41 *Opticks*, 1952, p. 102.

42 *Optical Papers*, vol. I, p. 283; "dioptrics" because an eye-lens is used and indeed (being of a very short focal length in Newton's telescope) effects a large part of the magnification. This tiny lens, though stopped down, necessarily introduced chromatic effects.

43 The rainbow is promised in the shorter version of the lectures but is not present.

44 In the lectures Newton made Descartes the discoverer of the cause of the rainbow (i.e. the refraction of light by falling raindrops). In *Opticks*, better informed but less generous, he gave the credit to "some of the Antients" and Antonio de' Dominis (*Opticks*, 1952, p. 169). Descartes's process involved tabulation of the angles of refraction produced by various angles

of incidence upon the surface of the drop. Earlier, about 1601, Thomas Harriot had discovered the determining condition for the primary bow analytically, in which feat he was followed by Huygens (1652, in complete ignorance of Harriot's work before him) and finally by Newton (in ignorance of both). Newton was the first to publish. (*Mathematical Papers*, vol. III, p. 507 n. 47.)

45 Newton adds the Sun's apparent diameter of 31' to make the breadth of the bow 2° 37'.

46 *Optical Papers*, pp. 73, n. (23) and 116, n. (4). Newton made a careful study of this book (J. Harrison, *The Library of Isaac Newton*, 1978, no. 713).

47 That Newton held this idea in 1672 is evident from his letter to Pardies (10 June 1672; *Optical Papers*, p. 116, n. 4; *Correspondence*, vol. I, p. 164, originally published in *Philosophical Transactions*, no. 85, 15 July 1672). For the following experiment see *Optical Papers*, vol. I, pp. 475–9.

48 *Correspondence*, vol. I, pp. 164, 175, 376.

49 The new introductory material includes the Definitions, Axioms and the (historically foremost) experiments on reflected colours. References in the later editions of *Opticks* to the *Lectures* were added after Newton's death and the printing of the lectures.

50 *Mathematical Papers*, vol. II, p. 278.

51 We may presume that Collins read neither Dutch nor German, nor perhaps even Latin with any fluency. Mercator (*Germanicè*, Kaufmann) a Dane from Schleswig-Holstein, spoke German, English, French and Latin. *Correspondence*, vol. I, p. 20.

52 Ibid., pp. 24, 30–1. Newton's "Observations" in *Mathematical Papers*, vol. II, occupy pages 364–445 (including the translation into English).

53 *Correspondence*, vol. I, pp. 32–8. Whiteside deals with all this matter in *Mathematical Papers*, vol. II, pp. 279–91.

54 *Correspondence*, vol. I, p. 36.

55 Ibid., pp. 43–4, 27 September 1670.

56 It was published in parts, 1673–4. Newton loyally promoted it in Cambridge. His own copy is Harrison no. 883.

57 *Correspondence*, vol. I, pp. 66, 68. This letter was written soon after a visit to Lincolnshire.

58 *Mathematical Papers*, vol. II, p. 288; vol. III, pp. 32–3.

59 Ibid., vol. II, p. 277 referring to the original article by J. C. Scriba in *British Journal for the History of Science*, 2, 1964, pp. 45–58. See also A. R.Hall, 'Newton's first book', *Archives Internationales d'Histoire des Sciences*, 13, 1960, pp. 39–61. I presume Pitts had paid Mercator for his labour.

60 Hall, 'Newton's first book', pp. 55–61; *Correspondence*, vol. I, p. 122.

61 H. W. Turnbull, *James Gregory*, 1939, pp. 218, 225; *Correspondence*, vol. I, pp. 119, 146–7. So far as is recorded, Newton did not lecture upon his method of series. *Optical Papers*, pp. 18–19. Shapiro's argument that the longer text was essentially complete by the end of February 1672 is not convincing.

62 *Correspondence*, vol. I, p. 161, 25 May 1672.

63 Reprinted material filles 115 tight pages in *Papers and Letters*; admittedly this is not all by Newton, but not all that he wrote was printed.

64 *Mathematical Papers*, vol. III, pp. 8–11.

65 That is, in Leibnizian terms differentiation and integration. Newton had written of 'flowing quantities' from the beginning.

66 Hall, *Philosophers at War*, 1980, p. 269. To be exact, Whiteside gives an equivalent passage from a draft of this printed paper: 'An Account of the Book entituled *Commercium Epistolicum . . .*', *Philosophical Transactions*, 16, 1715, pp. 173–224.

67 Ibid., p. 205: "the first ratios of nascent quantities". Note again Newton's desire to divorce himself from the mathematics of indivisibles, begun by Cavalieri and carried further by Wallis; in Newton's eyes Leibniz's differential was just another indivisible.

68 *Mathematical Papers*, vol. III, pp. 17, 72–3 and notes. The translation is Whiteside's, save that I have reverted to Newton's letters for the fluxions where, for convenience of reading the English, Whiteside had put ẋ, ẏ, ż etc., the notation introduced by Newton in 1691. After the passage I quote Newton again argues that fluxions are in the sound tradition of ancient geometry.

69 Ibid., p. 19. I have transposed two sentences.

70 Gregory's 'Tentamina' was obscurely published as appendices to Patrick Mather, *The great and new art of weighing vanity* (1672), which Newton would have been unlikely to come across unless it was sent to him. It was not in Newton's library. I think Newton's silence about the 'Tentamina' in letters to Collins is a weighty argument against his having received it. *Correspondence*, vol. I, p. 224 and n. (5) – rather confused; *Mathematical Papers*, vol. III, pp. 390–1, 421 n. (2). Whiteside is inevitably ambiguous about whether Newton's paper was written in 1672 or 1673.

71 Mechanically, Huygens's clockmakers relied upon an adaptation of the swiftly swinging balance of a table-clock; the pendulum-bob hung upon a thread 9–10 ins long, the arc of swing being some 60°. Hence the difference between a circular arc and a cycloidal arc was just considerable. After the English invention of the anchor escapement (?c.1671), to which a second's pendulum about one metre long was coupled with a swing of only a few degrees, the difference between a cycloidal and a circular arc was negligible. Huygens's cycloidal cheeks vanished.

72 *Correspondence*, vol. I, p. 284. Newton (17 September 1673) received another work on mechanics – "the little but ingenious tract of P. Pardies" – *La statique ou la science des forces mouvantes* by Ignace Gaston Pardies (1673). This also contains a demonstration of the cycloidal oscillation of a pendulum.

73 This point was suggested to me by Professor Whiteside.

74 *Correspondence*, vol. I, pp. 290–5.

75 D. T. Whiteside in *Mathematical Papers*, vol. III, p. 425, n. (15) infers that Newton's familiarity with this error came from James Gregory's 'Tentamina'.

76 Stillman Drake (trans.) *Galileo Galilei, Two New Sciences*, 2nd edn., 1989, p. 178.
77 *Mathematical Papers*, vol. III, pp. 420–3.
78 Huygens, *Oeuvres*, vol. XVIII, p. 489 quoted ibid., p. 423, n. 14.
79 *Mathematical Papers*, vol. III, pp. 420–3.
80 Newton's well-read copy is Harrison no. 1072.

Chapter 5

1 Oldenburg, *Correspondence*, vol. IX, 1973, p. 384 and note, pp. 488–98.
2 Ibid., vol. VIII, 1971, p. 292.
3 R. S. Westfall, *Never at Rest*, 1980, p. 233. *Tool*, in a special sense, here means the flat or curved metal surfaces upon which lenses and specula were ground to shape and polished. Newton inserted into *Opticks* a full account of his method of preparing the primary speculum of a telescope (1952, pp. 102–5).
4 Oldenburg, *Correspondence*, vol. I, 1965, pp. 72ff., 88. Hooke's reaction was to proposed his own 'highly considerable improvement of all sorts of optic' glasses by which 'whatever almost in notion and imagination [ever was] desired in optics, may be performed with great facility and truth' (18 January 1672; Birch, *History*, vol. III, p. 4). Nothing came of this.
5 Newton, *Correspondence*, vol. I, 1959, pp. 82–3.
6 Ibid., p. 95.
7 Both these points are lightly touched on (ibid., p. 98). But Newton's statement that "The Original or primary colours, are *Red, Yellow, Green, Blew, and a Violet-purple*, together with Orange, Indico, and an indefinite variety of Intermediate gradations", though correct in one sense, is ambiguous. Did he mean that there are five primaries, or seven, or an infinite number?
8 Ibid., p. 100.
9 Ibid., pp. 98, 102.
10 Ibid., pp. 96–7.
11 Ibid., p. 100.
12 T. Birch, *History of the Royal Society*, vol. III, 1757, pp. 9, 10–15; *Correspondence*, vol. I, pp. 107–8, 110–14.
13 Birch, *History*, vol. III, p. 11; *Correspondence*, vol. I, p. 111.
14 Birch, *History*, vol. III, pp. 14–15.
15 Ibid., p. 15.
16 *Correspondence*, vol. I, pp. 89–91, 285–6; Oldenburg, *Correspondence*, vol. IX, p. 14; Huygens, *Oeuvres*, vol. VII, pp. 151, 157, 159 etc.
17 Oldenburg, *Correspondence*, vol. IX, pp. 117–19. Huygens had received the printed letter from Oldenburg.
18 *Correspondence*, vol. I, pp. 212, 235–6.
19 Ibid., pp. 262, 264–6.

20 Ibid., pp. 209–10.
21 Newton's library as we know it contained Bacon's *Opuscula varia posthumu* (1658) which Newton had used. The 1706 *Essays*, the only other book by Bacon, seems unused (Harrison, *Library of Isaac Newton*, 1978, nos 108, 109).
22 I. A. Sabra, *Theories of Light from Descartes to Newton*, 1967, p. 248. With all respect to an excellent monograph, part of Sabra's argument here seems to me strained. J. A. Lohne, 'Newton's 'proof' of the sine law and his mathematical principles of colours,' *Archive for History of Exact Sciences*, 1, 1961, pp. 389–405; 'Experimentum crucis', *Notes and Records of the Royal Society*, 23, 1968, pp. 169–99.
23 As already pointed out, Newton did not perceive that spectra may be of different lengths if formed by glass prisms of different chemical composition, that is, that dispersion in glass is variable.
24 *Opticks*, 1952, pp. 113, 121.
25 See his 1678 letters to Lucas in *Correspondence*, vol. II, 1960, and especially his letter to Aubrey, pp. 266–8.
26 Newton to Oldenburg, 11 June 1672, ibid., vol. I, pp. 171–88; Birch, *History*, vol. III, p. 10. Newton's rejoinder was printed in *Philosophical Transactions*, 7, no. 88, pp. 5084–103.
27 *Correspondence*, vol. I, pp. 195–203.
28 Ibid., pp. 247–8.
29 Oldenburg, *Correspondence*, vol. IX, pp. 386–96, 427–30.
30 *Correspondence*, vol. I, p. 307.
31 Ibid., p. 358.
32 Ibid., p. 150.
33 Towneley to Oldenburg, 24 April 1673, ibid., p. 277.
34 The whole document is printed in Birch, *History*, vol. III, pp. 248–60, 262–9, 272–8, 280–95, 296–305. The "Hypothesis" only is printed in *Correspondence*, vol. I, pp. 362–85, followed by a summary of the 'Observations'.
35 Birch, *History*, vol. III, p. 269. Newton had vainly looked for such a refraction "in an Air Pump here at Christs College". Newton's relations with experimenters at Christ's, possibly still including the aged philosopher Henry More (1614–87) who had an interest in pneumatics, is one of the many small mysteries in his life; *Correspondence*, vol. I, p. 361. It is hardly necessary to insist that Newton's hypothesis was quite different from Hooke's.
36 It cannot be *later* than 1672, but I know no reason to fix it to 1670 with Westfall, *Never at Rest*, p. 216.
37 Parts I, II and III; *Opticks*, 1952, pp. 193–269.
38 Birch, *History*, vol. III, pp. 272–3; Hall, 'Further optical experiments', 1955, p. 31; note the likeness of the printed figures to the early MS drawing, p. 32. *Correspondence*, vol. I, pp. 404–6 (21 December 1675).
39 The construction in *Opticks* is modified from that of 1675 but the result is the same.

40 Birch, *History*, vol. III, p. 290. Newton altered the details of all this in *Opticks*, 1952, pp. 231–4. Even 'experimental' numbers he treated with a good deal of flexibility.

41 Birch, *History*, vol. III, p. 295.

42 Westfall, *Never at Rest*, pp. 217–18 in a more detailed examination of Newton's investigation than I have given.

43 Birch, *History*, vol. III, p. 294.

44 Ibid., p. 301.

45 Ibid., pp., 289, 301.

46 Ibid., pp. 181, 193, 194; see Hall, 'Beyond the fringe', 1990, p. 14.

47 Ibid., pp. 248–9. The analogy, though commonplace, is not really sound. Note that Newton here quotes from his own letter of 11 June 1672, his answer to Hooke's 'Considerations' (*Correspondence*, vol. I, p. 174).

48 I follow the text of the "Hypothesis" printed in Birch, *History*, vol. III, pp. 248–69, corrected according to Newton's later instructions.

49 "De Aere et Aethere" in *Unpublished Papers*, 1978, pp. 214–28. Westfall dates this document to 1679 (*Never at Rest*, p. 374) for reasons unclear to me. Newton's experiments on electric attraction and repulsion appear later in the "Hypothesis".

50 Birch, *History*, vol. III, pp. 250–1.

51 Ibid., p. 250.

52 The letter to Boyle was first printed by Thomas Birch in his 1744 edition of Boyle's *Works*; cf. Westfall, *Never at Rest*, p. 271.

53 Birch, *History*, vol. III, p. 254.

54 Ibid., p. 256.

55 Ibid., pp. 268–9. Hooke thought himself the first to discover diffraction, cf. Hall, 'Beyond the fringe', p. 14. The first discoverer was of course Francesco Maria Grimaldi, *Physico-mathesis de Lumine*, 1665, Cf. *Principia*, Book I, Section XIV.

56 *Correspondence*, vol. I, pp. 405–6.

57 Ibid., pp. 412–13.

58 Ibid., p. 416. On the metaphor, see Robert K. Merton, *On the Shoulders of Giants: A Shandean Postscript*, 1965. Hooke would have known that in the legend *pygmies* see further from the giants' shoulders!

59 A. R. Hall, 'Beyond the fringe', *Notes and Records of the Royal Society*, 44, 1990, pp. 13–15.

60 Birch, *History*, vol. III, p. 261, 14 December 1675.

61 *Correspondence*, vol. II, p. 268, perhaps a draft rather than a letter sent; Westfall, *Never at Rest*, pp. 274–9. I cannot explain the lacuna.

62 Ibid.

63 *Correspondence*, vol. I, p. 414, 25 January 1676. On all this see Hall, 'Newton's first book', *Archives Internationales d'Histoire des Sciences*, 13, 1960, pp. 49ff.

64 I. Bernard Cohen, 'Versions of Isaac Newton's First Published Paper', *Archives Internationales d'Histoire des Sciences*, 11, 1958, pp. 357–75. The sur-

viving pages (numbered 9 to 16) were found at Royston, Cambs., in the binding of an old book and therefore probably they were printed nearby.

65 *Correspondence*, vol. II, p. 239, 18 December 1677.

66 Ibid., pp. 220–21. Collins's letter survives only in a copy and Newton's answer to it is lost.

67 Westfall, *Never at Rest*, p. 227 and n. 119. The author of the tale about Newton's distress, Abraham de la Pryme, was an undergraduate long after (1692).

Chapter 6

1 *Correspondence*, ed. J. Edleston, 1850, p. lix, n. 95.

2 Ibid., p. xxv; *Correspondence*, vol. VII, 1977, pp. 371–2.

3 *Correspondence*, ed. J. Edleston, p. lviii, n. 90, records the erection of scaffolding so that a 'Mr Newton' could measure the 'fretwork of the staircase' of the New Library; surely *not* the Professor?

4 *Correspondence*, vol. VII, pp. 388–9; R. S. Westfall, *Never at Rest*, 1980, p. 481 and note.

5 *Correspondence*, ed. Edleston, p. lix, n. 95.

6 A. R. Hall, *Henry More*, 1990, p. 204.

7 *Correspondence*, vol. II, 1960, p. 415.

8 Ibid., pp. 288–95. The printing of this letter by Thomas Birch in Boyle's *Works* (1744) did much to strengthen the eighteenth-century taste for aetherial speculation. Cf. I. B. Cohen, *The Newtonian Revolution*, 1980, pp. 114–20.

9 Ibid., vol. I, p. 369. Cf. Westfall, *Never at Rest*, p. 373. Surprisingly, Westfall does not connect "sociability" with alchemy.

10 Cf. B. J. T. Dobbs, *The Foundation of Newton's Alchemy*, 1975, pp. 207–9.

11 'A geometrical exercise on the measure of figures', published at Edinburgh in 1684. It is the first *published* account of expansions by power series. The copy sent to Newton is lost.

12 *Correspondence*, vol. II, p. 396.

13 *Mathematical Papers*, vol. IV, 1971, pp. 413–16.

14 Ibid., pp. 417–19, 529. The translation is Whiteside's.

15 All this, with learned commentary, is to be found in *Mathematical Papers*, vol. IV, pp. 526–653. I hope my friend D. T. Whiteside will forgive my reproducing here his admirably terse analytical sentences:

'Using this storehouse [the 1676 letters to Leibniz] as a base and also borrowing from the chapters of his 1671 fluxional tract he would draft a new treatise on analysis which would effectively supplement Gregory's introduction and so render its intended sequel superfluous: at the same time he could set the historical record straight by stressing that David's uncle James had learnt the method of extracting infinite series (as [Newton] thought) from that of Newton's for the general circle zone. Such was the initial plan for the 'Matheseos Universalis Specimina'. But as he began to gather his material Newton's purpose subtly changed: the impulse to outpace a mathematical junior became transmuted into the need to reply, years after they were made, to the deeper, more penetrating criticisms of his contemporary in Hanover. And then in turn this latter compulsion, too, was sublimated and the 'Specimina' changed into an abstract technical treatise, 'De computo serierum' in which the names of neither David Gregory nor Leibniz appear. With the passion quenched, however, so was the fire and the latter tract was soon abandoned in the middle of its third chapter. Ibid., pp. 417–18.

16 *Correspondence*, vol. II, p. 484.

17 *Mathematical Papers*, vol. VI, 1974, p. xvi note; vol. VII, 1976, pp. 5–6; *Correspondence*, vol. II, p. 451.

18 *Correspondence*, vol. II, pp. 95–6, 5 September 1676; pp. 179–80, 5 November 1676.

19 *Correspondence*, vol. I, 1959 p. 309 and A. R. Hall, *Ballistics in the Seventeenth Century*, 1952, pp. 120–5. James Gregory had already criticized Anderson in his 'Tentamina' for his facile assumption; either Newton did not know this, or he had forgotten.

20 *Mathematical Papers*, vol. IV, pp. xv–xvi, 659.

21 Oldenburg, *Correspondence*, vol. IX, 1973, pp. 438–47, 3 February 1673. We do not know when this copy was sent to Newton. See J. E. Hofmann, *Leibniz in Paris*, 1974, ch. 20.

22 Oldenburg, *Correspondence*, vol. XI, 1977, pp. 42–6.

23 Wallis evidently found his name strange, writing to Collins (11 September 1676): 'Last year I was visited by a certain German named Tschirnhaus, that is (in both sound and sense) in English spelling Churn-house, that is, the building where the churning of the cream of milk into butter takes place.' (My translation is from S. P. Rigaud, *Correspondence*, vol. II, 1841, p. 591.)

24 Oldenburg, *Correspondence*, vol. XI, p. 141 (Latin pp. 139–40).

25 Ibid., pp. 253–62 (Collins's English draft), pp. 265–73 (Oldenburg's Latin letter, 12 April 1675, now in Hanover).

26 Ibid., p. 369 (14 June 1675) and vol. XII, p. 101, n. 12; see also *Correspondence*, vol. I, pp. 229–32.

27 Oldenburg, *Correspondence*, vol. XII, 1986, pp. 268–70. Newton later regarded this letter as significant and included an extract from it in his *Commercium Epistolicum* (1712).

28 Ibid., p. 269 and n. 2; *Correspondence*, vol. II, p. 198; Hall, *Philosophers at War*, 1980, pp. 64, 81. Leibniz's annotations from *De Analysi* are printed in *Mathematical Papers*, vol. II, pp. 248–59, his notes from the 'Historiola' in Leibniz's *Samtliche Schriften*, Dritte Reihe, Band I (1976), pp. 485–503. See also Hofmann, *Leibniz in Paris*, ch. 20.

29 *Correspondence*, vol. II, p. 20. The Latin part of the letter (for transmission to Leibniz) is of almost ten printed pages.
30 *Mathematical Papers*, vol. IV, pp. 666–70 gives a brilliant summary of the *First Letter*.
31 Oldenburg, *Correspondence*, vol. XII, p. 338; vol. XIII, pp. 1–17, 40–9 (English translation).
32 Ibid., vol. XIII, p. 140.
33 Owing to a mistake made in the initial printing of the *First Letter* by John Wallis, Newton supposed that it had been sent to Paris on 6 July (rather than the 26th). He therefore argued that by 17 August Leibniz had had plenty of time to put together this answer from the materials he had himself received from Collins.
34 *Correspondence*, vol. II, pp. 93–4, 110.
35 *Mathematical Papers*, vol. IV, pp. 671–3.
36 *Correspondence*, vol. II, pp. 212, 226, n. (1).
37 Ibid., vol. II, p. 179
38 Further details in *Mathematical Papers*, vol. IV, p. 672. n (54). The note in *Correspondence*, vol. III, p. 220, n. (4) is incorrect.
39 *Correspondence*, vol. IV, pp. 140–1. Compare Newton's autobiographical note dated 4 July 1699 in ch. 1, p. 19.
40 Ibid., vol. II, p. 239; 18 December 1677. Possibly Newton purposed an edition of the highlights of all his optical papers, as already suggested.
41 Ibid., vol. II, pp. 239, 240. Flamsteed was not involved, as Newton imagined. Hooke falsely accused Oldenburg of not recording it properly, he 'of late omitting all things done by me'.
42 Ibid., p. 240; T. Birch, *History of the Royal Society*, vol. III, 1757, p. 213.
43 *Correspondence*, vol. II, pp. 205–7, 21 April 1672.
44 Ibid., p. 264, 18 May 1678.
45 Ibid., pp. 300–1. It is hardly needful to remark that Newton's revulsion from philosophy was largely illusory, for Hooke's benefit.
46 Ibid., p. 436; Boas and Hall, 'Newton's chemical experiments', *Archives Internationales d'Histoire des Sciences*, 11, 1958, p. 121.
47 F. E. Manuel, *Portrait of Isaac Newton*, 1980, pp. 25, 27–8; cf. also p. 262 and *Mathematical Papers*, vol. IV, p. xiv, n. (15).
48 *Correspondence*, vol. II, pp. 246, 251, 253. Newton wrote to Lucas about the fire in a vanished letter of 2 February.
49 Ibid., p. 266, 8 June 1678. He wished not to expose himself to the ague (malaria) of the Fens. He proposed, but did not carry out, the measurement of a degree of latitude. Newton left Cambridge on 6 May 1678, returning on the 27th, presumably from Lincolnshire.
50 Ibid., pp. 297–8.
51 Ibid., p. 436, 20 June 1678.
52 As Newton next pointed out, his reasoning was opposite to that of the old anti-Copernicans who had supposed that in this experiment, if the Earth rotated beneath the falling body, it would be left far *behind* (to the west).

53 *Correspondence*, vol. II, pp. 301–3. "Fellows", presumably, of the Royal Society, not Trinity College. Nothing came of this.

54 Ibid., pp. 205–7, 21 April 1679. I wonder if Newton was the first to introduce the concept of phase into acoustics?

55 *Correspondence*, ed. Edleston, 1850 pp. 262–3, reprinted in *Correspondence*, vol. II, pp. 287–8 but spurned as a fake by Westfall.

56 *Correspondence*, vol. II, pp. 269–85. Storer was (it seems) living in Babington's rectory at Boothby [Pagnell] since he wrote thence.

57 Ibid., pp. 368–71, 387–93.

58 Ibid., p. 315. The treatment here of these letters is misleading. As Newton stated (p. 340) Flamsteed began to write to Crompton about the comet; the latter sent copied extracts of his letters to Newton. In one letter now lost Flamsteed made some favourable allusion to Newton which permitted the latter to intervene (via Crompton) on 28 February 1681.

59 Flamsteed to Halley, 17 February 1681, ibid., pp. 336–9. Though Newton never saw this letter, his paper for Flamsteed (pp. 340–7) shows knowledge of some similar one sent to Crompton.

60 Ibid., pp. 340–2; Francis Baily, *Account of the Rev. John Flamsteed*, 1835, reprinted 1966, pp. 29, 36. *Mathematical Papers*, vol. V, 1972, p. 32, n. (1).

61 Later Newton found his scholar to have mistaken the date.

62 *Correspondence*, vol. II, letter no. 254, pp. 360–1. This may be a draft for the "larger answer" to Flamsteed's letter of 7 March 1681 (ibid., pp. 348–53) mentioned in n. (1), p. 367. There would have been time for Newton to prepare this draft before leaving for ten days' absence on 15 March. In the event Newton sent Flamsteed letter no. 255 (16 April 1681) which contained little on the mechanics of cometary motion. It seems unlikely that Flamsteed saw a letter corresponding to no. 254.

63 Ibid., pp. 364–5.

64 Westfall, *Never at Rest*, pp. 394–5; *Correspondence*, vol. II, p. 419. Calculations of a rectilinear orbit for the comet(s) of 1680–1 in *Mathematical Papers*, vol. V, pp. 524–31; the method used was presented in the 'Lectures on Algebra', ibid., pp. 298–303.

65 Westfall, ibid., pp. 397, 499, 501; *Correspondence*, vol. II, pp. 373, 375–6; vol. III, pp. 357–68; vol. IV, pp. 93–4, 106, 133.

66 Burnet's letter answered by Newton (*Correspondence*, vol. II, pp. 329–34) is dated 13 January 1681. Thomas Burnet was not related to the Scot, Gilbert Burnet, bishop and historian of his own times.

67 An unexplained letter from Newton in 1691 (ibid., vol. III, p. 184) intimates his reluctance to become Master of the Charterhouse in succession to Burnet should the latter be made a bishop (he was now a royal chaplain). He was not.

68 Ibid., vol. II. pp. 377–8; Birch, *History*, vol. IV, pp. 136, 137.

69 *Correspondence*, vol. II, pp. 381–5, 417–18. The *New Theory* was appended to the reprinted *Ophthalmographia*.

70 *Mathematical Papers*, vol. III, pp. xviii-xix; *Correspondence*, vol. II, p. 335 (?1680).

71 In *Mathematical Papers*, vol. V, the text makes about 231 large printed pages.
72 Ibid., p. 5. D. T. Whiteside's introduction exhausts the generalities of the topic.
73 Ibid., pp. 13–14, 528–621.
74 At least this is true of the copies of the Latin 1722 and English 1728 editions in my possession.
75 *Mathematical Papers*, vol. V, pp. 16–18, 23–31.
76 Ibid., pp. 4, 508; *Correspondence*, vol. II, pp. 86–7, 187. The editorial reasons for dating these letters do not seem convincing to me, but they must be of this period.
77 *Mathematical Papers*, vol. V, p. 8 and n. (27). James E. Force, *William Whiston, Honest Newtonian*, 1985, ch. 1. It seems that Newton simply neglected his professorial duties from 1696 to 1700 (if not before).
78 This material occupies pp. 526 to 653 in vol. IV of *Mathematical Papers*.
79 Ibid., p. 11; the texts, pp. 116–67. Wharton took orders, became chaplain to Archbishop Sancroft, and published books on the Church of England.
80 D. T. Whiteside has traced an Oxford graduate of this name living not far from Cambridge, born perhaps about 1628, who lived to take a second wife in 1681; ibid., pp. 12, 168–9, n. (1). Two notebooks bearing Hare's name were in Newton's possession, but the trigonometry text is lost.
81 Ibid., pp. 218–25, 274–353.
82 Ibid., pp. 354–405; *Enumeratio curvarum tertii ordinis* was published with *Opticks* in 1704.
83 See ibid. pp. 409–518; my quotation is from p. 423, the editor's translation slightly modified.
84 Ibid., p. 427.

Chapter 7

1 Shaw, Peter (trans.) *A New Method of Chemistry . . . Translated from . . . Dr. Boerhaave's Elementa Chemiae*, London, 1753, p. 173, note *r*. (This is the third revised edition of a book with a different title.) See also I. Bernard Cohen, *Franklin and Newton*, 1956, p. 223.
2 Marie Boas, *Robert Boyle and Seventeenth Century Chemistry*, Cambridge, 1958.
3 B. J. T. Dobbs, *The Foundations of Newton's Alchemy*, 1975, pp. 93, 121. The date of Oxford Bodleian Library MS Don b.15 was presumably derived from the handwriting. Another and much larger dictionary (Keynes MS 30) is dated in K. Figala, 'Newton as Alchemist', *History of Science*, 15, 1977, p. 109 to *post* 1690.
4 Dobbs, *Foundations*, p. 94.

5 Ibid. p. 97. The 'fact' that Nidd was a practising alchemist, a recent dogma, is based on his owning Glauber's *Philosophical Furnaces* (1648, 1651) and, perhaps, an iron retort. Cf. note 13 below.

6 Ibid., pp. 99, 102.

7 Ibid., p. 100.

8 Ibid., ch. 4 *passim*. To be fair, Dr Dobbs quotes one or two instances – there are many – of More's contempt for chemists and their 'slibber-sauce' experiments but without attending to their meaning.

9 A. Rupert Hall, *Henry More*, 1990. A similar view is taken in F. E. Manuel, *The Religion of Isaac Newton*, 1974, p. 72.

10 Dobbs, *Foundations*, p. 122. At this point I venture to differ from this authority only in her supposition that the dictionary necessarily implies considerable chemical experience on Newton's part.

11 *Correspondence*, vol. I, 1959, pp. 10–11.

12 Dobbs, *Foundations*, p. 122.

13 *Correspondence*, vol. I, pp. 345, 356 (extracts). The letters may be found in H. W. Turnbull, *James Gregory*, 1939, pp. 310–11 and S. P. Rigaud, *Correspondence*, 1841, vol. II, p. 280. In reality, Newton had written mathematical letters to Collins as recently as 24 July and 27 August 1675. It is obvious that Collins does not mean to link Barrow with Newton in the 'Chimicall Studies', though some historians have written as though he did so.

14 Boas and Hall, 'Newton's chemical experiments', *Archives Internationales d'Histoire des Sciences*, 11, 1958, pp. 119–21. Though MSS 3973 and 3975 have been used by other scholars, notably Dobbs and Westfall, they have not yet been published in their entirety.

15 I have not ascertained, and perhaps one could not, which editions of these books Newton read. The 1660 edition of the former is Harrison 269, the latter is not listed among Newton's books.

16 Boas and Hall, 'Newton's chemical experiments', p. 121.

17 Ibid., pp. 125–6.

18 On fol. 81 of the MS Newton carefully describes the preparation of a 'regulus' by melting a metal with a greater quantity of antimony (i.e. stibnite, the sulphide) purified with saltpetre. "The better your proportions are the brighter and britler will the Regulus bee and the darker the scoria and the easier will they part. And also the more perfect the star." (ibid., p. 125, n. 22).

19 'Basil Valentine', *The Triumphal Chariot of Antimony*, trans. A. E. Waite, reprinted 1962, p. 175.

20 Boas and Hall, 'Newton's chemical experiments', pp. 123–4. The first paragraph of this quotation is a loose translation from 'Philalethes' [George Starkey], *Secrets Reveal'd*, 1669: 'Learn what *Diana's Doves* are, which do vanquish the *Lion* by asswaging him; I say the Green *Lion*, which is in very deed the *Babylonian Dragon*, killing all things with his Poyson: Then at length learn to know the *Caducean Rod* of *Mercury*, with which he worketh Wonders' (quoted in Dobbs, 1975, p. 68). The Green Lyon is

perhaps stibnite, so ibid. p. 184, yet green would suggest a compound of copper, which might well be highly poisonous.

21 Dobbs, *Foundations*, p. 90 from King's College MS Keynes 29, fol. lv, Newton's notes on a work of Michael Maier.
22 Ibid., p. 151. The passage is in 'Basil Valentine' (*Triumphal Chariot*, n. 19), pp. 176–7. 'Basil' goes on to describe the making of the star, claiming that in it the alchemist has a 'hot and ignitable substance, in which wonderful [medical] possibilities are latent'.
23 Boas and Hall, 'Newton's chemical experiments', pp. 128, 129–30, 141. *Mercurius dulcis* reappears in *Opticks*, Query 31 (1952, p. 385)
24 Ibid., p. 147.
25 Some (in an obvious analogy) spoke of mercury as the spirit of metal, sulphur its soul, and salt its body. In transmutational operations the first was all-important. Cf. Dobbs, *Foundations*, pp. 134–6.
26 Ibid., pp. 136–7.
27 *Opticks*, Query 31; 1952, p. 382, first printed in *Optice*, 1706.
28 Dobbs, *Foundations* p. 139. As the process was only published in this edition (Harrison no. 255), Newton's probably is not earlier than 1669.
29 Ibid. quoting Boyle, *Certain Physiological Essays*, 1669, pp. 202–3.
30 Boas and Hall, 'Newton's chemical experiments', pp. 124–5; *Opticks*, 1952, p. 381.
31 Ibid., p. 138; I do not know what Newton means by 'spar' (normally a fissile, crystalline mineral of various sorts).
32 Dobbs, *Foundations*, pp. 144–5 and note. The suggestion made here that Newton was already (in 1675, say) interpreting alchemical teaching in atomistic terms, though it may be true, seems without documentary support. Starkey's *Ripley reviv'd* (1677–78; J. Harrison, *The Library of Isaac Newton*, 1978, no. 1407).
33 Dobbs, *Foundations*, pp. 144–5.
34 Ibid., p. 160.
35 Ibid., p. 163.
36 Ibid., p. 170.
37 So Harrison, *Library*, and also Dobbs, *Foundations*, p. 131. *Mathematical Papers*, vol. II, p. xiii takes the other view.
38 Harrison, *Library*, p. 59. The grand total is 1752 titles.
39 Ibid., p. 11 and nos 150, 938, 954.
40 *Sotheby Catalogue*, pp. 58–9.
41 R. S. Westfall, *Never at Rest*, 1980, p. 361; Dobbs, 1975, p. 8.
42 A list of the MSS as sold is in *Sotheby Catalogue* pp. 1–19 (Lots 1–120), revised with present locations in Dobbs, 1975, Appendix A.
43 A. N. L. Munby, 'The Keynes Collection of the Works of Sir Isaac Newton at King's College, Cambridge', *Notes and Records of the Royal Society*, 10, 1952, pp. 42–4.
44 Dobbs, *Foundations*, p. 111.
45 Ibid., pp. 112–21.
46 *Sotheby Catalogue*, p. 8, Lot 45; Dobbs, *Foundations*, p. 240.

47 The earliest of Newton's historical researches to be published (anonymously) was his study of the ancient knowledge of gravitation, printed by David Gregory in his *Elementa astronomiae physicae et geometricae* (1702); see P. Casini, 'Newton: The Classical Scholia', *History of Science*, 22, 1984, pp. 1–58.

48 F. Sherwood Taylor, 'An alchemical work of Sir Isaac Newton', *Ambix*, 5, 1956, pp. 59–84. Boas and Hall, 'Newton's chemical experiments', p. 116. The MS in question is Keynes 38 (*Sotheby Catalogue*, Lot 57).

49 Dobbs, 1975, p. 178; Westfall, *Never at Rest*, p. 370.

50 William Newman, 'Newton's *Clavis* as Starkey's *Key*', *Isis*, 78, 1987, pp. 564–74. That 'Newton's' *Clavis* was by Starkey was first suggested in Figala, 'Newton as alchemist', 1977, p. 107.

51 Quoted in Westfall, *Never at Rest*, p. 371.

52 *Correspondence*, vol. II, 1960, pp. 1–2. Clearly Newton felt no embarrassment about informing Oldenburg of his familiarity with the Hermetic writers.

53 Locke had assisted Boyle in preparing his *History of the Air* for the press, and saw the book into print after Boyle's death (30 December 1691; Maddison, *Life of Boyle*, 1969, p. 182). He had grown very intimate with Newton since his return to England after the Revolution.

54 *Correspondence*, vol. III, 1961, pp. 218–19.

55 Westfall, *Never at Rest*, p. 298.

56 Boas and Hall, 'Newton's chemical experiments', p. 151.

57 Ibid., pp. 151–2. Cf. F. E. Manuel, *The Religion of Isaac Newton*. 1974, pp. 44–6: the alchemists studied Nature but presented their results in disguise. Manuel notes that to Newton Rosicrucians were imposters.

58 *Opticks*, 1952, p. 381; K. Figala, 'Newton as alchemist', *History of Science*, 15, 1977, p. 105.

59 Figala, 'Newton as alchemist', pp. 118, 119.

60 *Opticks*, 1952, pp. 268–9.

61 R. de Villamil, *Newton*, [1931?], p. 50. Those who have made a special study of Newton's (al)chemical researches (including B. J. T. Dobbs, K. Figala, W. Newman, P. M. Rattansi and R. S. Westfall in recent years) differ considerably in their views of the dating, significance and interpretation of individual MSS in the Newtonian corpus, and also in their treatment of alchemy itself. It is impossible here to enter into these detailed and often speculative questions.

Chapter 8

1 *The Preliminary Manuscripts for Isaac Newton's 1687 'Principia': 1684–85*, ed. D. T. Whiteside, 1989; *idem*, 'The Prehistory of the *Principia* from 1664 to 1686', *Notes and Records of the Royal Society*, 45, 1991, pp. 11–61.

2 It should be noted here that Newton seems to have made two assumptions unstated in his letter: (1) the tower, the globe and the observer all rotate together; (2) the body falls under the action of a *constant* gravity.

3 *Correspondence*, vol. II, 1960, pp. 304–6. In a curious passage Hooke praised Newton for not being a Drudge or Devotion [?] and went on 'Covetousness Slavery or Supersticon act them and they produce nought but Molas or chymeras sume what without life or Sole' (p. 305). The first printing of this letter by Alexandre Koyré in 1952 (*Isis*, 43, pp. 312–37) was the foundation of the recent interest in the origins of the *Principia*.

4 Ibid., pp. 307–8, 13 December 1679 – allowing little enough time for Newton to have examined the question.

5 Ibid., pp. 309–10. The opening sentence of Hooke's letter suggests that he might have made an experimental check on Newton's curve, for example by rolling a ball inside a large tundish.

6 Ibid., p. 309. Algebraically, Hooke posits

$$f\alpha \frac{1}{d^2} \rightarrow v\alpha \sqrt{f} \ [\text{from Huygens}] \rightarrow v\alpha \frac{1}{d}$$

7 The finest expression of Hooke's qualitative concept of universal gravitation, and of its effect in forming the planetary orbits, is to be found in his *Attempt to prove the Motion of the Earth* (1674); see R. T. Gunther, *Early Science at Oxford*, vol. VIII, 1931, pp. 27–8.

8 *Correspondence*, vol. II, pp. 433, 435, 27 May and 20 June 1686. Newton asserted that, meeting Wren in London in 1677, his knowledge of the inverse-square law was already evident. Newton's own recognition of it, he assured Halley, went back before 1673 (ibid., pp. 446–7).

9 Ibid., p. 438, 20 June 1686.

10 Ibid., p. 433, 27 May 1686.

11 Ibid., p. 444, 14 July 1686.

12 *De motu* (in brief, the various drafts are variously titled) has been printed several times since its first publication in S. P. Rigaud, *Historical Essay on the first publication of Sir I. Newton's* Principia, 1838. See, for example, *Unpublished Scientific Papers*, 1962, pp. 243–92 (text and English translation of final state) and *Mathematical Papers*, vol. VI, 1974, pp. 30–74 (first draft, with translation).

13 Until this time, the expression 'laws of motion' signified the rules of collision between ideal particles. The modern, wider meaning was introduced by Newton. It ought to be added that Newton did not enter a new world of celestial force mechanics because no alternative offered: Huygens, Fatio de Duillier (his friend) and especially Leibniz all demonstrated, about 1690, each to his own satisfaction, that there was a preferable, neo-Cartesian aetherial alternative.

14 See Halley's first-hand account to Newton, 29 June 1686, *Correspondence*, vol. II, pp. 441–2.

15 Cf. Patri J. Pugliese, 'Robert Hooke and the Dynamics of Motion in a curved Path', in M. Hunter and S. Schaffer (eds.) *Robert Hooke: New Studies*, 1989, pp. 203–4.

16 Halley asked Newton for an answer to what later came to be called the *inverse* problem of central force, that is, given the force, what curve(s) may result? Newton had already solved the *direct* problem, that is, given the curve, what force is required to produce it? Therefore he could not yet fully answer Halley's question by replying 'a conic'. Newton's main investigations were all to be directed to the solution of the direct problems.

17 *Mathematical Papers*, vol. VI, p. 17, n. (52).

18 We gave this title to the paper when printing it in *Unpublished Scientific Papers*, 1962, pp. 293–301. It has been printed several times, before and since, both from Locke's copy and from the slightly different version that Newton retained (CUL MS Add. 3965.1). See Whiteside, 'Prehistory of the *Principia*', pp. 28 and n. 87, 89.

19 *De motu corporum in gyrum*, as sent to Halley, was later entered in the Royal Society's Register, then sent to Flamsteed and lost. Halley had made a copy of his own, later returned to Newton, and in part surviving. Cf. note 12 above.

20 *Mathematical Papers*, vol. VI, pp. 92–187, *Correspondence*, vol. II, pp. 403–5. Part of Newton's concern was with the satellites of Jupiter and Saturn.

21 T. Birch, *History of the Royal Society*, vol. IV, 1757, p. 347; my reconstruction is confirmed by Paget's still having the (November) *De Motu* in January 1685 (*Correspondence*, vol. II, p. 412) and Newton's not thanking Aston for having it copied into the Royal Society's Register until 23 February 1685 (ibid., p. 415). It explains also why Newton's friends, regarding *De Motu* (rightly) as a preliminary sketch, kept quiet about it pending the arrival of Newton's 'treatise'.

22 Not only Hooke but also John Wallis at Oxford seems to have lacked detailed knowledge of *De Motu*.

23 Note that Newton's initial formulations are imperfect. The fact that in Theorem I the body revolves around a centre *of force* only appears incidentally. Much improvement was effected by the final version of the tract, *De Motu . . . in fluidis* (see n. 12).

24 *Mathematical Papers*, vol. VI, p. 49; *Unpublished Scientific Papers*, p. 277.

25 *Mathematical Papers*, vol. VI, pp. 58–9; *Unpublished Scientific Papers*, p. 283.

26 Ibid., vol. VI, pp. 63–74; ibid., pp. 287–92. For an analysis of the *De Motu*'s contents in relation to those of the *Principia* see I. B. Cohen, *Introduction to Newton's 'Principia'*, 1971, pp. 59ff.

27 Cohen, *Introduction*, pp. 79, 80; A. R. Hall, *Philosophers at War*, 1980, p. 296. The writer in both quotations is of course Newton himself.

28 D. T. Whiteside, 'The Mathematical Principles underlying Newton's *Principia Mathematica*', *Journal for the History of Astronomy*, I, 1970, p. 119. This is the essential study of Newton's mathematical techniques in his masterpiece.

29 *Mathematical Papers*, vol. VI; Cohen, *Introduction*, pp. 82–115.

30 *Correspondence*, vol. II, p. 415.

31 Ibid., p. 413.

32 Ibid., p. 437. Already in January 1685 Newton intended to "determin the lines described by the Comets of 1664 & 1680" according to the principles of motion observed by the planets, hoping for Flamsteed's help in establishing the places (ibid., p. 413).

33 Cohen, *Introduction*, pp. 68–9. By Edleston's record (p. lxxxv) Newton was absent from Cambridge for about three weeks in April/May 1685, then not till March 1687 (indefinite absence).

34 Cohen, *Introduction*, pp. 83–92; the early 'lectures' draft is printed in *Mathematical Papers*, vol. VI, pp. 92–187, the later draft in pp. 229–409; both are also reproduced in the *Preliminary Manuscripts* (see n. 1).

35 *Correspondence*, vol. II, p. 437. "Yours" because Halley had been authorized by the Royal Society to publish the *Principia* as a work receiving its imprimatur (signed by Pepys), at his own expense; Newton obviously surrendered to Halley any lien that he might have had on author's profits (apart from the right to receive a few presentation copies).

36 Ibid., p. 443, 29 June 1686.

37 Ibid., pp. 444 (14 July 1686), 473 (5 April 1687). Book III left Cambridge about the end of March. In its revised version, Newton referred to its first state: "I had indeed composed the third book in a popular method, that it might be read by many. But afterwards considering that such as had not sufficiently entered into the principles, could not easily discern the strength of the consequences, nor lay aside the prejudices, to which they had been many years accustomed; therefore to prevent the disputes that might be raised upon such accounts, I chose to reduce the substance of that book into the form of propositions (in the mathematical way) which should be read by those only, who had first made themselves masters of the principles established in the preceding books." *Principia*, 1687, p. 386; F. Cajori, *The Mathematical Principles*, 1946, p. 397.

38 *Correspondence*, vol. II, p. 470, 1 March 1687.

39 Ibid., p. 481. Halley was selling calf-bound copies in London at nine shillings each – a great price.

40 *Principia*, 1687, preface. Roger Cotes is said to have been highly praised by Newton for his abilities, but he received no word of public thanks for his pains in editing the great second edition of the *Principia*. There is no note of the publication of the first edition in the records of the Royal Society.

41 A summary of the whole work, proposition by proposition, is given in W. W. Rouse Ball, *An Essay on Newton's 'Principia'*, 1893, pp. 75–112.

42 Besides gravity, Newton examined in the *Principia* two forces: (1) in Book I, §14, small particles are attracted by great bodies with a force varying inversely with the distance; (2) in Book II, Proposition 23, particles in a fluid repel each other with a force varying inversely as the distance. The phenomena in the first case are similar to those of light, in the second case to those of elastic fluids (gases) in that they obey Boyle's Law.

43 Clifford Truesdell, 'Rational Fluid Mechanics 1687–1765', editor's introduction to *Euleri Opera Omnia Series II*, vol. 12, 1954, p. XII. Many 1687 results were corrected in later editions.

44 In 1687 Newton found the speed of sound by measurement to be between 870 and 1270 feet/sec and judged his computed value of 968 feet/sec verified. By the second edition Sauveur had convinced him that 1142 feet/sec was correct, whereupon by strange manipulations of a first derived value of 979 feet/sec he arrived at exactly this figure. More than a century later Laplace showed that Newton's computed speed was too low because he was unaware of the heating effect of corpuscular vibrations. (*Principia*, 1687, pp. 370–1; 1713, pp. 343–4.)

45 Heat may be a mode of motion, but Newton intuitively understood that by diffusion heat is, effectively, lost (*Opticks*, 1952, p. 398).

46 Ibid., pp. 399–400.

47 Ibid., p. 402. The passage was first printed in slightly variant form in *Optice*, 1706, p. 345.

48 *Principia*, 1713, p. 482; cf. *Opticks*, 1952, pp. 369–70.

49 [Samuel Clarke, (ed.)] *A Collection of Papers which passed between the late learned Mr Leibnitz and Dr Clarke in the years 1715 and 1716*, 1717, p. 15.

50 *Opticks*, 1952, p. 376.

51 Ibid., pp. 369–70. The continuation of this passage is well-known but merits transcribing once more: ". . . and not only to unfold the Mechanism of the World, but chiefly to resolve these and such like Questions. What is there in places almost empty of Matter, and whence is it that the Sun and Planets gravitate towards one another, without dense Matter between them? Whence is it that Nature doth nothing in vain; and whence arises all that Order and Beauty which we see in the World? To what end are Comets, and whence is it that Planets move all one and the same way in Orbs concentrick, while Comets move all manner of ways in Orbs very excentrick; and what hinders the fix'd Stars from falling upon one another? How came the Bodies of Animals to be contrived with so much Art, and for what ends were their several Parts? Was the Eye contrived without Skill in Opticks, and the Ear without Knowledge of Sounds? How do the Motions of the Body follow from the Will, and whence is the Instinct in Animals? And these things being rightly dispatch'd, does it not appear from Phaenomena that there is a Being incorporeal, living, intelligent, omnipresent, who in infinite Space, as it were in his Sensory, sees the things themselves intimately, and throughly perceives them, and comprehends them wholly by their immediate presence to himself . . . And though every Step made in this Philosophy brings us not immediately to the Knowledge of the first Cause, yet it brings us nearer to it, and on that account is to be highly valued." The passage was first printed in *Optice*, Query 20, pp. 314–15.

52 Perhaps on this point I differ from I. Bernard Cohen who in various writings (e.g. *The Newtonian Revolution*, 1980, pp. 72–4) seems to suppose

Newton to have allowed more ontological reality to possible mechanical explanations of forces than I think probable.

53 W. G. Hiscock, *David Gregory*, 1937, p. 30.
54 *Opticks*, 1952, pp. 365, 369.
55 See *Unpublished Scientific Papers*, pp. 302–8 (a long draft preface) and pp. 321–47 (a draft conclusion); material of the same kind was transferred to the Queries.
56 *Mathematical Papers*, vol. VI, pp. 425–6, n. (10).

Chapter 9

1 D. Brewster, *Memoirs*, 1855, pp. 91, 92, 96, 98. The book mentioned was David Gregory, *Astronomiae physicae et geometricae elementa*, 1702.
2 Christopher Wordsworth, *Scholae Academicae*, 1877, p. 331.
3 Brewster, *Memoirs*, vol. II, pp. 92–4.
4 *Correspondence*, vol. II, 1960, p. 484, 2 September 1687. This letter makes Gregory the first known reader of the *Principia* after Halley.
5 Ibid., p. 487, [? September 1687].
6 Ibid., p. 501, 29 December 1687. Craige's letter to Campbell of 30 January 1689 is of interest as detailing the help he received from Newton in 1685 (ibid., vol. III, pp. 8–9).
7 A. E. Shapiro, 'Newton's "achromatic" dispersion law: theoretical background and experimental evidence', *Archive for History of Exact Sciences*, 21, 1979, p. 97.
8 *Opticks*, 1952, p. 103.
9 *Correspondence*, vol. II, pp. 467–8. Nothing else is known of this letter.
10 Newton was to refuse a possible appointment as Master of the Charterhouse a few years later; the current Master (who carried the motion for disobedience by his casting vote) was Thomas Burnet, author of *Telluris Theoria Sacra* (1681), with whom Newton had corresponded about this book in 1680–1.
11 R. S. Westfall, *Never at Rest*, 1980, p. 475 and note.
12 Ibid., pp. 477–8; Brewster was the first to print this story in *Memoirs*, vol. II, pp. 107–8.
13 Westfall, *Never at Rest*, pp. 478–9.
14 Brewster, *Memoirs*, vol. II, p. 113; *Correspondence*, vol. III, 1961, p. 20 (no. 335).
15 *Correspondence*, vol. III, p. 21.
16 Ibid., vol. II, p. 415.
17 The first surviving paper is of August 1689; see *Correspondence*, vol. III, p. 25.
18 The accounts of the Huygens brothers of the comings and goings do not tally (Westfall, p. 480 notes) but the three certainly had an audience.

19 *Correspondence*, ed. J. Edleston, 1850, p. lix, n. 96. *Victoria County History, Cambridge*, vol. III, 1959, pp. 397–8. Newton was trebly disqualified from becoming Provost.

20 Westfall, *Never at Rest*, p. 516 from *Mathematical Papers*, vol. VII, p. 79 (draft of *De quadratura*, 1691). See also Whiteside's note here on Fatio's real originality in mathematics.

21 *Correspondence*, vol. III, p. 187.

22 W. G. Hiscock, *David Gregory*, 1937, pp. 6–8.

23 *Correspondence*, vol. III, p. 45. Fatio's letters were probably destroyed. Newton here stated a rather brusque opinion of Robert Boyle, "[who] has divers times offered to communicate & correspond with me in these [alchemical?] matters but I ever declined it because of his [?sociability] & conversing with all sorts of people & being in my opinion too open & too desirous of fame". Newton's choice of Fatio for intimacy was to prove less than judicious.

24 *Correspondence*, vol. VII, 1977, pp. 390–1, 17 April 1690.

25 Ibid., vol. III, p. 168, 8 September 1691. Newton was absent from Cambridge 12–19 September.

26 Ibid., p. 191, 28 December 1691 and cf. n. 27 below.

27 Ibid., p. 69, 24 February 1690. I presume that by 'facility' Fatio meant something like 'amiability' or 'good temper'. With this letter he sent his essay 'De la cause de la pesanteur' (presented to the Royal Society on 27 June 1688); Fatio proposed a mechanical, aetherial explanation of gravity which (he said) Newton favoured. See B. Gagnebin, *Notes and Records of the Royal Society*, 6, 1949, pp. 106–60.

28 *Correspondence*, vol. III, p. 193. Gregory gave a full account of this idea of Newton's in his Preface to *Astronomiae . . . elementa*, 1702.

29 Ibid., p. 230 and p. 231, 21 November 1692.

30 Ibid., p. 241 (24 January 1693), pp. 267–70 (18 May 1693).

31 James L. Axtell, 'Locke's Review [1688] of the *Principia*', *Notes and Records of the Royal Society*, 20, 1965, pp. 152–61. Apart from Halley's account in the *Philosophical Transactions* (1687), originally prepared for James II (reprinted in Cohen 1958, pp. 405–11), the only other notices of the book in journals were a full account in the *Acta Eruditorum* (1688) and in the same year a terse, hostile notice in the *Journal des Sçavans* (cf. I. B. Cohen, *Introduction to Newton's 'Principia'*, 1971, pp. 145–56).

32 *Correspondence*, vol. III, p. 391. Damaris Cudworth, who married Sir Francis Masham, was witty and intelligent, the best woman philosopher of the age. It is difficult not to believe that she and Locke were devoted lovers. Later, Newton visited Locke at Oates on a number of occasions.

33 *Correspondence*, vol. III, pp. 71–6; *Unpublished Scientific Papers*, 1962, pp. 293–301. Cf. *Mathematical Papers*, vol. VI, p. 553 and n. (33).

34 Ibid., vol. III, pp. 76–7, quoting *On Education*, 1693.

35 Ibid., pp. 390–1.

36 Ibid., p. 79, n. 1.

37 Throughout his fellowship Newton's periods of absence from Cambridge were short until the time of his mother's death, when he was away fourteen weeks (1679) and fifteen and a half weeks (1680). Afterwards he was as before, rarely away, missing only one week of residence from 1684 to 1687. Then, during his political involvement, he was absent fifty-five weeks in two years. From 1691 to 1695 once more he spent nearly all his time in Cambridge, but only half of 1696 (*Correspondence*, ed. J. Edleston, 1850, p. lxxxii).

38 *Correspondence*, vol. III, pp. 83–122; the MS is an autograph, in English.

39 The English Authorized or King James's Bible of 1611 (which I quote) follows the Vulgate version, regarded by Newton and modern scholars as corrupt; the Revised Version of the Bible prints a reading close to Newton's preferred text.

40 *Correspondence*, vol. III, p. 109.

41 Ibid., p. 82. Presumably this method of publication was proposed by Locke.

42 Ibid., pp. 129–42.

43 Ibid., pp. 123, n. (1), 147.

44 *Sotheby Catalogue*, pp. 70–1. There is also an unfinished treatise on the Apocalypse (110,000 words), p. 64. Besides Brewster, *Memoirs*, vol. II, ch. XXIV, and Westfall, *Never at Rest*, pp. 310–30 etc. see H. McLachlan, *Theological Manuscripts*, 1950, and F. E. Manuel, *The Religion of Isaac Newton*, 1974.

45 Westfall, *Never at Rest*, pp. 310–11. The notebook is *Sotheby Catalogue*, Lot 235.

46 Newton certainly read Jerome and his omission from the library is puzzling. See J. Harrison, *The Library of Isaac Newton*, 1978.

47 *Principia*, concluding *Scholium Generale* (1713).

48 Edward Gibbon, *The Decline and Fall of the Roman Empire*, ch. 21 (Everyman edn. 1910, vol. II, pp. 265–72).

49 Newton was particularly bitter about the story that Arius, whose fortunes at this time were rising, died (perhaps of poison?) in a privy in 336.

50 Westfall, *Never at Rest*, p. 314 quoting the Jerusalem MSS.

51 *Sotheby Catalogue*, Lot 268. Brewster, *Memoirs*, vol. II, pp. 342–6 states the 16 questions.

52 Westfall, *Never at Rest*, pp. 315–16.

53 Brewster, *Memoirs*, vol. II, p. 354.

54 Ibid., pp. 347–8; a passage from "A short scheme of the true religion" is very close to one printed in *Opticks*, Query 31 (1952, pp. 402–3).

55 *Correspondence*, ed. J. Edleston, pp. xxxvi, lxx, n. (144); *Correspondence*, vol. IV, pp. 405–6. Newton's letter states firm reasons for rejecting Locke's interpretation of one Pauline phrase.

56 See among other places *Correspondence*, vol. III, pp. 184–6, 214, 216–18.

57 Fatio wrote seven extant letters to Newton in 1693, before leaving temporarily for Switzerland in (?) June. It may be that Newton visited him

in London just before he left. Newton may also have visited (in the autumn of 1693) at Brigstock in Northants his half-sister Hannah Barton, whose husband was dying (*Correspondence*, vol. III, pp. 278–9; *Correspondence*, ed. J. Edleston, p. xc).

58 *Correspondence*, vol. III, p. 380. Newton referred of course to Locke's *Essay concerning Human Understanding*, of which the second edition was about to appear (1694).

59 Brewster, *Memoirs*, vol. II, p. 142, 13 September 1693. The letters to and from Pepys seem to have vanished.

60 Ibid., pp. 143–6. These few letters prove, once again, how little is known of Newton's relations with other men. But for the breakdown, the closeness of his relation to Pepys – not to say to his Magdalene contemporary Millington – would have been unsuspected.

61 *Correspondence*, vol. III, pp. 283–4, 5 October 1693.

62 Ibid., p. 284. Perhaps Newton suffered from a 'summer dysentery', a common water-borne infection until recent times.

63 Ibid., vol. IV, p. 282. At least one preceding letter from Locke to Newton is missing, possibly several.

64 The reasons for the rupture – other than Fatio's (temporary) desire to return to Geneva – are unknown. Westfall remarks that its effects upon Fatio were more severe than those upon Newton; however, he did become tutor to the Duke of Bedford's children, surely a normal, respectable career move (*Never at Rest*, pp. 538–9).

65 Papers in *Notes and Records of the Royal Society*, 34, 1979, by L. W. Johnson and M. C. Wolbarrsht, and P. E. Spargo and C. A. Pounds, pp. 1–32.

66 Ibid., 35, 1980, pp. 1–16.

67 At this time Newton delayed answers to letters from Leibniz and his friend Menke for six months. He made the excuse to both that he had mislaid their letters.

68 Richard Bentley, *A Confutation of Atheism*, 1693. See Perry Miller in I. B. Cohen's edition of Isaac Newton's *Papers and Letters*, 1978, pp. 271–394; Westfall, *Never at Rest*, pp. 498, 590.

69 Brewster, *Memoirs*, vol. I, pp. 465–9; *Correspondence*, vol. III, pp. 364–6. Bentley's protégé William Wotton made the enquiry of Craige, who replied to him.

70 *Correspondence*, vol. III, pp. 152, 155–6 (June/July 1691).

71 Ibid., p. 233, 10 December 1692.

72 Ibid., pp. 234–5.

73 Ibid., p. 239.

74 Ibid., p. 240. Here Newton seems to accept 'Plato's' fancy, if only the gravitating power of the Sun be doubled at the instant of conversion from descent to orbital motion. Later (p. 255) he rightly maintained that no such point of origin of all the planetary motions can be found. See I. Bernard Cohen, 'Galileo, Newton, and the divine order of the solar system' in Ernan McMullin (ed.) *Galileo: Man of Science*, 1967, pp. 207–31. In the

'classical scholia' of the early 1690s Newton abortively proposed to outline the ancient antecedents of the idea of universal gravitation.

75 *Correspondence*, vol. III, p. 240.

76 Ibid., pp. 253–4.

Chapter 10

1 Leibniz's letters were dated 11 June and 12 July (*Correspondence*, vol. II, 1960, pp. 212–34; Oldenburg's letter is of 9 August 1677 (ibid., p. 235, extract). This was a cordial letter of scientific news.

2 A. R. Hall, *Philosophers at War*, 1980, pp. 74–6.

3 G. W. Leibniz, 'Nova methodus pro maximis et minimis', *Acta Eruditorum*, October 1684, pp. 467–73; 'De geometria recondita et analysi indivisibilium atque infinitorum', ibid., June 1686, pp. 292–300.

4 Hall, *Philosophers*, pp. 77–8.

5 Ibid., pp. 31–4; *Principia*, 1687, pp. 250–4. The Scholium was much changed in subsequent editions.

6 Hall, *Philosophers*, p. 33. I should point out here that Wallis, having received copies from Collins of Newton's 1676 letters to Leibniz, in his English *Treatise of Algebra* (1685) excerpted from them liberally (pp. 330–46). Needless, perhaps, to repeat that though there was here rich material on series, there was nothing on fluxions (*Mathematical Papers*, vol. VII, 1976, pp. 8–9, especially n. (27)).

7 *Correspondence*, vol. II, p. 238.

8 Hall, *Philosophers*, pp. 40–3. The literal rendering of the encoded message of the *Second Letter* was longer than Newton made it in the Scholium.

9 *Mathematical Papers*, vol. VII, pp. xi, xxiii. Whiteside finds no immediate degeneration caused by Newton's illness in 1693.

10 *Correspondence*, vol. III, 1961, pp. 154–5, 27 July 1691. After some delay Gregory was appointed to the chair.

11 Barely published by Archibald Pitcairne in 1688, Gregory's work was now to be set out in some state and illustrated by examples.

12 Gregory's letters in *Correspondence*, vol. III, pp. 165–6, 169–81; Newton's draft response, pp. 181–3 and in *Mathematical Papers*, vol. VII, pp. 21–3.

13 Ibid., vol. VII, pp. 24–48.

14 Ibid., p. 35.

15 Compare ch. 4, at the end.

16 *Mathematical Papers*, vol. VII, pp. 64–5. The Leibnizian equivalent is $x.dx = y.dy$.

17 Ibid., pp. 49–130. The text ends in the middle of a problem which (the editor points out) offered no difficulties in its completion.

18 Ibid., p. 113. Newton defines the curvature at any chosen point of a curve as the diameter of a circle with the same curvature.

19 Ibid., pp. 123–9.

20 The brachistochrone is the curve along which a body falls in its swiftest possible descent from one point to another not vertically below; see *Mathematical Papers*, vol. VIII, 1981.

21 Ibid., vol. VII, pp. 507–9, nn. (1)–(3). The third version of the text was copied by David Gregory during his stay in Cambridge in May 1694.

22 Ibid., vol. VII, pp. 196–7; *Correspondence*, vol. III, pp. 385–6.

23 Henry Pemberton, *A View of Sir Isaac Newton's Philosophy*, 1728, preface.

24 *Mathematical Papers*, vol. VII, pp. 251, 253, slightly modified.

25 Ibid., p. 255, n. (21). I have Englished 'constructio sine Analysi inventa'.

26 Ibid., p. 566.

27 Ibid., pp. 567–78, 579 n. (1). For Gregory see *Correspondence*, vol. IV, p. 277 and for Cotes see ibid., vol. V, p. 279 (26 April 1712).

28 Ibid., pp. 170–80.

29 Ibid., p. 181, n. (26). From this time the letters exchanged between Leibniz, Huygens and the Bernoullis are filled with references to Newton's mathematics.

30 Ibid., quoting the *Commercium Epistolicum* of Leibniz and Bernoulli, Lausanne and Geneva, 1745, vol. I, pp. 190–1 (15 August 1696). See Hall, *Philosophers*, p. 117. This was the first suggestion – made privately – that someone had cheated in the duplicate invention of calculus.

31 Newton had proved (*Principia*, Book II, Scholium to Proposition 53) that a normal fluid vortex could not meet the astronomical requirements.

32 Varignon was an indirect pupil in calculus of Johann Bernoulli, but was not a strong Leibnizian party-man. The mistake was easily remedied. See E. J. Aiton, *Leibniz*, 1985, pp. 153–8 and his *Vortex Theory of Planetary Motions*, 1972.

33 The truth about Leibniz's reading of the *Principia* was established by D. Bertoloni Meli in an unpublished Cambridge Ph.D. thesis, 1988. See his 'Leibniz's Excerpts from the *Principia Mathematica*', *Annals of Science*, 45, 1988, pp. 477–504 and 'Public Claims, Private Worries: Newton's *Principia* and Leibniz's theory of Planetary Motion', *Studies in History and Philosophy of Science*, 22, 1991, pp. 415–49.

34 For several years the British seemed to have paid little heed to the publications of the Leibnizians in the continental journals. The first mention of the 'Tentamen' in Newton's correspondence is by Cotes (vol. V, 1975, p. 389, 10 March 1713) and Newton seems to have known nothing of this paper till near this date. *Mathematical Papers*, vol. VIII, 1981, p. 494, n. (86) lists drafts by Newton concerned with it.

35 *Correspondence*, vol. III, pp. 257–9.

36 This work, then unpublished, was part of the background to the 1689 conversations and letters between Huygens and Newton, already mentioned. The *Traité* was a skilful geometrical presentation of the wave theory of light, including an ingenious explanation of double refraction (to which Newton responded in *Opticks*, Query 25). With the *Traité* Huygens

published in *Discours sur la Cause de la Pesanteur* his hypothesis of an aetherial cause of gravity, first framed in 1669. Huygens's inability to accept Newton's physics was not new to him nor did it much disturb him.

37 He probably received the letters from Leibniz and Mencke in June (*Correspondence*, vol. III, p. 291), well before his collapse. The latter alone does not explain why his replies were put off until October and November.

38 Ibid., pp. 285–7.

39 Ibid., pp. 308–9. We need not necessarily imagine Fatio heedless of the opinions expressed by Newton to Bentley. For Newton opposed the inherence of force in matter *per se*, not a possible divine coalescence of matter and force.

40 Hall, *Philosophers*, pp. 110–14. *Correspondence*, vol. IV, 1967, p. 24, n. (5); pp. 237–8 (Wallis to Newton quoting Leibniz's words).

41 Hall, *Philosophers*, pp. 105–6, 118–21; *Mathematical Papers*, vol. VIII, pp. 3–12 is a full account of Newton's part in the challenge affair. Bernoulli's challenge and Newton's answer are printed in *Correspondence*, vol. IV, pp. 220–7. *Ex ungue leonem*, a common tag, does not attribute a leonine stature to Newton (*Mathematical Papers*, vol. VIII, pp. 9–10, n. (21) to (22)). The personal incidents of Newton's solution came from Catherine Conduitt, Newton's niece. He gained his results by fluxional analysis but only a synthetic proof of them survives (ibid., pp. 76–9).

42 Hall, *Philosophers*, p. 106. Newton's abbreviation of the *Investigatio* is printed in *Mathematical Papers*, vol. VIII, pp. 86–90. It is further evidence of a continuing relationship between Newton and Fatio de Duillier. Fatio also printed in his pamphlet a partially successful solution to Newton's problem of the solid of least resistance (*Principia*, Book II, Proposition 35, Scholium) – see *Mathematical Papers*, vol. VI, p. 466, n. (25).

43 Hall, *Philosophers*, pp. 120–6; *Mathematical Papers*, vol. VIII, pp. 471–2.

44 Hall, *Philosophers*, pp. 131–4, quotation p. 132; Hiscock, *David Gregory*, 1937, pp. 17, 19, 20, 23, 25.

45 Hall, ibid., pp. 132, 133.

46 Ibid., p. 15; *Mathematical Papers*, vol. VII, p. 20.

47 Hall, ibid., pp. 137–8; *Mathematical Papers*, vol. VII, p. 29.

48 Hall, ibid., pp. 143–5; *Philosophical Transactions*, 26, 1708, p. 185. The periodical being in arrears, the volume for 1708 was issued in 1710.

49 Hall, ibid., pp. 143–69; quotation p. 145. *Correspondence*, vol. V, pp. 96–7 (21 February 1711), pp. 115–17.

50 William Jones (ed.) *Analysis per Quantitatum Series, Fluxiones, ac Differentias* (1711) contained Newton's *De Analysi, De Quadratura curvarum, Enumeratio, Methodus Differentialis*, and correspondence.

51 *Correspondence*, vol. V, pp. 132–49.

52 Ibid., pp. xxiv–xxv, 212–14.

53 Ibid., pp. xxv–xxvi. Newton's various drafts are fully printed in *Mathematical Papers*, vol. VII, pp. 539–60.

54 *Mathematical Papers*, vol. VII, p. 538.

55 Hall, *Philosophers*, pp. 180–3.
56 That is, his *Opticks* (*Correspondence*, vol. III, p. 164). Newton asked whether satellites at the instant of occultation showed a flash of colour, as on some emission theories of light they should do.
57 Ibid., pp. 199–203. Cf. I. B. Cohen, *Introduction to Newton's 'Principia'*, 1971, pp. 172–7.
58 Ibid., vol. IV, pp. 7–8, 12–13.
59 Ibid., pp. 24–5 (7 October 1694), 34. Later (p. 46) Newton thought well of Caswell's experiments and said they should be published. Newton could sometimes be peevish and repent of it later.
60 Ibid., p. 46 (17 November 1694). Newton had a sizar working for him at this time (name unknown) since he wrote: "My servant has lately learnt Arithmetick . . . "
61 Ibid., p. 62 (20 December 1694).
62 Ibid., p. 106.
63 Ibid., p. 132 (14 June 1695, to Nathaniel Hawes).
64 Ibid., pp. 134–43; quotation p. 143.
65 Ibid., p. 171 (19 September 1695).
66 Ibid., pp. 151–2 (20 July 1695); see on all this D. T. Whiteside, 'Newton's Lunar Theory: From High Hope to Disenchantment,' *Vistas in Astronomy*, 19, 1975–6, pp. 317–24.
67 In Book III, Scholium to Proposition 35; see Whiteside, *loc. cit.* pp. 323–4. The MS is printed in *Correspondence*, vol. IV, pp. 322–6 (without title).
68 The two men had been in touch again at least since September 1697, when Flamsteed sent Newton a correction to data supplied earlier. They met in December 1698, and Newton obtained further observations – again, corrected by Flamsteed later. No letter from Newton about these affairs survives.
69 *Correspondence*, vol. IV, pp. 296–7 (6 January 1699), 302–3 (10 January 1699).
70 Ibid., pp. 331–2 (10 May 1700), 333–46 (18 June 1700).
71 Ibid., p. 311. Newton cannot have been pleased by Flamsteed's rather absurd criticism of Newton's *Opticks*, pp. 424–5 (1952 edn.). Flamsteed claimed to see that a star was no more than 1" of arc ($\approx 5.10^{-5}$ ins.) in diameter; Newton made the least visible point seen through a telescope five times as great. He went down to Greenwich to hear Flamsteed's objections to his book.
72 Ibid., pp. 430, 436.
73 In June 1709 (ibid., pp. 538–9) Newton drew up an account showing £166 . . 12s paid for printing 400 copies of *Historia Coelestis*, £125 paid to Flamsteed, £30 to John Machin for checking calculations. If Gregory was ever involved, as Flamsteed imagined, he was paid nothing. See also pp. 462–3, 480, 515.
74 *Correspondence*, vol. V, pp. 99, 120 (19 April 1711).
75 Ibid., pp. 80, 131. The phrase 'conditions permitting' was omitted!

76 Ibid., pp. 165–6. Flamsteed was also angered because Halley printed lunar and planetary observations imparted to Newton in confidence years before (p. 194). The only book in Newton's known library bearing Flamsteed's name is this one of 1712.

77 Ibid., pp. 209–10. Flamsteed's side of his disputes with Newton is told at length in his autobiography (in Francis Baily, *An Account of the Revd. John Flamsteed*, 1835–7, reprinted 1966). Newton's side can be seen only in his correspondence.

78 For the anti-Flamsteed, post-Baily view see *Correspondence*, ed. J. Edleston, n. 118, pp. lxiv–lxvii; for the modern qualification, *Mathematical Papers*, vol. VII, pp. xxiv–xxv, and D. T. Whiteside, 'Newton's lunar theory', n. 66 above, pp. 323–4, and n. (42).

Chapter 11

1 W. G. Hiscock, *David Gregory*, 1937, p. 15; *Opticks*, 1952, p. cxxii (misnumbered cxxiv). George Cheyne, *Fluxionum methodus inversa* had appeared in 1703.

2 Hiscock, *Gregory*, 1937, p. 14, 15 November 1702. Note again the presence of Fatio among Newton's friends, and the indication that the Captain's name was pronounced 'Hawley'.

3 *Correspondence*, vol. IV, 1967, pp. 336, 338–9.

4 Ibid., pp. 100, 115, 117, 130, 238; Hiscock, *Gregory*, 1937, p. 8.

5 Cambridge University Library, MS Add. 3970.3, fol. 393–426. A. E. Shapiro, 'Newton's achromatic dispersion law', *Archive for History of Exact Sciences*, 21, 1979, p. 97. The 1687 date is confirmed by an allusion in *Opticks*, 1952, p. 103. The Latin text went beyond this point.

6 *Opticks*, 1952, p. cxxi.

7 Ibid., pp. cxxi–ii, 339; *Principia*, 1687, pp. 231–2.

8 A. Rupert Hall, 'Beyond the Fringe: Diffraction as seen by Grimaldi, Fabri, Hooke and Newton', *Notes and Records of the Royal Society*, 44, 1990, p. 14.

9 T. Birch, *History of the Royal Society*, vol. III, 1757, pp. 181, 193, 194; *Diary of Robert Hooke*, ed. H. W. Robinson and W. Adams, 1935, pp. 145, 152, 153; Newton, *Correspondence*, vol. I, 1959, pp. 383–4.

10 *Opticks*, 1952, pp. 336–7.

11 They were renumbered 25 to 28 in later English editions, in consequence of the still newer material adjoined in Queries 17 to 24.

12 *Erasmi Bartholini Experimenta Chrystalli Islandici Dis-Diaclastici quibus mira & insolita REFRACTIO detegitur*, Copenhagen 1669, was given a seven-page article by Oldenburg (*Phil. Trans.* vol. V, January 1671, pages 2041–8 of the second pagination).

13 Huygens visited England from 1 June to 14 August 1689 and met Newton several times. At the Royal Society on 12 June 'Mr Huygens of Zulichem

being present gave an account that he himself was about publishing a Treatise concerning the cause of Gravity and another about refractions given [*sic*!] the reasons of the double refracting Island Chrystall . . . Mr Newton considering a piece of the Island Chrystall did observe, that of the two Species wherewith things do appear through that body, the one suffered no refraction when the visuall Ray came parallel to the Oblique sides of the parallelopiped; the other as is usual in all other transparent bodies suffered none, when the beam came perpendicular to the planes through which the Object Appeared' (Journal Book Copy). Oldenburg had been sent a piece of the spar (calcite) by Bartholin, which Newton perhaps borrowed. (*Phil. Trans.* vol. V, p. 2040).

14 *Opticks*, 1952, pp. 360–1 (Query 26).
15 Cambridge University Library MS Add. 3970.3, fol. 337ff; see Zev Bechler, 'Newton's search for a mechanistic model of colour dispersion; a suggested interpretation', *Archive for History of Exact Sciences*, 11, 1973, pp. 20–33; *Mathematical Papers*, vol. VI, 1974, p. 424, n. (3); R. S. Westfall, *Never at Rest*, 1980, pp. 520–2.
16 Westfall, ibid., p. 521. Newton went on that the strangeness of such notions to philosophers had caused him to expunge it from the *Principia* lest it be accounted an "extravagant freak" and prejudice the reader against his book.
17 *Unpublished Scientific Papers*, pp. 339–42.
18 See the suppressed "Conclusion" to *Opticks* (Cambridge University Library MS Add. 3970.3, fol. 337v–338v) quoted in *Mathematical Papers*, vol. VI, pp. 425–6, n. (10).
19 *Correspondence*, vol. III, 1961, pp. 164, 202. Flamsteed replied to Newton's question of 10 August 1696 only on 24 February 1692. Bechler, 'Newton's Search', pp. 22–3, n. 15,
20 *Opticks*, 1952, pp. 38–9.
21 Ibid., pp. 130–1.
22 As Simon Schaffer has pointed out, the *experimental* definition of a homogeneous ray by colour analysis through a prism may be circular, since if a critic claims to have divided a 'homogeneous' ray by a further refraction, the retort can be made against him that his ray was evidently not homogeneous in the first place! (Simon Schaffer, 'Glass Works: Newton's Prisms and the uses of experiment', in David Gooding, Trevor Pinch, and Simon Schaffer (eds.), *The Uses of Experiment*, 1989, p. 69.) However, the homogeneous ray was to Newton also a mathematical concept, indeed that which alone permitted the mathematical theory of colour.
23 See Bechler, 'Newton's Search', pp. 22–3.
24 *Opticks*, 1952, pp. 45–6, 64–73, 122. Cf. Schaffer *loc. cit.* above and J. A. Lohne, 'Experimentum crucis' in *Notes and Records of the Royal Society*, 23, 1969, pp. 169–97, especially, p. 187.
25 *Opticks*, 1952, p. 339.

26 Ibid., pp. 154–8 (Book I, Part II, Proposition 8). This is a highly imaginative and theoretical presentation of 'Newton's colour circle'. Newton did not paint such a circle with pigments, nor did he rotate it to compound the colours into an approximate white.

27 Ibid., p. 208. The words are copied (thoughtlessly?) without change from the "Observations" of December 1675 (Birch, *History*, vol. III, p. 277) and may not really reflect Newton's ideas of 1687, 1691 or 1704.

28 Ibid., Book II, Part III, Propositions 12 to 20 and Part IV.

29 Ibid., p. 348.

30 The third edition of *Opticks*, 1721, was the last supervised by its author. The fourth is said to have been printed from a copy of the third bearing further emendations in Newton's hand.

31 I do not mean that Newton was invariably and totally objective. His distinction between the spectral bands indigo and violet (for the sake of making seven primary colours) was clearly subjective and highly dubious.

32 *Correspondence*, vol. I, p. 174, 11 June 1672.

33 These experimental tools were as decisive for the development of physics down to the nineteenth century as were the spectroscope and the cyclotron in later times; cf. Schaffer, 'Glass Works', n. 22 above.

34 Book II, Part III, Proposition 11. Since Newton possessed a nearly correct idea of the diameter of the Earth's orbit, it gave him a close measure of the velocity of light at $\pm 18.10^7$ miles/sec. All he *needed* was certainty that this velocity is less than infinite.

35 *Opticks*, 1952, p. 25.

36 The translator was an immigrant Huguenot, Pierre Coste, his work first appearing at Amsterdam in 1720. Pierre Varignon took charge of the Paris edition, modified Coste's translation to his own satisfaction, and corresponded amicably with Newton. J. P. Marat retranslated the book later.

37 George Cheyne, *Philosophical Principles of Religion*, 1715, p. 85.

Chapter 12

1 *Correspondence*, vol. III, 1961, p. 279, 13 September 1693.

2 Ibid., p. 280, 16 September 1693.

3 Ibid., p. 79, 28 October 1690. Charles Mordaunt (1658–1735), Earl of Monmouth, later as Earl of Peterborough commander of the ill-starred British forces in Spain (1705–6), had resigned from the Treasury on 18 March 1690.

4 Ibid., pp. 152, 30 June 1691, and 185–6, 13 December 1691. As it turned out, Newton's friend Thomas Burnet was not promoted from the Mastership of the Charterhouse.

5 Ibid., p. 192. and vol. IV, 1967 p. 193.

6 Ibid., vol. IV, pp. 188, 189, 195. Montague had secured King William's agreement to Newton's nomination which he would not have done if there was a chance that Newton might decline.

7 Ibid., p. 200; pp. 207–8 (Newton's note on the state of the Mint in 1696); *Correspondence*, ed. J. Edleston, 1850, p. xxxvi; James E. Force, *William Whiston, Honest Newtonian*, 1985, pp. 11–14.

8 *Correspondence*, ed. Edleston, lxiv, n. (113). The story is from Abraham de la Pryme.

9 *Correspondence*, vol. IV, pp. 196–8, after 3 March 1696.

10 "Scala graduum caloris", *Philosophical Transactions*, 22, no. 270, March–April 1701, pp. 824–9. The *Journal Book* (vol. IX, p. 260) records that it was read on 28 May at the Royal Society. It was reprinted by Castiglioni in his edition of Newton's *opuscula*, 1744.

11 The notes are in CUL MS Add. 3975, fol. 45–6. On all that follows see J. A. Ruffner, 'Reinterpretation of the Genesis of Newton's "Law of Cooling"', *Archives for History of Exact Sciences*, 2, 1964, pp. 138–52.

12 Newton supposed the cooling effect to be proportional to the difference between the temperature of the body and that of its surroundings – another way of stating the law.

13 The paper was first printed in John Harris, *Lexicon Technicum*, vol. II, 1710, introduction; it is reprinted in *Letters and Papers*, 1978, pp. 256–8. A version with Pitcairne's comments is in *Correspondence*, vol. III, pp. 205–12.

14 Cohen, *Introduction*, 1977, pp. 177–98.

15 Ibid., pp. 188–9; P. Casini, 'Newton: The Classical Scholia', *History of Science*, 22, 1984; McGuire and Rattansi, 'Newton and the "Pipes of Pan"' *Notes and Records of the Royal Society*, 21, 1966. I agree with Casini that the significance of this Renaissance antiquarianism in Newton's thought is easily exaggerated. It exercised no ascertainable effect upon his scientific writings.

16 *Correspondence*, vol. IV, p. 276.

17 W. G. Hiscock, *David Gregory*, 1937, pp. 13, 16, 19, 36.

18 The Hamburg book was Johann Groening, *Historia cycloeidis*, 1701; it contained a list of corrections to the *Principia* found in Huygens's papers, therefore supposed to be by him but in reality sent by Fatio to Huygens in 1689–90 from Newton's early corrected copy. See Cohen, *Introduction* 1971, pp. 184–7 and *Principia*, ed. Koyré and Cohen, 1972, II, Appendix IV.

19 Comparison may be made with the aid of *Principia*, 1972, vol. II, pp. 707–65.

20 *Correspondence*, vol. IV, pp. 165, 169, 171–2.

21 Ibid., pp. 180–1, Newton to Halley 17 October 1695.

22 Ibid., vol. III, pp. 260–2.

23 Ibid., vol. IV, p. 190 (undated).

24 *Principia*, 1713, p. 480.

25 Samuel Clarke (1675–1729) defended a Newtonian thesis at his act for the BA (1695). His tutor was Newton's friend John Ellis. During Newton's last

year at Cambridge Clarke was a Fellow of Caius College. Another link was that he followed Whiston as Bishop Moore's chaplain at Norwich. It is likely that he met Newton in Cambridge. After his move to London he translated *Opticks* into Latin at Newton's request. His theological irregularity forced him to give up preaching in 1714.

26 *Correspondence*, vol. IV, 1967, pp. 377–80; vol. VI, 1976, p. 381 (1717); vol. VII, 1977, p. 50 (1719).

27 24 August 1693; ibid., vol. III, 1961, pp. 278–9.

28 See ch. 10, p. 263. Catherine's report is the source of details about the solving of the problem by Newton.

29 A. De Morgan, *Isaac Newton: his Friend and his Niece*, 1885, made a curious mistake about this (p. 21), supposing the twenty years to have extended to Newton's death, leaving about ten years unaccounted for. Cf. *Correspondence*, vol. VII, p. 74, 16 November 1719.

30 *Correspondence*, vol. IV, p. 349, 5 August 1700. Pudlicote, the place to which Newton sent his letter, is a small isolated seat near the tiny village of Chilson, a few miles from Charlbury. I have failed to identify Mr Gyre, unless this is another spelling of Gayer (cf. letter 760).

31 *Correspondence*, vol. V, 1975, p. 201, undated.

32 Ibid., pp. 199, 345; D. Brewster, *Memoirs*, vol. II, 1855, pp. 396–7 and note; Conduitt was his authority. Baydon is south-east of Swindon. Katherine (Greenwood) Barton's daughter *en secondes noces* married Cutts Barton, a grandson of Robert Barton the elder by his first marriage. This family died out in the nineteenth century.

33 Bushey Park passed to Halifax's heir, his nephew George, the new earl. So, seemingly did Apscourt Manor, near Weybridge.

34 *Correspondence*, vol. V, pp. xliv–xlv; the anonymous biography of Halifax, in which Newton thought Steele had a hand, by an easy confusion, printed 'widow' for 'sister'. Katherine (Greenwood) Barton was the wife of the late Robert. Newton noted a mistake in his copy about Mint medals, nothing about Catherine (Harrison, *Library*, 1978, no. 733 and p. 15).

35 I translate from De Morgan, *Newton*, 1885, p. 3. His attribution to the *Lettres* is mistaken and I follow Westfall, *Never at Rest*, 1980, p. 596, n. 154 who says it was first printed in 1757.

36 *Correspondence*, vol. VI, p. 232, 9 July 1715.

37 For a review, see Westfall, *Never at Rest*, pp. 596–600.

38 *Correspondence*, vol. VI, p. 225, 23 May 1715.

39 Newton succeeded Lord Somers, who followed Halifax. For the instrument see Cohen in *Papers and Letters*, 1978, pp. 237–8; in an improved form it is known as Hadley's sextant. The incident must have confirmed Newton in his resolution not to publish anything on optics while Hooke lived.

40 *Correspondence*, vol. V, p. 45. Woodward was eccentrically absurd: he had tried to fight a duel with Richard Mead, a fellow-physician and a friend of Newton's, on the steps of the College of Physicians!

41 Ibid., p. l, n. 29. Sir Henry Lyons, *History of the Royal Society*, 1944, pp. 130ff.

42 Ibid., pp. 127–8, 137.

43 Ibid., p. 137.

44 *Correspondence*, vol. VI, pp. 33, 69–70, 255–6, 294 (letter 1189, n. (1)), 315, 333–4, 23 April 1716. For Flamsteed's account of his conflagration see William Cudworth, *Life and Correspondence of Abraham Sharp*, 1889, pp. 122–3.

45 So James Gregory, David's brother, professor of mathematics at Edinburgh. *Correspondence*, vol. VI, pp. 332, 380, 14 February 1717; vol. IV, p. 311, 29 May 1699; Hall, 'Newton in France', 1975.

46 William Young (ed.), *Contemplatio philosophica* [by Brook Taylor], privately printed, 1793, p. 94; R. de Villamil, *Newton: the Man*, [n.d.], pp. 52–3.

47 S. Sorbière, *A Voyage to England*, 1709, pp. 36–7; Westfall, *Never at Rest*, pp. 681–2. The orders are hardly 'imperial' (his word); they codify the English way of running a meeting. W. Stukeley, *Memoirs of Sir Isaac Newton*, 1930, pp. 78–9, 80.

48 A. Rupert Hall 'Sir Isaac Newton's Steamer', *History of Technology, Tenth Annual Volume*, 1985, pp. 17–29. In these paragraphs I have borrowed freely from Marie Boas Hall, *Promoting Experimental Learning*, 1991.

49 The interest on Copley's legacy of £100 was paid to Desaguliers for presenting experiments to the Royal Society; the annually presented medal began in 1736.

50 *Opticks*, Query 8; 1952 edn. p. 341.

51 H. Guerlac, *Essays and Papers*, 1977, p. 123.

52 Ibid., pp. 123–4. Desaguliers placed two similar, cool thermometers in similar closed glass vessels, of which one was exhausted of air. When exposed to a hot fire, the mercury in both thermometers rose similarly. Radiant heat was not yet understood. Newton supposed that the transmission of heat demanded a material continuum.

53 On all this see A. R. Hall, *Philosophers at War*, 1980, ch. 9, quotation p. 195; *Correspondence*, vol. V, pp. 348–50; *Mathematical Papers*, vol. VI, pp. 72–91. Both Johann and Nikolaus Bernoulli claimed the detection of Newton's mistake as their own.

54 *Correspondence*, vol. V, pp. 347, 361. There was a precedent for Newton's discourtesy to Bernoulli in that the latter had printed a Newtonian central force theorem, imparted to him by de Moivre, as his own discovery.

55 Leibniz was living in Vienna at this time. The fly-sheet, dated 18 July 1713, was printed and issued for him by Christian Wolf. See *Correspondence*, vol. VI, pp. 15–21, quotation pp. 18–19.

56 Ibid., vol. V, pp. 133–52.

57 Ibid., vol. VI, p. 62 (5 February 1714), pp. 80–95 (undated), pp. 106–9 (19 and 20 April 1714).

58 Newton's primary demonstration in the *Principia* is that, if a body moves in a conic section about some centre of force, that force varies as the inverse square of the distance from the centre; he then argues inversely that given that the force is inversely as the square of the distance, the orbit

must necessarily be a conic. The validity of this inverse theorem, denied by Bernoulli, has lately been questioned again and defended by D. T. Whiteside.

59 *Correspondence*, vol. V, pp. 389–93, 400. Perhaps Newton was disingenuous for he had some grudge against Cotes, and wished him to take the blame if he should choose to stick his neck out.

60 Hall, *Philosophers*, pp. 226–31; quotations p. 230. The "Account" is there reproduced in an appendix.

61 *Correspondence*, vol. VII, pp. 69–71 (29 September 1719), 80 (n.d.). Bernoulli supposed that Newton had caused the Royal Society to deprive him of his fellowship; it was not so.

62 Samuel Clarke, *A Collection of Papers which passed between the late Learned Mr. Leibnitz and Dr. Clarke, In the Years 1715 and 1716 . . . 1717*. See H. G. Alexander, *The Leibniz–Clarke Correspondence*, Manchester, 1956.

63 Newton is said to have given Clarke the fantastic present of £500 as a reward for the Latin translation of *Opticks*. At the other extreme Cotes received nothing for his labours on the *Principia* (1713). For a hint of the social relations between Clarke and Newton see *Correspondence*, vol. VII, p. 180 (14 September 1721).

64 See Alexander, *Leibniz-Clarke Correspondence*; Alexandre Koyré and I. Bernard Cohen, 'Newton and the Leibniz-Clarke Correspondence', *Archives Internationales d'Histoire des Sciences*, 15, 1962, pp. 64–126; quotation pp. 66–7; F. E. L. Priestley, 'The Leibniz-Clarke Controversy' in Robert E. Butts and John W. Davis, *The Methodological Heritage of Newton*, 1970. Westfall, *Never at Rest*, p. 778 cites evidence that Newton received material directly from the Princess.

65 See *Correspondence*, vol. VI, *passim* especially pp. 285–90 (Newton to Conti, 26 February 1716) also considered by Koyré and Cohen, *loc. cit.* pp. 69–80. Newton added his later correspondence with Leibniz to unsold copies of Raphson's *History of Fluxions* (1715) which were re-issued in 1717.

66 Koyré and Cohen, *loc. cit.* n. 64, pp. 63, 75–6.

67 Des Maizeaux, a naturalized Huguenot, published a *Recueil de Diverses Pièces sur la Philosophie, la Religion Naturelle, l'Histoire, les Mathématiques, etc par Mrs Leibniz, Clarke, Newton et autres Auteurs célèbres*, Amsterdam, 1720, 1740. For his relations with Newton see *Correspondence*, vols VI and VII (especially VI, pp. 454–62) and Cohen, *Introduction*, 1971. In some drafts Newton developed the false claim that fluxions had been essential to his celestial mechanics. There is also important material in *Mathematical Papers*, vol. VI.

Chapter 13

1 *Correspondence*, vol. IV, 1967, pp. 382–3 (early 1702, addressee unknown). On these elections see Westfall, *Never at Rest*, 1980, pp. 623–6.

2 *Correspondence*, ed. J. Edleston, 1850, p. lxxiv, n. (153). *Correspondence*, vol. IV, p. 439, one of Halifax's few known letters to Newton, is about this election and the Queen's possible 'help' to him.

3 Yet there is not a little evidence to suggest that despite his Unitarianism and his philosophic conception of the Deity, conventional piety came easily enough to Newton. In (?1694) David Gregory noted two graces before meals used by Newton: Good Lord bless what is here offered us for use, and us for thy service; For this, and all other mercies, the Lord be blest through His Son Jesus Christ. (*Correspondence*, vol. IV, p. 18, n. 1)

4 Ibid., pp. 201, 204–5, 205–6. The other sureties (for £1000 each) were Halifax, Newton's relative by marriage Thomas Pilkington, and (perhaps) Peter Floyer, later Prime Warden of the Goldsmiths' Company. Newton bound himself for £2000.

5 John Craig, *Newton at the Mint*, 1946, pp. 8–9; Westfall, *Never at Rest*, pp. 553–6. Also among the eight consultants were Christopher Wren and John Wallis. For more detail on these and all Mint matters see Craig's study and his history of the Royal Mint. I concentrate upon Newton's personal concern in its business. In the recoinage only those who paid coins to officials (by way of taxes, customs etc.) were allowed face value for their old coins; others, when the coins went out of circulation, had to sell them at a loss as bullion. The estimated value of the silver money in 1695 was £12 million; a century later it was reduced to £3 million – an enormous deflation, partly compensated by the greater circulation of gold.

6 D. Brewster, *Memoirs*, vol. II, 1855, pp. 192–3. Craig points out that an assay furnace and touchstone were then still preserved at the Mint as Newton's, but makes it clear that they were never used officially. 'He cared not a whit for scientific advances towards obtaining complete purity of metals or measuring their fineness in absolute terms' (p. 22).

7 *Correspondence*, vol. IV, pp. 213–14, 229–31, 246, 254 (November 1696– December 1697). Halley had been appointed at Chester by the Comptroller, James Hoare, seemingly at Newton's suggestion.

8 Craig, *Newton at the Mint*, pp. 20–1; *Correspondence*, vol. IV, p. 210.

9 Craig, *Newton at the Mint*, pp. 17–19; *Correspondence*, vol.IV, pp. 231–2, 259–60, 305, 307–8.

10 F. E. Manuel, *Portrait of Newton*, 1980, p. 244. Not long before ceasing to be Warden Newton requested a refund of £120 for "various small expenses in coach-hire & at Taverns & Prisons & other places" visited in the cause of his war on crime. *Correspondence*, vol. IV, p. 317.

11 Ibid., pp. 306–7, 242–5; on Mint practices, pp. 207–8, 233–5, 255–8.

12 Ibid., pp. 253, 265, 314 (the first letter on longitude), 319–20.

13 Ibid., vol. VII, pp. 172–4, 330–2.

14 Ibid., vol. VI, pp. 77, 160–3. Westfall, *Never at Rest*, pp. 834–5 is too kind to Newton, whose paper quite failed to address the precise issues.

15 The MS is dated 27 February 1700 (*Correspondence*, vol. IV, pp. 322–6); Gregory's text is slightly longer towards the end. In the English edition of his book (1715) the text is at pp. 563–71 of vol. II. See D. T. Whiteside, 'Newton's Lunar Theory: From high hope to disenchantment', *Vistas in Astronomy*, 19, 1976, pp. 317–28.

16 Craig, *Newton at the Mint*, pp. 34–5; *Correspondence*, vol. IV, pp. 352–3, 388–90.

17 Craig, ibid., p. 34.

18 Ibid., pp. 42–3.

19 Ibid., p. 44. *Correspondence*, vol. IV, pp. 388–90 (20 September 1701). Newton's comprehensive table of coins was first published in Arbuthnot's *Tables*, (?1707) and again in [Thomas Prior], *Observations*, 1720.

20 *Correspondence*, vol. V, pp. 245–6, 309–12 (Newton to Oxford, 3 March and 23 June 1712). These letters were first printed in [Thomas Prior], *Observations*, pp. 22–6.

21 Ibid., pp. 418, 419, Newton to the Treasury, 21 September 1717, first printed in a newspaper, then by Prior at Dublin.

22 *Correspondence*, vol. IV, pp. 399–400, 407, 508 (Union medal); Craig, *Newton at the Mint*, pp. 51–6.

23 *Correspondence*, vol. IV, pp. 409–11, 466–7; vol. V, pp. 11–12; Craig, pp. 57–61.

24 Trial-plates were pieces of metal of standard purity with which samples of the currency coined at the Mint were compared, from time to time. The samples were hoarded in boxes called 'pyxes' hence the test was the 'Trial of the Pyx' – see Craig, *Newton at the Mint*, pp. 70–75; *Correspondence*, vol. IV, pp. 371–3.

25 *Correspondence*, vol. IV, pp. 485–7, 491–500 (especially). The mopping up operation figures largely at the end of vol. IV and the beginning of vol. V, 1975.

26 Craig, *Newton at the Mint*, pp. 76, 79–81; *Correspondence*, vol. V, pp. 123–4 (Newton's calculation of the values of Britannia (9854 parts silver/1000), sterling (9250 parts silver/1000), "old" and inferior plate, 25 April 1711), 126, 175–86 *passim*, Newton's narratives.

27 Craig, *Newton at the Mint*, pp. 77–9; *Correspondence*, vol. IV, pp. 371–3; vol. V, pp. 82–8.

28 *Correspondence*, vol. V, p. 85.

29 Craig, *Newton at the Mint*, pp. 95–101; *Correspondence*, vol. V, pp. 81–2, 357–60, 405–8, 514–16.

30 *Correspondence*, vol. VI, pp. xxii–iii, 55–9, 75–6, 404–5, 442, 451–4.

31 Ibid., vol. VI, pp. 451–4; vol. VII, pp. 217–18. *Prose Works of Jonathan Swift*, ed. T. Scott, vol. VI, 1902, appendices. I have found no explanation of Swift's use of 'drapier' rather than the usual 'draper'.

32 Swift, *loc. cit.* pp. 35, 39.

33 *Correspondence*, vol. VII, 1977, pp. xliii–iv, 97, 349, 351.

34 Ibid., pp. 181–2, 247, 352 (6 September 1726). Haynes (1672–1749) had served in the Mint since about 1687 and was highly regarded by Newton who promoted him to be weigher and teller in 1701. He became assay master in 1723.

35 A. Rupert Hall, 'Isaac Newton's Steamer', *History of Technology, Tenth Annual Volume*, 1985, pp. 17–19; Westfall, *Never at Rest*, p. 684; Guerlac, *Essays and Papers*, 1977, pp. 107–30. Hauksbee was never an official curator.

36 *Correspondence*, vol. IV, pp. 446–7, 14 September 1705 [the editor's date] – the year might be later. Westfall, *Never at Rest*, 1980, pp. 488, 581. I do not know the authority for the introduction story before Rouse Ball, 1893, p. 116.

37 *Correspondence*, vol. VI, p. 332; vol. VII, pp. 315–16 (29 April 1725).

38 Ibid., vol. IV, pp. 138, 468, 518–19. Antecedent negotiations between Bentley and Newton are lost. Newton said that Bentley was covetous and so he let him make money. There is no reason to believe that Bentley had long before attended Newton's Lucasian lectures.

39 *Correspondence*, vol. V, p. 28; vol. VI, pp. 406–7.

40 As all students of Newton must be, I am much indebted in what follows to Manuel's *Newton*, 1963. Newton believed that the earliest Greek writing was in verse and therefore not to be trusted as to facts; *Chronology of Ancient Kingdoms*, p. 45.

41 Manuel, *Newton*, pp. 17, 71. David Gregory by October 1695 had heard of Newton's 'papers about the mythologies; & Christian religion'; W. G. Hiscock, *David Gregory*, 1937, p. 4.

42 F. E. Manuel, *Isaac Newton, Historian*, 1963, pp. 95–7. Though brought up as an orthodox Anglican and never formally quitting his Church, the tendency of Newton's mind was strongly anti-Roman and pro-Puritan. His opinion of Jewish pre-eminence is clearer in MSS quoted by Manuel than in the *Chronology*.

43 *Chronology*, pp. 26–7, 115.

44 Ibid., pp. 14–15, 25.

45 Manuel, *Newton*, pp. 92–4. Josephus lived from AD 38 to 100, Clement from 150 to 216.

46 *Chronology*, pp. 83–92. Newton drew much information about ancient astronomy from Denis Petau, especially *Opus de doctrina temporum* (Antwerp, 1703 = Harrison no. 1284). He owned four books by this scholar.

47 *Correspondence*, vol. VII, pp. xl–xli, 279 (30 April 1724), 311 (9 March 1725), 322 (27 May 1725); Manuel, *Newton*, pp. 22–6; *Phil. Trans.* 33, no. 389, July–August 1725, pp. 315–21.

48 Manuel, *Newton*, pp. 264–5, n. 45; *Chronology* dedication. R. de Villamil, *Newton: The Man* [1931?], p. 55.

49 Manuel, *Newton*, p. 10.

50 I omit here Newton's special investigation of the structure of Solomon's

Temple, *Chronology*, ch. V and last, as of very specialized interest.

51 McGuire and Rattansi, 'Newton and the "Pipes of Pan"', *Notes and Records of the Royal Society*, 21, 1966, p. 108. The allusion to Pappus is to Book VIII of his *Collection*. Newton took the neo-Platonists to be great enemies of Christian truth.

52 P. Casini, 'Newton: The Classical Scholia', *History of Science*, 22, 1984, p. 1 (= F. Cajori, *The Mathematical Principles*, 1946, p. 549).

53 *Correspondence*, vol. III, pp. 193 (translated), 338. McGuire and Rattansi, pp. 109–10, make the dubious proposition that Newton wished his idea to be conveyed to Huygens. More likely, Fatio was boasting of his intimacy with Newton. By a quotation from Macrobius Newton wrote in explanation: "The philosophers loved to modify their mystic utterances in this way, so that they might unfittingly set commonplace things before the crowd to amuse it, and conceal the truth beneath this kind of utterance" (Casini, 'Newton: The Classical Scholia', 1984, p. 11).

54 Casini, 'Newton: The Classical Scholia', 1984, pp. 31–2, 55–8.

55 Ibid., p. 10.

56 Ibid., p. 11; 'Ten Books of mythology, or an explanation of legends'. Conti's book is in J. Harrison, The *Library of Isaac Newton*, 1978 (no. 439).

57 Cassini, 'Newton: The Classical Scholia', p. 16.

Chapter 14

1 I do not know whether Newton wrote the final form of the new material added in 1706 in Latin, or Clarke translated it.

2 See *Optice*, 1706, sig. a, "Errata, Corrigenda & Addenda"; cf. *Opticks*, 1952, pp. 99–100, 110–11, 122 for English equivalents. Another long passage on the orders of corpuscles in the structure of matter was added at the end of Book II, Part III, Prop. 8 (ibid., pp. 267–9).

3 *Opticks*, 1952, pp. 339, 345–6.

4 *Optice*, 1706, pp. 309–10, 313. On all these changes see Koyré. 'Les Queries de l'Optique', *Archives Internationales d'Histoire des Sciences*, 13, 1960, pp. 15–29.

5 *Optice*, 1706, pp. 314–15; cf. the revision in *Opticks*, 1952, pp. 369–70. In translating from *Optice* I have used the words of Newton's English text, so far as possible.

6 *Optice*, pp. 315ff; cf. *Opticks*, 1952, pp. 370ff.

7 *Optice*, pp. 320, 322; cf. *Opticks*, pp. 375–6.

8 *Optice*, pp. 327, 335, 340–1, 343; cf. *Opticks*, pp. 381, 389, 397, 399. On these aspects of Newton's thought see J. E. McGuire, 'Body and Void in Newton's De Mundi Systemate: Some new sources', *Archive for History of Exact Sciences*, 3, 1966, pp. 206–48; and idem, 'Force, Active Principles, and Newton's Invisible Realm', *Ambix*, 15, 1968, pp. 154–208.

9 *Optice*, p. 346; cf. *Opticks*, p. 403.

10 W. G. Hiscock, *David Gregory*, 1937, pp. 29, 35.
11 N. R. Hanson, *Patterns of Discovery*, 1958, ch. 1. H. Guerlac, *Essays and Papers*, 1977, pp. 120–30. The experiment on the flow of heat was made by Desaguliers for Newton in 1716 (p. 124).
12 To be consciously tautological, if a causal chain implies a mechanism, it must itself be of mechanical form.
13 *Optice*, p. 321; Zev Bechler, 'Newton's law of forces which are inversely as the mass . . .', *Centaurus*, 18, 1974, pp. 184–222. The same scholar has published other excellent papers on Newton's dynamics of light.
14 *Mathematical Papers*, vol. V, 1972, pp. 52–68; vol. VIII, 1981, p. xiii, n. (8).
15 William Jones, *Analysis per Quantitatum series, fluxiones, ac differentias* (1711). I presume that Collins's MSS used by Newton in *Commercium Epistolicum* (1712) and since then in the possession of the Royal Society were given by Jones to Newton. The former was also an industrious collector of Newton's own MSS or copies of them, and compiled the first narrative of Newton's mathematical discoveries (ibid., vol. VIII, pp. xxi–xxiii).
16 *Mathematical Papers*, vol. VIII, pp. xix, 425–41, 603–24; *Correspondence*, vol. VI, 1976, pp. 254, n. 2, 440, n. 1 (etc.); Pierre des Maizeaux, *Recueil de diverses pièces sur la Philosophie, la Religion . . .* , 1720.
17 *Mathematical Papers*, vol. VIII, pp. 68–70, 442–59, 527–8, 625–75.
18 *Correspondence*, vol. VIII, 1977, pp. 153, 199, 206–7, 214; the vignette is reproduced in plate 1.
19 Ibid., pp. 104–6.
20 Ibid., vol. V, 1975, p. 8, 20 October 1709. This portrait is reproduced in R. S. Westfall, *Never at Rest*, 1980, fig. 14.1. Bentley chided Newton because the picture was slow in coming to Cambridge, but Newton supported the schemes for changes in Trinity College introduced by Bentley.
21 Ibid., vol. V, pp. 4–5 (n. 4), 5; I. B. Cohen, *Introduction*, 1971, pp. 216–23. It is obvious that Newton must have placed his final corrected text of the first part of the book in Bentley's hands before June 1708. Whilst Whiston carried the MS Clarke carried the letter: hints of Newton's everyday contacts now forever lost.
22 Ibid., pp. 5, 7, Cotes's letter of 18 August 1709, Bentley's of 20 October. Newton told Cotes not to check every demonstration, but to correct by his copy amending obvious slips. Cotes insisted on understanding what was printed.
23 Ibid., p. 24.
24 Ibid., pp. 24–6. Cotes stopped at p. 224 of the new edition.
25 Ibid., p. 35, at pp. 254–8 of the new edition. Newton was now residing at Chelsea. Cohen, *Introduction*, p. 231.
26 Ibid., pp. 73–4, 5 October 1710 and *ante*; 75, 103–4, 107–110, 152–3, 155, 156–9, 164.
27 Ibid., pp. 215, 226, 16 February 1712, and 233, 28 February 1712. See also Richard S. Westfall in *Science*, 179, 1973, pp. 751–8.
28 Ibid., pp. 315–16, 20 July 1712.

29 Ibid., p. 347, 14 October 1712. The substitution comprises pp. 476–81 of the second edition. For the mistake in Proposition 10, see *Mathematical Papers*, vol. VIII, pp. 312–419.

30 Newton was much concerned with the pay to be given to the British troops in Dunkirk: a Mint letter about this was signed on 7 October – when perhaps the Prop. 10 crisis was over.

31 *Correspondence*, vol. V, pp. 347–50, 361–9; Newton was not pleased when Johann Bernoulli later published his correction of Newton's error, with some pomposity. Absurdly, and quite wrongly, he traced Newton's mistake in Proposition 10 to his misuse of second differentials. Newton's mistake was a slip in geometry.

32 Ibid., pp. 361, 384, 386 etc.

33 Ibid., pp. 397, 412. The phrases were added to the General Scholium.

34 Ibid., pp. 107, 114; vol. VI, pp. 48–9. Some of the amendments did eventually pass into the third edition, 1726. Newton may have made some sort of apology to Cotes (vol. VI, pp. 54–6), now lost. See A. Rupert Hall, 'Newton and his editors', *Notes and Records of the Royal Society*, 29, 1974, pp. 44–5.

35 *Correspondence*, vol. V, pp. 113–14, 279. *Mathematical Papers*, vol. VIII, pp. 46–8, 258–311, quotation p. 259. Cf. the more detailed prefatorial drafts, ibid., p. 442–57. The translation here is Whiteside's.

36 *Mathematical Papers*, vol. VIII, pp. 420–4, quotation p. 420, n. (1).

37 Correspondence, vol. V, pp. 361–9, 6 January 1713, 384, 2 March 1713; *Unpublished Scientific Papers*, pp. 348–64; Cohen, *Introduction*, pp. 240–5.

38 I do not know if Halley's copy has been found. *Correspondence*, vol. VI, pp. 151–2; vol. V, p. 413 and appendix.

39 Hall, 'Newton and his editors', pp. 46–8; Cohen, *Introduction*, pp. 65–79; *idem*, 'Pemberton's translation of Newton's *Principia*', *Isis*, 54, 1963, pp. 319–51; Henry Guerlac and M. C. Jacob in *Journal of the History of Ideas*, 30, 1969, pp. 307–18.

40 Evidence does not justify the opinion of older writers that the Conduitts made their London home with Newton in St Martin's Street. *Correspondence*, vol. VII, p. 321, n. 1; pp. 325, 336.

41 Ibid., vol. V, p. 409; vol. VII, pp. 180, 321. *Correspondence*, ed. J. Edleston, 1850, p. lxxiv, n. (156).

42 *Correspondence*, vol. VII, pp. 329, 336–9, August–November 1725; Newton supported Maclaurin's application for a post in Edinburgh, and offered funds to help support it.

43 The Newton–Varignon correspondence extended from 1713 – the latter's thanks for a presentation copy of the second *Principia* – to 1722, the year of Varignon's death. It is pleasant to record that in the last year of his life Varignon so far succeeded in his aim as a peacemaker that Newton agreed to his sending three copies of the *Traité d'Optique* edited by Varignon (Paris, 1722) to Johann Bernoulli, his son and his nephew (*Correspondence*, vol. VII, pp. 218–21).

Chapter 15

1 W. Stukeley, *Memoirs*, 1936, pp. 82–3; E. Turnor, *History of Grantham*, 1806, pp. 165–6, 172–3; *Sotheby Catalogue*, 1936, pp. 54–5. Conduitt was in fact mistaken about Newton's final attendance at the Royal Society: it was on 2 March. He had previously presided on 16 February 1727.

2 D. Gjertsen, *Newton Handbook*, 1986, p. 28; Turnor, *Grantham*, p. 165.

3 *Sotheby Catalogue*, p. 59.

4 *Principia*, 1713, p. 482 (my translation).

5 F. E. Manuel, *The Religion of Isaac Newton*. 1974, p. 61.

6 Ibid., p. 62; D. Brewster, *Memoirs*, vol. II, 1855, pp. 349–50.

7 Gjertsen, *Handbook*, pp. 396–9; R. S. Westfall, *Never at Rest*, 1980, pp. 319–21. The seventy weeks were allowed to the people for repentance: Daniel, 9: 24–27.

8 Westfall, *Never at Rest*, p. 319, n. 114; cf. *Correspondence*, vol. VII, 1977, p. 387, letter no. X.132. The manuscripts on prophecy were listed summarily in *Sotheby Catalogue* pp. 64–9, lots numbered 227–31, 242–8.

9 The 'Common Place Book' on theology (described in *Sotheby* no. 235 as 'slightly water-stained throughout, edges of some leaves rotted and a number of passages illegible') is thought to have been begun well before 1670.

10 *Sotheby Catalogue* nos 227(2), 231; Manual, *Religion*, pp. 122–5; on the text see Newton's correspondence with Dr John Mill, 1693–4, in *Correspondence*, vol. III, 1961.

11 Apart from historical objections against post-Nicene Christianity Newton objected to the complexity of its theology. Did Christ send his Apostles to preach metaphysics to the unlearned common people and to their wives and children? he asked (Brewster, *Memoirs*, vol. II, pp. 532–4).

12 Gjertsen, *Handbook*, p. 274; Westfall, *Never at Rest*, pp. 325–8; Manuel, *Religion*, pp. 94–8. I do not understand Newton's argument.

13 Manuel, *Religion*, pp. 126–36, quotation p. 132; *Sotheby Catalogue* no. 244.

14 Manuel, *Religion*, pp. 100–1; Brewster, *Memoirs*, vol. II, p. 354.

15 Brewster, *Memoirs*, vol. II, pp. 351–2, 526–31; *Sotheby Catalogue* no. 240.

16 *Correspondence*, vol. VII, pp. 338, 344–6, 347, 349, 355 (4 February 1727).

17 Useful references are: *Correspondence*, vol. II, pp. 502–4; vol. III, p. 393; vol. VII, pp. 365, 368–9.

18 *Correspondence*, vol. VII, appendix II; R. de Villamil, *Newton: The Man*, [1931?], pp. 57–61. The six cousins – all descendants of Newton's Smith half-siblings – were Hannah Clark and Thomas Pilkington (siblings); Benjamin Smith, Newton Smith and Hannah Tompson (siblings) and Catherine Conduitt. George Pilkington, Thomas's brother, was also involved in the legal settlement. The eighth surviving cousin, Margaret Warner, sister to Catherine Conduitt, apparently was not. Her husband John appears as a debtor to Newton, and was involved. Newton had years before bought Margaret an annuity and in 1726 he gave their son Isaac

Warner rents worth £100. Hannah Clark, who signed the inventory, is said to be the same woman as Mary Pilkington in the genealogy.

19 Villamil, *Newton*, p. 35; Stukeley, *Memoirs*, p. 84.
20 Villamil, *Newton*, p. 60; *Correspondence*, vol. VII, pp. 96–7.
21 Turnor, *Grantham*, p. 167. A full account of the family wind-up appears in Westfall, *Never at Rest*, pp. 870–4.
22 Voltaire, *Lettres Philosophiques*, Lettre XIV. Francis Haskell, 'The Apotheosis of Newton in Art' in Robert Palter (ed.), *The Annus Mirabilis of Sir Isaac Newton 1666–1966*, 1970, pp. 303–5.

Bibliography

Aiton, E. J., *Leibniz: A Biography*, Bristol and Boston, 1985.

Alexander, H. G., *The Leibniz-Clarke Correspondence*, Manchester, 1956.

Arbuthnot, John, *Tables of Grecian, Roman and Jewish Measures*, London, 1707.

Axtell, James, 'Locke's Review of the *Principia*', *Notes and Records of the Royal Society*, 20, 1965, 152–61.

Baily, Francis, *An Account of the Rev. John Flamsteed*, London, 1835, reprint 1966.

Baird, K. A., 'Some influences upon the young Isaac Newton', *Notes and Records of the Royal Society*, 41, 1987, 169–79.

Barrow, Isaac, *Optical Lectures 1667*, trans. and ed. H. C. Fay, A. G. Bennett and D. F. Edgar, London, 1987.

Bechler, Zev, 'Newton's search for a mechanistic model of colour dispersion: a suggested interpretation', *Archive for History of Exact Sciences*, 11, 1973, 20–33.

Birch, Thomas, *The History of the Royal Society of London*, London, 1756–7, 4 vols.

Boas, Marie and Hall, A. Rupert, 'Newton's chemical experiments', *Archives Internationales d'Histoire des Sciences*, 11, 1958, 113–52.

Brewster, Sir David, *Memoirs of . . . Sir Isaac Newton*, Edinburgh, 1855.

Cajori, Florian, *The Mathematical Principles of Natural Philosophy*, trans. A. Motte, revised edition, Berkeley, CA, 1946.

Casini, Paolo, 'Newton: The Classical Scholia', *History of Science*, 22, 1984, 1–58.

Cohen, I. Bernard, *Introduction to Newton's 'Principia'*, Cambridge, 1971.
—— *The Newtonian Revolution*, Cambridge, 1980.
—— 'Versions of Isaac Newton's first published paper', *Archives Internationales d'Histoire des Sciences*, 11, 1958, 357–75.
Craig, Sir John, *Newton at the Mint*, Cambridge, 1946.
Curtis, Mark H., *Oxford and Cambridge in Transition, 1558–1642*, Oxford, 1959.
Dobbs, Betty Jo Teeter, *The Foundations of Newton's Alchemy or "The Hunting of the Greene Lyon"*, Cambridge, 1975.
Feingold, Mordechai (ed.), *Before Newton: The Life and Times of Isaac Barrow*, Cambridge, 1990.
—— *The Mathematicians' Apprenticeship, 1560–1640*, Cambridge, 1984.
Figala, Karen, 'Die exakte Alchemie von Isaac Newton', *Verhandlungen der Naturforschenden Gesellschaft Basel*, 94, 1984, 157–228.
—— 'Newton as alchemist', *History of Science*, 15, 1977, 102–37.
Force, James E., *William Whiston, Honest Newtonian*, Cambridge, 1985.
Gjertsen, Derek, *The Newton Handbook*, London, 1986.
Greenstreet, W. J. (ed.), *Isaac Newton, 1642–1727*, London, 1927.
Guerlac, Henry, 'Amicus Plato and other friends', *Journal of the History of Ideas*, 39, 1978, 627–33.
—— *Essays and Papers in the History of Modern Science*, Baltimore, MD and London, 1977.
Hall, A. Rupert, 'Beyond the fringe: diffraction as seen by Grimaldi, Fabri, Hooke and Newton', *Notes and Records of the Royal Society*, 44, 1990, 13–23.
—— 'Further optical experiments of Isaac Newton', *Annals of Science*, 11, 1955, 27–43.
—— *Henry More: Magic, Religion and Experiment*, Oxford, 1990.
—— 'Newton in France: a new view', *History of Science*, 13, 1975, 233–50.
—— 'Newton on the calculation of central forces', *Annals of Science*, 13, 1957, 62–71.
—— 'Newton's first book', *Archives Internationales d'Histoire des Sciences*, 13, 1960, 39–61.
—— *Philosophers at War: The Quarrel between Newton and Leibniz*, Cambridge, 1980.
—— 'Sir Isaac Newton's Notebook, 1661–65', *Cambridge Historical Journal*, 9, 1949, 239–50.
—— 'Sir Isaac Newton's steamer', *History of Technology, Tenth Annual Volume*, 1985, 17–29.
Harrison, John, *The Library of Isaac Newton*, Cambridge, 1978.
Haskell, Francis, 'The apotheosis of Newton in art' in Robert Palter (ed.), *The Annus Mirabilis of Sir Isaac Newton 1666–1966*. Cambridge, MA and London, 1970, 302–21.
Hendry, John, 'Newton's theory of colour', *Centaurus*, 23, 1980, 230–51.
Herivel, J. W., *The Background to Newton's Principia*, Oxford, 1965.
—— 'Interpretation of an early Newton manuscript', *Isis*, 52, 1961, 410–16.
—— 'Newton's discovery of the law of centrifugal force', *Isis*, 51, 1960, 546–53.

—— 'Sur les premières recherches de Newton en dynamique', *Revue d'Histoire des Sciences*, 15, 1962, 105–40.

Hiscock, W. G., *David Gregory, Isaac Newton and their Circle*, Oxford, 1937.

Hofmann, J. E., *Leibniz in Paris, 1672–76*, Cambridge, 1974.

Huygens, Christiaan, *Oeuvres complètes*, The Hague, 1888–1950, 22 vols.

Koyré, Alexandre, 'Les Queries de l'Optique', *Archives Internationales d'Histoire des Sciences*, 13, 1960, 15–29.

Koyré, Alexandre and Cohen, I. Bernard, 'Newton and the Leibniz-Clarke Correspondence', *Archives Internationales d'Histoire des Sciences*, 15, 1962, 64–126.

Lohne, J. A., 'Experimentum crucis', *Notes and Records of the Royal Society*, 23, 1968, 169–99.

—— 'Newton's "proof" of the sine law and his mathematical principles of colours', *Archive for History of Exact Sciences*, 1, 1961, 389–405.

McGuire, J. E., 'Body and void in Newton's De Mundi Systemate: some new sources', *Archive for History of Exact Sciences*, 3, 1966, 206–48.

McGuire, J. E. and Rattansi, P. M., 'Newton and the "Pipes of Pan"', *Notes and Records of the Royal Society*, 21, 1966, 108–43.

McGuire, J. E. and Tamny, M., *Certain Philosophical Questions: Newton's Trinity Notebook*, Cambridge, 1983.

McMullin, Ernan, *Newton on Matter and Activity*, Notre Dame, IN and London, 1978.

Maddison, R. E. W., *The Life of the Honourable Robert Boyle*, London, 1969.

Mamiani, Maurizio, *Isaac Newton Filosofo della Natura*, Florence, 1976.

—— *Il Prisma del Newton*, Rome and Bari, 1986.

Manuel, Frank, E., *Isaac Newton, Historian*, Cambridge, 1963.

—— *A Portrait of Isaac Newton*, London, 1968, 2nd edn, 1980.

—— *The Religion of Isaac Newton*, Oxford, 1974.

Meli, D. Bertoloni, 'Leibniz's excerpts from the *Principia Mathematica*', *Annals of Science*, 45, 1988, 477–504.

— — 'Public claims, private worries: Newton's *Principia* and Leibniz's theory of planetary motion', *Studies in History and Philosophy of Science*, 22, 1991, 415–49.

More, Louis Trenchard, *Isaac Newton*, New York, 1934.

Munby, A. N. L., 'The Keynes Collection of the works of Sir Isaac Newton at King's College, Cambridge', *Notes and Records of the Royal Society*, 10, 1952, 42–4.

Newman, William, 'Newton's *Clavis* as Starkey's *Key*', *Isis*, 78, 1987, 564–74.

Newton, Isaac, *The Chronology of Ancient Kingdoms Amended*, London, 1728.

—— *The Correspondence of Isaac Newton*, ed. H. W. Turnbull, J. F. Scott, A. Rupert Hall and Laura Tilling, Cambridge, 1959–77, 7 vols.

—— *Correspondence of Sir Isaac Newton and Professor Cotes*, ed. Joseph Edleston, London, 1850, reprint 1969.

—— *The Mathematical Papers of Isaac Newton*, ed. D. T. Whiteside, Cambridge, 1967–81, 8 vols.

—— *The Optical Papers of Isaac Newton*, ed. Alan E. Shapiro, vol. I,

Cambridge, 1984.

—— *Opticks, or a Treatise of the Reflections, Inflections and Colours of Light*, London, 1936 (based on the 4th edn. 1730); reprinted New York, 1952, with an introduction by I. Bernard Cohen.

—— *Papers and Letters on Natural Philosophy*, ed. I. Bernard Cohen, Cambridge, MA, 1958, 2nd edn, 1978.

—— *Philosophiae Naturalis Principia Mathematica*, 3rd edn (1726) with variant readings, ed. Alexandre Koyré and I. Bernard Cohen, Cambridge, 1972.

—— *Preliminary Manuscripts for Isaac Newton's 1687 Principia, 1684–1685*, ed. D. T. Whiteside, Cambridge, 1989.

—— *Unpublished Scientific Papers of Isaac Newton*, ed. A. Rupert Hall and Marie Boas Hall, Cambridge, 1962, 2nd edn, 1978.

Oldenburg, Henry,*The Correspondence of Henry Oldenburg*, ed. A. Rupert Hall and Marie Boas Hall, London and Madison, WI, 1965–86, 13 vols.

Priestley, F. E. L., 'The Leibniz-Clarke controversy' in Robert E. Butts and John W. Davis (eds), *The Methodological Heritage of Newton*, Oxford, 1970.

[Prior, Thomas], *Observations upon Coin in general*, Dublin, 1720 and London, 1730.

Raven, C. E., *John Ray, Naturalist*, Cambridge, 1950.

Rigaud, S. P., *Correspondence of Scientific Men of the Seventeenth Century . . .* Oxford, 1841, 2 vols.

—— *Historical Essay on the First Publication of Sir I. Newton's Principia*, Oxford, 1838.

Rouse Ball, W. W., *An Essay on Newton's 'Principia'*, London, 1893.

Ruffner, J. A., 'Reinterpretation of the genesis of Newton's "Law of Cooling"', *Archive for History of Exact Sciences*, 2, 1964, 138–52.

Sabra, I. A., *Theories of Light from Descartes to Newton*, London, 1967.

Scriba, J. C., 'Mercator's Kinckhuysen translation in the Bodleian Library at Oxford', *British Journal for the History of Science*, 2, 1964, 45–58.

Shapiro, Alan E., 'Newton's "achromatic" dispersion law: theoretical background and experimental evidence', *Archive for History of Exact Sciences*, 21, 1979, 91–128.

Smith, D. E., 'Two unpublished documents of Sir Isaac Newton' in W. J. Greenstreet (ed.), *Isaac Newton, 1642–1727*, London, 1927, 16–34.

Sotheby & Co., *A Catalogue of the Newton Papers sold by order of Viscount Lymington . . . 13–14 July 1936*, London, 1936.

Stukeley, William, *Memoirs of Sir Isaac Newton's Life*, ed. A. Hastings White, London, 1936.

Taylor, F. Sherwood, 'An alchemical work of Sir Isaac Newton', *Ambix*, 5, 1956, 59–84.

Turnbull, H. W., *James Gregory Tercentenary Memorial Volume*, London, 1939.

Turnor, Edmund, *Collections for the History . . . of Grantham containing authentic memoirs of Sir Isaac Newton*, London, 1806.

Villamil, R. de, *Newton: The Man*, London, [1931?].

Wesbster, Charles, *The Great Instauration*, London, 1975.

—— 'Henry More and Descartes: some new sources', *British Journal for the History of Science*, 4, 1969, 359–77.

Westfall, Richard S., 'The foundations of Newton's philosophy of Nature', *British Journal for the History of Science*, 1, 1962, 171–82.

—— 'Isaac Newton's Coloured Circles twixt two Contiguous Glasses', *Archive for History of Exact Sciences*, 2, 1965, 181–96.

—— *Never at Rest: A Biography of Isaac Newton*, Cambridge, 1980.

—— 'Short-writing and the state of Newton's conscience, 1662', *Notes and Records of the Royal Society*, 18, 1963, 10–16.

Whiteside, D. T., 'The mathematical principles underlying Newton's *Principia Mathematica*', *Journal for the History of Astronomy*, 1, 1970, 116–38

—— 'Newton's lunar theory: from high hope to disenchantment', *Vistas in Astronomy*, 19, 1975–6, 317–24.

—— 'The prehistory of the *Principia* from 1664 to 1686', *Notes and Records of the Royal Society*, 45, 1991, 11–61.

Index

Dates have been added to the names of some members of Newton's family to clarify the distinctions between them.